E&M Endocrinology and Metabolism

Progress in Research and Clinical Practice

Margo Panush Cohen   Piero P. Foà

Series Editors

Endocrinology and Metabolism
Progress in Research and Clinical Practice

Margo Panush Cohen   Piero P. Foà
  Series Editors

Cohen and Foà (eds.): Hormone Resistance and Other Endocrine
  Paradoxes (Vol. 1)

Forthcoming volumes:

Jovanovic (ed.): Controversies in Diabetes and Pregnancy (Vol. 2)

Cohen and Foà (eds.): The Brain as an Endocrine Organ (Vol. 3)

Ginsberg-Fellner and McEvoy (eds.): Autoimmunity and the
  Pathogenesis of Diabetes (Vol. 4)

Margo Panush Cohen   Piero P. Foà
Editors

Hormone Resistance and Other Endocrine Paradoxes

With 41 Figures

Springer-Verlag
New York Berlin Heidelberg
London Paris Tokyo

Margo Panush Cohen, M.D., Ph.D.
Professor of Medicine
University of Medicine and
  Dentistry of New Jersey
Newark, New Jersey 07103
Director, Institute for
  Metabolic Research
University City Science
  Center
Philadelphia, Pennsylvania
  19104
USA

Piero P. Foà, M.D., Sc.D.
Professor Emeritus of Physiology
Wayne State University
Chairman Emeritus
Department of Research
Sinai Hospital
Detroit, Michigan 48202
Mailing address:
  2104 Rhine Road
  West Bloomfield, Michigan 48033
  USA

Library of Congress Cataloging-in-Publication Data
Hormone resistance and other endocrine paradoxes.
  (Endocrinology and metabolism series ; v. 1)
  Includes bibliographies and indexes.
  1. Hormone resistance. 2. Endocrine glands—
Diseases. I. Cohen, Margo P. II. Foà, Piero P.
(Piero Pio), 1911– . III. Series: Endocrinology and
metabolism series (Springer-Verlag) ; v. 1. [DNLM:
1. Hormone—physiology. WK 102 H8118]
RC664.H67 1987   616.4   87-12799

Typeset by TC Systems, Shippensburg, Pennsylvania.
Printed and bound by R.R. Donnelley & Sons, Harrisonburg, Virginia.
Softcover reprint of the hardcover 1st edition 1987

9 8 7 6 5 4 3 2 1

ISBN-13: 978-1-4612-9143-5   e-ISBN-13: 978-1-4612-4758-6
DOI: 10.1007/978-1-4612-4758-6

Preface

Endocrinology and Metabolism: Progress in Research and Clinical Practice is a new series that has been designed to present timely, critical reviews of constantly evolving fields; to provide practical and up-to-date guidance in the solution of pertinent clinical problems; to offer an alternative to the laborious search of the literature (and the often frustrating reading of highly technical articles); and to translate the language of the laboratory into that of the practice of medicine.

We think that this volume and those to come will prove useful to physicians (and to physicians in training), as well as to investigators in a wide variety of specialties; in short, to anyone who seeks answers to questions in endocrinology and metabolism.

The first chapter of this volume could well serve as a general introduction to the entire series. It points out how our growing understanding of the molecular basis of biologic communication has led to the discovery of a growing number of clinical syndromes, as well as to the realization that phenotypically similar diseases may have radically different pathogenetic mechanisms and thus may require radically different therapeutic stratagems.

Endocrinology and metabolism deal with complex systems comprising innumerable components responsible for signal transmission, rapid adjustment and fine-tuned responses, and many layers of "fail-safe" protection. Sometimes one component of the system misfires, masquerading as a teleologically counterproductive or "paradoxical" event. Our hope is that this volume and those that follow will provide at least some explanation for such events, teaching us that paradoxes do not exist in nature, but only in the minds of the uninformed.

New York, New York
Detroit, Michigan

Margo P. Cohen
Piero Foà

Contents

Contributors

MARGARET JOHNSON BIA, M.D.
Associate Professor of Medicine, Associate Director, Dialysis and Transplant
Unit, Yale University School of Medicine, New Haven, Connecticut, USA

LEWIS E. BRAVERMAN, M.D.
Director, Division of Endocrinology and Metabolism, Professor of Medicine
and Physiology, Acting Chairman, Department of Nuclear Medicine, University of Massachusetts Medical School, Worcester, Massachusetts; Lecturer on
Medicine, Harvard Medical School, Boston, Massachusetts, USA

ARNOLD S. BRICKMAN, M.D.
Professor of Medicine, University of California at Los Angeles School of Medicine, Los Angeles, California; Chief, Mineral Metabolism Unit, Sepulveda
Veterans Administration Medical Center, Sepulveda, California, USA

TERRY R. BROWN, PH.D.
Associate Professor of Pediatrics, Division of Endocrinology, Johns Hopkins
University School of Medicine; Assistant Director, Pediatric Endocrine Laboratories, Johns Hopkins Hospital, Baltimore, Maryland, USA

HAROLD E. CARLSON, M.D.
Professor of Medicine, Department of Medicine, State University of New York
at Stony Brook, Stony Brook, New York; Chief, Endocrinology Section,
Northport Veterans Administration Medical Center, Northport, New York,
USA

MARGO PANUSH COHEN, M.D., PH.D.
Professor of Medicine, University of Medicine and Dentistry of New Jersey,
Newark, New Jersey; Director, Institute for Metabolic Research, University
City Science Center, Philadelphia, Pennsylvania, USA

PIERO P. FOÀ, M.D., SC.D.
Professor Emeritus of Physiology, Wayne State University; Chairman Emeritus, Department of Research, Sinai Hospital, Detroit, Michigan; Mailing address: 2104 Rhine Road, West Bloomfield, Michigan 48033, USA

JOHN G. HADDAD JR., M.D.
Professor of Medicine, Chief, Endocrine Section, Department of Medicine, University of Pennsylvania School of Medicine, Philadelphia, Pennsylvania, USA

JAMES F. MCLEOD, M.D.
Fellow in Endocrinology, Department of Medicine, University of Pennsylvania School of Medicine, Philadelphia, Pennsylvania, USA

CLAUDE J. MIGEON, M.D.
Professor of Pediatrics, Division of Endocrinology, Johns Hopkins University School of Medicine; Director, Pediatric Endocrine Clinic and Laboratories, Johns Hopkins Hospital, Baltimore, Maryland, USA

MARK D. OKUSA, M.D.
Nephrology Fellow, Yale University School of Medicine, New Haven, Connecticut, USA

MARJORIE SAFRAN, M.D.
Assistant Professor of Medicine, Department of Endocrinology and Metabolism, University of Massachusetts Medical Center, Worcester, Massachusetts, USA

MORRIS SCHAMBELAN, M.D.
Professor of Medicine, Department of Medicine, University of California School of Medicine; Chief, Division of Endocrinology, Program Co-Director, General Clinical Research Center, San Francisco General Hospital, San Francisco, California, USA

ANTHONY SEBASTIAN, M.D.
Professor of Medicine, Department of Medicine, University of California School of Medicine; Program Co-Director, General Clinical Research Center, Moffitt Hospital, University of California, San Francisco, California, USA

HOWARD S. TAGER, PH.D.
Louis Block Professor and Chairman, Department of Biochemistry and Molecular Biology, The University of Chicago, Chicago, Illinois, USA

1
Introduction: The Journey of the Endocrine Signal: *A Paradigm of Murphy's Law*

Piero P. Foà

"The harmony of life," wrote Claude Bernard in 1866 (1), depends on the integrity of two types of "organic elements," one type represented by muscle and nerve fibers, which functions through direct anatomical connections, and one type represented by the "organs of internal secretion," which influences distant structures through "peculiar substances" introduced into the blood. After more than a century, this concept continues to serve us well, even as we know that "peculiar substances" can be produced by neurons and that, as in paracrine and autocrine relationships, they need not be released into the circulation. Indeed, intercellular communication through chemical signals starts during the earliest stages of embryologic development, when cells begin to segregate toward the formation of specific tissues and well before the appearance of a cardiovascular system. From this beginning and from the embryogenesis of the endocrine system, the journey of the signal molecules takes them through many steps including biosynthesis, release and transport, recognition by the target cell, transduction of the message, and, finally, extinction of the signal through counterregulation, metabolism, and excretion.

The purposes of this chapter are to review selected experiments, biologic models, and clinical syndromes that illustrate these steps and their malfunction and to speculate about future developments, thus introducing the other chapters of this volume, in which some of these clinical syndromes are discussed in detail.

I. Chemical Control of Morphogenesis

Normal morphogenesis requires the orderly succession of cell division, movement, adhesion, differentiation, and death (2). The crucial role of chemical messengers in this process was recognized many years ago by the French physiologist Gley (3), who, perhaps inspired by Claude Bernard's "harmony of life," suggested that some endocrine products be called "harmozones" (from the Greek, meaning "I create an orderly union''), citing among others the hormones of the thyroid, for their role in amphibian metamorphosis. To Gley's list, we could add the Müllerian inhibiting substance, the testis-determining factor, and the gonadal steroids for their role in sexual development and the many peptides that stimulate differentiation and growth of specific tissues (4–13). Among these peptides are fibronectin, ubiquitin, the cell adhesion molecules (CAM), and other widely distributed substances that promote mutual attraction between cells of similar ancestry and thus the formation of different organs and tissues (2,14,15).

Once contact between cells has been established through these or other mechanisms, portions of adjacent cell membranes fuse to form "tight junctions," or structures capable of creating separate compartments by sealing off intercellular spaces, whereas other portions form transmembrane channels or "gap junctions," providing the structural means for paracrine functions, such as the traffic of ions, nutrients, metabolic regulators, and electrical charges between cells (16–21). Although the degree of opening or "gating" of the gap junctions is regulated by membrane potential and by the concentration of hydrogen and other ions and can thus be modified by pharmacologic agents (22; see Section VIII), specific abnormalities of CAM, ubiquitin, or gap junctions have not been described. Nevertheless, the changes in the level of intercellular communication following exposure to prostaglandins (23), tumor promoters, or viruses that cause unregulated growth (24), or following the phosphorylation of gap junction proteins (25) and, finally, the functional deficiencies caused by the experimental disruption of intercellular connections (26–28), suggest that a pathology of these molecules may indeed exist.

By contrast, the developmental defects that lead to hypo- or hyperplasia of specific endocrine tissues are well known (4) and need not be discussed here, except for two recently described animal models. The first is a mutant mouse in which the absence of preoptic nuclei leads to a deficiency of gonadotropin releasing hormone (GnRH) and consequent gonadal failure, a syndrome that can be corrected by transplants of normal fetal hypothalamic tissue (29) and that in some ways resembles idiopathic hypothalamic hypogonadism in humans (30). The second is another mutant mouse with an inherited form of thyroid hypoplasia, in which the presence of low or undetectable levels of thyroxine (T4) and increased secretion of thyroid-stimulating hormone (TSH) resembles human congenital primary hypothyroidism (31).

II. Hormone Biosynthesis

The biosynthesis of the signal molecules by normally developed endocrine tissues depends on the structure and expression of genes encoding peptide hormone precursors (prohormones) and enzymes that either regulate the posttranslational conversion of prohormones into active hormones or the synthesis of hormones from other molecules, such as amino acids, fatty acids, and cholesterol (Fig. 1.1). Missing prohormone genes or defective gene transcription may result in endocrinopathies such as growth hormone (GH)- or growth hormone–releasing hormone (GHRH)-deficient dwarfism (32–35), isolated corticotropin (ACTH) deficiency with hypoglycemia (36,37), or GnRH-deficient hypogonadism (30). Alternatively, endocrinopathies may arise from defects in gene structure. Such may be the case of dwarfism due to the secretion of a growth hormone variant with normal immunologic properties but decreased biologic activity, a syndrome which, as may be expected, responds to treatment with "normal" recombinant human GH (38).

Other structural abnormalities may lead to the synthesis of inactive TSH (39), of proinsulin molecules that cannot be converted to insulin, and of abnormal insulin variants with decreased biologic activity (40). Theoretically, it should be possible to correct these deficiencies by introducing the missing gene, a feat that so far has been attempted with some success only for the gene of growth hormone, using a viral vector (41,42), for that of insulin using liposomes (43), and by the above-mentioned transplant of normal hypothalamic tissue into GnRH-deficient mice.

In cases where the missing or abnormal gene may be one that normally encodes a prohormone-processing enzyme, the result is either a defective production of the active hormone or a change in the relative amounts of different peptides that derive from the same precursor. Although since the discovery of proinsulin (44), many prohormones have been identified (45–54), and although it is known that different tissues process them in different ways, producing specific peptides in characteristic amounts,

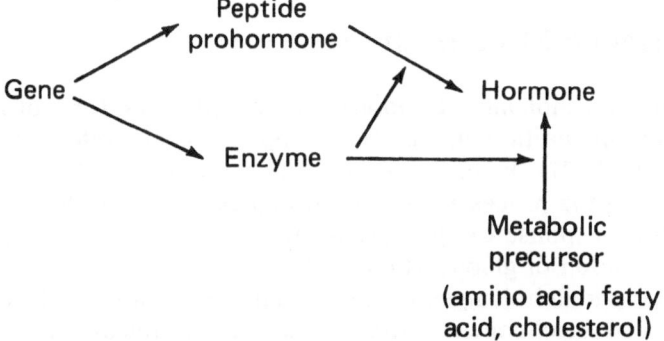

FIGURE 1.1. Schematic representation of hormone biosynthesis through posttranslational processing of peptide prohormones or through enzymatic conversion of metabolic precursors.

only relatively few examples of conversion defects are known to exist. Among them are familial hyperproinsulinemia (55), the abnormal production of glucagonlike peptides in animals and patients with glucagon-producing tumors and other disorders (56–59), a form of pituitary dwarfism due to posttranslational polymerization of growth hormone (60), and the production of a high-molecular-weight TSH with impaired biologic activity in a euthyroid man (61).

Still other enzymatic deficiencies may block the conversion of metabolic precursors into hormones. Such is the case of defective steroidogenesis from cholesterol leading to adrenal hyperplasia and to male pseudohermaphoditism (4,62–64) and of the impaired neurotransmitter biosynthesis associated with autonomic disorders (65) and loss of memory (66). Finally, the structure and the processing of genes may be normal, but their expression may occur in "nonendocrine" tissues, such as the central nervous system (67,68), or it may occur "ectopically," contributing to the clinical syndromes due to hormone-producing tumors (69,70) and, possibly, to the growth of the tumor itself (71).

Finally, the synthesis of a hormone may be modified by environmental factors. Thus, iodine availability limits the synthesis of thyroid hormones (Fig. 1.2), the diurnal and seasonal cycles of light and darkness regulate the synthesis of pineal melatonin (72), and the abundance of polyunsaturated fatty acids (fish oil) in the diet may direct the synthesis of relatively inactive thromboxane A3 (TXA3) at the expense of the more active TXA2, decreasing the risk of thromboembolic disease (73). Endocrinopathies due to ectopic hormone secretion are well known (70) and need not be described here except to mention the additional diagnostic difficulties created by prohormones and by abnornmal posttranslational products when they react in presumably "specific" radioimmunoassay systems (38,56,74,75).

III. Hormone Storage and Release

Peptide and catecholamine hormones are stored in the form of granules which, after maturation in the Golgi apparatus, are released through exocytosis (4,76,77). In the case of peptide hormones, this maturation coincides with the processing of prohormones whose products are then coreleased in response to appropriate stimuli. Thus, adrenalectomy and the administration of glucocorticoids, known modifiers of ACTH secretion, induce parallel changes in the secretion of POMC-derived endorphin (45), and insulinogenic stimuli result in the release of equimolar amounts of insulin and C peptide, providing a convenient tool for measuring B cell function in patients with circulating insulin antibodies (78).

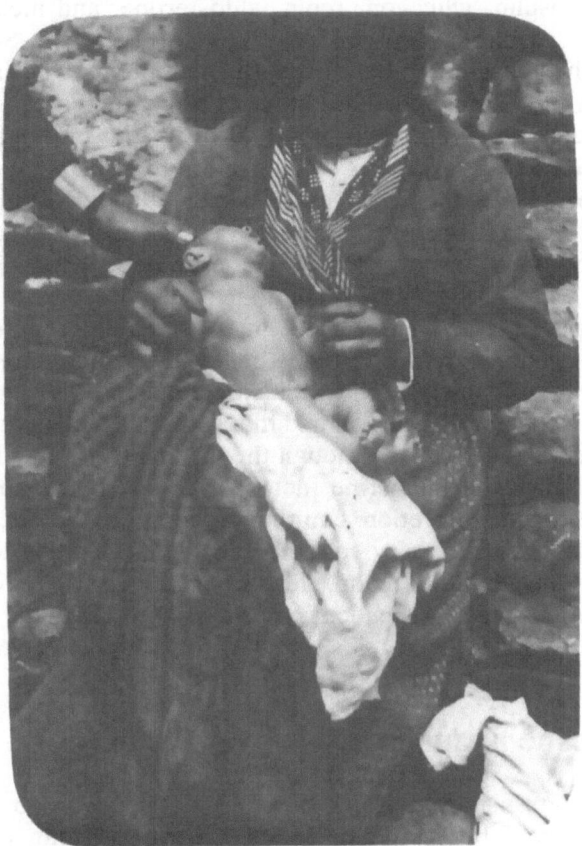

FIGURE 1.2. Congenital goiter in a 2-day-old offspring of a cretin suffering from chronic dietary iodine insufficiency. Failure to synthesize thyroid hormones caused the uninhibited secretion of TSH and thyroid growth. Courmayeur, Italy, ca. 1910. (Photograph by U. Cerletti.)

The release of hormones, whether in the form of granules or not, is under the control of neural, metabolic, and endocrine factors that regulate available energy, membrane polarization, and the concentration of calcium ions in the secretory cell. Although most of these factors are exhaustively discussed in many textbooks of endocrinology (4) and in other chapters of this volume (79,80), a few will be cited here either because they have been the object of renewed interest or because they have hypothetical or proven pathogenetic and therapeutic implications (see Section VIII).

One example is the inhibition of insulin secretion by islet cell surface antibodies (81). Another is the role of biogenic amines in the secretion of numerous hormones, including prolactin and luteinizing hormone (LII),

GnRH, GH, insulin, glucagon, renin, aldosterone, and melatonin, and hence in the regulation of gonadal activity (4,82–90). Still another example is the pulsatile, rather than the continuous, release of many hormones (91–95), a phenomenon that apparently increases such diverse biologic effects as those of glucagon and of insulin on hepatic glucose production (96,97) and of GnRH when used in the treatment of hypothalamic amenorrhea (98) or delayed puberty (99). Finally, one should mention the "enteroinsular axis" whereby "incretins" (100) and other B cell–regulating hormones, such as the glucose-dependent insulinotropic peptide (GIP), are released following a meal.

The pathophysiologic role of these insulin-modulating intestinal peptides is not fully understood, but it is believed that they are the reason why food or an oral glucose load elicits a greater insulinogenic response than does an intravenous glucose load that results in a comparable degree of hyperglycemia (101–104). Although the mechanisms of action of these neural, metabolic, and endocrine factors may differ, the common final pathway in stimulus-secretion coupling appears to be the uptake of calcium ions by the secretory cell and subsequent binding to a specific protein such as calmodulin (calm), leading to the activation of Ca-calm-dependent and/or cAMP-dependent protein kinases and to the phosphorylation of the cytoplasmic proteins required for exocytosis (76,105–108).

IV. Hormone Transport

Hormones are transported from the site of production to their target tissues either free or bound to blood cells, to serum albumin, or to specific binding proteins (4,109). These different hormone compartments are in a state of dynamic equilibrium that determines the amount of free hormone available for biologic activity and for metabolic clearance. Thus, an increase in progesterone-binding protein may lead to hypogonadism and a decrease in androgen-binding protein to hirsutism (110,111), whereas estrogen-binding α-fetoprotein protects the fetus from the high maternal levels of estrogen and plays an essential role in the development of the sexual phenotype.

Carrier proteins for cortisol (4), for insulinlike growth factors (IGF) I and II (112), and for thyroxine (T4; 113) have also been described. Indeed, several inherited variants of thyroxine-binding protein, some with increased and some with decreased binding capacity, have been found in a number of individuals with otherwise normal thyroid function and may be a source of diagnostic difficulties (114,115) complicating the evaluation of thyroid function in malnourished rats (116) and in patients with severe nonthyroid illness (117). Finally, one of the functions of the binding proteins may be to aid in the transport of hormones through the target cell membrane (118) or through the vascular endothelium and the blood-brain barrier (119–121), toward their sites of action and metabolic disposal (see Section VII).

V. Hormone Binding and Recognition

The recognition of a hormone by its target cell is a function of the receptors and depends on their number and their affinity and specificity for the hormone (122–127). The number of available receptors, in turn, depends on the equilibrium between synthesis, translocation to the site of action, degradation, and, in the case of membrane receptors, internalization and recycling (128,129). Thus, the absence of a gene or a gene defect may lead to a receptor deficiency syndrome, such as vasopressin-resistant diabetes insipidus, PTH-resistant pseudohypoparathyroidism, insulin-resistant acanthosis nigricans (n.) and leprechaunism, GH- and GHRH-resistant dwarfism, androgen-resistant infertility, male pseudo-hermaphoditism and testicular feminization, vitamin D–resistant rickets (5,30,130–136), and a form of cortisol resistance characterized by sodium retention, hypokalemic alkalosis, and hypertension (137).

In this syndrome, the anterior pituitary, lacking cortisol receptors, escapes the inhibitory feedback and secretes increased amounts of ACTH. The adrenal cortices are stimulated, the high levels of corticosterone and deoxycorticosterone result in the characteristic electrolyte imbalance, and the absence of cortisol receptors in the peripheral tissues prevents the development of signs and symptoms usually associated with Cushing's syndrome. In other circumstances, the genes for the receptors are normal, but their expression is modified by endocrine, dietary, or other factors. For example, the number of insulin receptors may be decreased by glucagon (138), the number of glucagon receptors may be increased by insulin (139), LH-releasing hormone (LHRH) increases estrogen binding (140), and GH induces the synthesis of its own receptors (141).

The list of examples may continue. The number of oxytocin receptors in the uterus and the mammary gland increases sharply at the end of gestation, triggering labor and lactation (142); estrogens control progesterone binding in human breast cancer (143); the angiotensin receptors in the adrenal cortex respond to changes in water and electrolyte balance (144,145); and those of triiodothyronine and insulin respond to obesity and to changes in the nutritional state (146–149). In still other circumstances, the number of receptors may be decreased for unknown reasons, as in the case of the muscarinic receptors in patients with Alzheimer's disease (150) or because of improper translocation to their site of action, as when the insertion of the insulin receptors into the cell membrane is defective, causing insulin resistance in patients with Rabson-Mendenhall syndrome (151).

Another control mechanism is "down-regulation," the decrease in the number of receptors with increasing concentrations of the hormone (122,123,152,153). Although down-regulation is a mechanism for the physiologic control of the endocrine signal, when exaggerated by the presence of excessive amounts of hormone, it may lead to clinically significant resistance. Such is the case of insulin-resistant obesity, acromegaly, type 2 diabetes, and other hyperinsulinemic states (130,152) and the refractoriness to adrenergic agonists that develops in rats with pheochromocytoma (154) or in patients treated with β-adrenergic drugs.

The opposite phenomenon, receptor "up-regulation," may also occur when the concentration of a hormone is low or its binding is prevented, a phenomenon that may explain the hypersensitivity to endogenous catecholamines and the occurrence of cardiovascular reactions upon sudden withdrawal of β-adrenergic blockers (126,127,155,156). Finally, the number of available receptors may be normal, but their effectiveness may be decreased by inhibitors of binding such as those found in the plasma of patients with ataxia telangiectasia (157) or by receptor antibodies such as those against the acetylcholine receptor in patients with myasthenia gravis (158,159), against the insulin receptor in type B acanthosis n. (160), or against TSH in women with primary myxedema (161).

Paradoxically, though receptor antibodies may block the binding of a hormone, they may also mimic its action, as in the case of antibodies against TSH (4,162,163) or against the insulin receptor (164). Perhaps this phenomenon occurs because the receptors contain different immunogenic determinants. Thus, antibodies against the binding site may promote receptor internalization or activation and have a hormonelike effect, whereas antibodies against the catalytic moiety may block its activity (164–166).

Affinity, the second essential property of the receptor, is a function of the molecular structure of the hormone (167–171), of the receptor itself (172), and of the degree of "occupancy" by the hormone. Thus, increasing concentrations of a hormone decrease not only the number of available receptors (down-regulation) but also their affinity, a phenomenon called negative cooperativity (122,123).

The third essential property of the receptor is specificity, and it is not always absolute. Indeed, when sufficient amino acid homologies exist either between the receptors or between the hormones, the hormones may "cross over" and bind to each other's receptors. Similarly, if one of the hormones is present in high concentration or if binding to its receptors is blocked, "spillover" to the other receptor may occur. Thus IGF1, acting through the insulin receptor, may stimulate glucose uptake and lipogenesis, whereas insulin, acting through the IGF receptors, may stimulate growth, explaining, for example, the increased growth of the placenta in insulin-treated rabbits (173), the development of organomegaly and macrosomia in the hyperinsulinemic fetus of diabetic mothers (174) and, possibly, the growth-promoting effect of insulin on blood vessels (175–178).

All receptors identified thus far are protein molecules, often composed of two or more moieties whose biologic functions, physicochemical and immunologic properties, and, in some cases, chromosomal localization and gene structure have been partially or completely clarified. Among them are the receptors for estrogen, progesterone, glucocorticoid, acetylcholine, glucagon, insulin, several growth factors (179–189), and some oncogene products that closely resemble the growth factor receptors and share their growth-promoting properties (190–193).

One of the better known is the insulin receptor, a protein derived from a precursor molecule whose gene is located on chromosome 19 (194). This receptor is composed of two subunits containing the insulin-binding sites and two subunits with a tyrosine-kinase, an ATP-binding, and a calmodulin-binding domain (195,196). According to the best evidence, insulin binding to the α subunits activates the tyrosine kinase domain of the β subunits, catalyzing the autophosphorylation of the receptor itself, as well as the phosphorylation of other kinases and phosphatases involved in the transduction of the insulin signal with ATP acting as the phosphate donor (197,198).

Among these phosphorylation reactions, that of the β subunit at the serine and threonine rather than at the tyrosine residues decreases both the activity of the tyrosine kinase and the binding affinity of the α subunit, thus decreasing the intensity of the insulin signal (see Section VI). It is believed that structural defects of the insulin receptor may contribute to the decreased binding and kinase activity observed in cases of obesity, in patients with type II diabetes or type A insulin resistance, and in rats with streptozotocin diabetes, and to insulin hypersensitivity and increased kinase activity in obese rat adipocytes (148,149,199,200).

A second cell membrane receptor whose structure has been clarified is adenylate cyclase. This receptor is composed of a hormone-binding moiety coupled with a nucleotide-binding regulatory protein called N or G, which may stimulate (Ns or Gs) or inhibit (Ni or Gi) the third or catalytic component of the system. Occupancy of the receptor by the hormone controls the activity of the N proteins and thus the catalytic activity and the production of cyclic AMP (cAMP). In particular, occupancy of the receptor by β-adrenergic agonists, glucagon, ACTH, TSH, FSH, LH, PTH, or secretin activates the Ns protein and increases cAMP formation, whereas occupancy by α-adrenergic or muscarinic agonists, opioid peptides, or somatostatin activates the Ni protein and inhibits the formation of cAMP (126,127,201).

Several abnormalities of the adenyl cyclase receptors have been described. Thus, their number appears to decrease with age (202), while cholera toxin, the enterotoxin of *E. coli,* and vasoactive intestinal peptide (VIP) lead to persistent stimulation of Ns and increased production of cAMP, causing the diarrhea and electrolyte loss characteristic of cholera, "traveler's enteritis", and VIP-secreting tumors (pancreatic cholera). On the other hand, the toxin of *B. pertussis,* acting on the Ni moiety, leads to a generalized inhibition of α-adrenergic responses, prevents the vasoconstrictive effect of the catecholamines released by counterregulation, and thus allows the toxin-induced hypotension to go unchecked. In the pancreatic islets, this effect reduces the α-adrenergic inhibition of insulin release, and the toxin behaves as an "islet-activating protein" (IAP; 203).

Other interesting defects of the cyclase system are found in mutant cells of lymphoma S49 that lack Ns, in the erythrocytes of the toad *Xenopus levis* that lack the catecholamine-binding moiety, in the maturing human reticulocyte and the aging rat myocardium that show a loss of the catalytic subunit, and, finally, in two types of pseudohypoparathyroidism in which PTH is ineffective owing to a lack of binding moiety and a lack of Ns, respectively. As one would expect, patients of the first type are resistant only to PTH, whereas those of the second are resistant also to other hormones whose action is mediated through Ns, such as the β-adrenergic agonists. A defect of N protein function has also been linked to retinitis pigmentosa and to some animal models of inherited retinal degeneration (126,127,204–206).

Finally, there is evidence that the glucocorticoids increase the number of β-adrenergic receptors and the abundance of the Ns protein in cultured osteosarcoma cells (207), an effect that, if applicable to other tissues, could explain the permissive action of these hormones on other endocrine functions. To my knowledge, no defects in the amino acid sequence of cytosolic steroid hormone receptors have been described, although it has been shown that the molecular mass of the estradiol receptors in the rat uterus and the structure of the progesterone receptors in human breast cancer may be altered by treatment with estrogen and progestin, respectively (143,208), and that end-organ resistance to steroid hormones, such as androgens, progesterone, aldosterone, vitamin D, and the glucocorticoids, may be associated with decreased receptor affinity (124,125,136).

VI. Transduction of the Endocrine Signal

The transduction of the endocrine signal into appropriate biologic responses depends on the interaction of the hormone-receptor complex with cytoplasmic and nuclear effector systems. In the case of the iodothyronines whose receptors are within the cell nucleus, this complex reacts directly with chromatin acceptor sites; in the case of the steroid hormones whose receptors are in the cytoplasm, transduction of the signal requires activation of the hormone-receptor complex and translocation to the nucleus. In contrast, in the case of hormones whose receptors are part of the cell membrane, transduction requires internalization of the complex and/or the formation of an intracellular second messenger. The nuclear events that follow thyroid hormone binding have been described (122,123,146,147,209). Similarly, a great deal is known about the mechanisms regulating steroid hormone binding, activation, translocation, and effect on the genome, even though we do not have a full understanding of the molecular events leading to postreceptor end-organ resistance (4,124,125,136,210,211) or of the mechanisms whereby estrogen stimulates DNA synthesis and tissue growth through the intervention of unknown growth factors (212) and in tissues that lack estrogen receptors (213). As stated above, plasma membrane receptors such as those of insulin and of some growth factors may act as their own transducers through internalization, through the kinase activity of the receptor itself (214), or, as in the case of the cholinergic nicotinic and the GABA receptors, acting as ion channels (126,160,183,215) which, in turn, bring about the activation of Ca-calm-dependent kinases (105,216–218).

Other hormones, such as the β-adrenergic agonists glucagon and PTH, stimulate the cyclase system and the production of cAMP, the typical "second messenger," and their effects depend on the activity of cAMP-dependent kinases (219,220). Finally, transduction of the signal may occur through the activation of kinase C by metabolites of phosphoinositides(221–227). Thus phosphorylation of cytoplasmic and nuclear proteins by specific kinases appears to be the major molecular mechanism for the synthesis, activation, and deactivation of the enzymes involved in signal transduction (185,186,195,197,228–232). Clearly, a deficiency of any of these target enzymes may decrease the effectiveness of a hormone. For example, a decreased activity of adenyl cyclase may explain the impaired effectiveness of PTH in patients with vitamin D deficiency (233), a decreased activity of cAMP phosphodiesterase may account for some of the metabolic difficulties of diabetes (234), and a deficiency of glucose-6-phosphatase, debranching enzyme, or phosphorylase explains why glucagon and epinephrine cannot stimulate hepatic glycogenolysis in some patients with glycogen storage disease (235).

On the other hand, the role of these kinases in the generation of pyruvate dehydrogenase, the putative intracellular mediators of insulin action (236) and in its alterations following exercise (237) or dexamethasone treatment (238), is not fully understood. Nor is their role in other insulin-mediated phenomena such as the expression of enzyme genes (226,239) or the synthesis of the recently discovered glucose transporter and its changes in exercise, in diabetes, and possibly in the remission of diabetes during the "honeymoon" period (240–244). Defects of these and other postreceptor mechanisms that contribute to insulin resistance in such diverse clinical states as hyperinsulinism, pregnancy, obesity, ketoacidosis, acromegaly, type 2 diabetes, or type A acanthosis n. (245–249) are discussed in another chapter of this volume (250). The role of the polyamines in the growth-stimulating activity of insulin, insulinlike growth factors, and oncogene products that resemble the IGF receptor and that of adenosine in the regulation of cardiovascular, respiratory, central nervous system, and other functions have been reviewed elsewhere (251–256).

VII. Extinction of the Endocrine Signal

The endocrine signal may be attenuated in many ways, including reduced secretion, reuptake of the hormone by the secretory cell, desensitization of the target tissue, action of antibodies and other inhibitors, secretion of counterregulatory hormones, or metabolic inactivation and excretion of the signal molecule. A discussion of mechanisms regulating hormone secretion is beyond the scope of this chapter. Well-known examples of their clinical importance are the feedback regulation of hypothalamic, anterior pituitary, and other hormones, a mechanism commonly exploited for diagnostic purposes and responsible for the inhibition of ACTH secretion in patients treated with glucocorticoids that makes gradual withdrawal of the steroid mandatory.

Reuptake by the secretory cell as a means of controlling the effective concentration of a hormone is characteristic of certain neurotransmitters and is discussed in many endocrinology and pharmacology textbooks (4,233,257). Desensitization of the endocrine cell that develops upon persistent stimulation has been described for several hormones (75,258,259). Indeed, desensitization of the gonads and of the anterior pituitary following prolonged exposure to LHRH agonists (260) is under investigation for the treatment of idiopathic precocious puberty (261). The effectiveness of a hormone may also be reduced by desensitization of the target tissue through a reduction of the number or affinity of the receptors, through the uncoupling of the hormone binding from the catalytic moiety, or through postreceptor mechanisms (122,123,262,263).

Thus, while desensitization of the secretory cells to stimulation may explain the cyclic nature of hormone production, desensitization of the target cells may explain why the pulsatile secretion of a hormone is more effective and less likely to induce resistance than its continuous administration (96,97,264). The action of a hormone may also be blunted by antibodies, by circulating inhibitors (see above and 265,266) or by the action of counterregulatory hormones. The latter is best illustrated by the interplay of insulin and ''antiinsulin'' hormones in the regulation of blood glucose concentration (267–269), an interplay that has been found wanting in insulin-dependent diabetes (270,271), in idiopathic hypoglycemia

(272), and in posthypoglycemic hyperglycemia (273). Finally, extinction of the signal may occur through metabolic clearance, a process that may require binding of the hormone to a specific serum protein or to the receptor as the means of transporting it through the vascular endothelium and through the cell membrane to its site of degradation and excretion (274–279). Metabolic clearance may be modified by the nature of the molecule, as in the case of renin (280); by the presence of receptor antibodies, as in the case of insulin (281); and, of course, by hepatic or renal disease (4,233).

VIII. Drug-Induced Changes of the Endocrine Signal

Many steps in the path of the endocrine signal can be modified by pharmacologic agents for experimental, diagnostic, or therapeutic purposes. To begin with, the secretory tissue may be destroyed by means of toxic agents or antibodies, with the production of experimental and clinical syndromes, such as alloxan and streptozotocin diabetes, goldthioglucose hypothalamic hyperphagia, monosodium glutamate–induced destruction of the arcuate nucleus with GH deficiency, stunted growth and obesity (282), neuronal death due to nerve growth factor antibodies (7,8), and numerous well-known autoimmune endocrinopathies (4,223), including diabetes (283) and the polyglandular autoimmune syndrome (284). Other drugs may alter the Gap junctions (22); still others may influence hormone biosynthesis. Such is the action of antithyroid agents (4,233,257), of nonsteroidal antiinflammatory agents on the synthesis of the eicosanoids (285), and of aminoglutethimide on steroidogenesis (286), a potential method of "chemical adrenalectomy" in patients with metastatic breast carcinoma (287) or with hyperaldosteronism (79).

Other examples are the production of male pseudohermaphroditism by a 5α-reductase inhibitor (288), the modulation of human endometrial estrogen synthetase (aromatase) by contraceptive agents (289,290), the treatment of precocious puberty in girls with the McCune-Albright syndrome with testolactone (an aromatase inhibitor) (29), the inhibition of proinsulin synthesis and processing by tris-(hydroxymethyl)-aminomethane (292), and the stimulation of proinsulin conversion by tolbutamide (293). Hormone secretion may also be modified pharmacologically. Well-known examples are the inhibition of ACTH secretion by glucocorticoids and the block of the cortisol feedback by metyrapone (4,233,257); the modulation of growth hormone, prolactin, gonadotropin, aldosterone, insulin, and glucagon secretion by monoamine receptor agonists and antagonists (82–86, 294–289); that of acetylcholine by ouabain and of acetylcholine, melatonin, T3, and T4 by calmodulin antagonists, such as trifluoroperazine and chlorpromazine (299–301); the selective inhibition of growth hormone and ectopic hormones by somatostatin analogues used in the therapy of acromegaly and of tumor-induced endocrinopathies (302,303); and the control of melatonin secretion and neurotransmitter release and reuptake by psychomimetic and neurotropic drugs (4,233,257).

The examples of drugs that interfere with receptor binding and signal transduction are equally numerous and include hormone antagonists that bind to the receptor without stimulating the normal postreceptor events; modified hormones with decreased, increased, prolonged, or more selective action; and receptor antibodies. Antagonists have been described for several steroid hormones, for the catecholamines, glucagon, ACTH, PTH, vasopressin, and LHRH (304–308). Some antagonists, such as the β-adrenergic blockers, are used commonly in clinical practice (4,233,257). Other inhibitors of receptor binding include snake venom, curarelike drugs, receptor antibodies and local anesthetics that act on the acetylcholine receptor (159), cholecystographic agents and sulfobromophthalein that interfere with T3 and T4 binding (309), and certain plant products that inhibit the binding of Graves's immunoglobulins (310). Still other drugs stimulate the activity of the receptor. Among these are many well-known adrenergic and cholinergic agonists and forskolin, a plant product that stimulates adenyl cyclase and has therapeutic potential in glaucoma and other diseases (311).

Finally, there are drugs that act at the postreceptor level. Among them are the modifiers of the intracellular concentration of calcium, such as chelators, ionophores, calcium channel blockers, and agents such as the phenothazines that inhibit the action of the Ca-calm complex (312–314); new compounds, such as methyl-palmoxyrate, which inhibits carnitine-palmitoyl transferase and and ketone body production (315); and old drugs, such as the sulfonylureas, for which new mechanisms of action have been proposed (222,316).

IX. Summary

The path of the endocrine signal from its origin to its extinction is long, complex, and hazardous (Table 1.1). A detailed discussion could easily have turned this chapter into a textbook of endocrinology rather than an introduction to a volume on hormone resistance and other endocrine paradoxes. I hope that the reader will consider my effort a useful guide to additional reading. For this purpose, I have added a bibliography which is necessarily extensive even though limited largely to recent papers and to a few papers of historic interest.

TABLE 1.1. Defects in the path of the endocrine signal. Biologic and experimental models, clinical syndromes.

I. Chemical control of morphogenesis
    A. Defective organogenesis
        1. Missing or defective genes
        2. Morphogenic peptides and their defects
        3. Environmental injury
    B. Hormone-secreting tumors derived from embryonic remnants
II. Hormone biosynthesis
    A. Quantitative defects
        1. Environmental factors
        2. Missing genes
        3. Inadequate or excessive gene expression
        4. Ectopic gene expression
    B. Qualitative defects
        1. Defective prohormone-encoding genes
        2. Defective enzyme-encoding genes
            a. Defective prohormone processing
            b. Defective conversion of metabolic precursors
III. Hormone release
        1. Defective feedback systems
        2. Autonomous secretion (tumors)
        3. Desensitization of the secretory cell
IV. Hormone transport
        1. Quantitative changes in transport proteins
        2. Qualitative changes in transport proteins
        3. Circulating hormone antibodies
V. Hormone binding and recognition
        1. Changes in the number of receptors
            a. Genetic defects
            b. Acquired defects
            c. Excessive down- and up-regulation
        2. Changes in affinity
            a. Defects of receptor structure
            b. Defects of hormone structure
            c. Negative cooperativity
        3. Defects due to limited receptor specificity
            a. Excessive crossover
            b. Excessive spillover
        4. Inhibition of binding
            a. Competitive inhibitors
            b. Receptor antibodies
        5. Changes in structure
            a. Genetic defects
            b. Defects caused by environmental factors
VI. Transduction of the endocrine signal
        1. Defective production of messengers
        2. Defects of effector enzymes
VII. Extinction of the endocrine signal
        1. Decreased secretion
            a. Feedback inhibition
            b. Desensitization of the secretory cell

TABLE 1.1. *Continued*

  2. Hormone reuptake
  3. Desensitization of the target cell
  4. Secretion of counterregulatory hormones
  5. Metabolic clearance
  6. Antibodies
 VIII. Drug-induced changes of the endocrine signal
  1. Destruction of the secretory tissue
  2. Disruption of paracrine relationships
  3. Block of hormone synthesis
  4. Altered hormone storage and release
  5. Altered hormone binding
  6. Increased or decreased postreceptor enzymatic activities

Addendum

No sooner had I submitted this manuscript to my coeditor for her constructive and much appreciated comments than several significant observations related to the journey of the endocrine signal came to my attention. Among the papers published before the arbitrary cutoff date of December 31, 1986, are a few dealing with the control of morphogenesis. One paper reviews the role of hormones in sexual development (317); a symposium on the interaction of developing neurons with their target tissues (318) discusses the role of neurotrophic factors in axonal growth and that of axonal contact with the target membrane in the survival of the neuron itself and in the development of the capacity to synthesize neurotransmitters and receptors; still another paper describes evidence that the adhesion of isolated ovine oligodendrocytes to a polylysine substratum activates the protein kinase C–mediated phosphorylation of myelin basic protein, a possible early step in myelin formation (319).

In other experiments, the hypogonadism and infertility of mice lacking a complete GnRH gene were corrected by "gene therapy" consisting of the microinjection of DNA fragments containing the missing gene into fertilized eggs (320). The hormonal regulation of gene expression was studied by examining the effect of estrogen on prolactin synthesis (321) and the role of calcium ions and cyclic AMP in regulating the level of proopiomelanocortin mRNA in rat pituitary (322). The tissue specificity of the posttranslational processing of prohormones was again underlined by the description of a pancreatic cell line that produces proglucagon-derived peptides that resemble those produced by the gut rather than those produced by the normal pancreatic A cells (323).

The role of protein kinase C activators and of arachidonic acid metabolites in the secretion of LHRH was discussed (324), and it was suggested that diacylglycerols may modulate insulin binding and receptor internalization and thus play a role in cell differentiation and proliferation (325). New evidence for receptor crossover was provided by the demonstration that somatomedin C can substitute for insulin in promoting the growth of cultured mammary epithelial cells (326). Two papers underline the significance of structure-function relationships by describing the synthesis of superactive glucagon analogues (327) and the difference in metabolic clearance of pituitary and serum TSH derived from euthyroid and hypothyroid rats, possibly due to changes in the carbohydrate moiety of the molecule (328). One paper illustrates the role of calmodulin as a mediator of hormone action by demonstrating that this calcium-binding protein influences the activity of the guanyl nucleotide–dependent adenylate cyclase of rat thymocytes, thus explaining the calcium-dependent nature of the stimulatory effect of T3 on this enzyme activity (329). Finally, one paper reports that a 20-h infusion of insulin resulting in plasma concentrations of approximately 30 mU/1 reduces the sensitivity of normal human subjects to the hormone, probably through a postreceptor mechanism (330).

References

1. Bernard C (1866) Leçons sur les Propriétés des Tissus Vivants. Germer Baillière, Paris.
2. Edelman GM (1985) Cell adhesion and the molecular process of morphogenesis. Annu Rev Biochem 54:136–169.
3. Gley MEE (transl. Fishberg M) (1918) The Internal Secretions. Wm Heineman, London.
4. Wilson JD, Foster DW (1985) (eds) Williams Textbook of Endocrinology. Saunders, Philadelphia.
5. Brown TR, Migeon CJ (1987) Androgen insensitivity syndromes: Paradox of phenotypic feminization with male genotype and normal testicular androgen secretion. In Cohen MP, Foà PP (eds): Hormone Resistance and Other Endocrine Paradoxes. Springer-Verlag, New York, pp 157–203.
6. Levi-Montalcini R, Booker B (1960) Excessive growth of the sympathetic ganglia evoked by a protein isolated from mouse salivary glands. Proc Natl Acad Sci USA 46:373–384.
7. Levi-Montalcini R, Booker B (1960) Destruction of the sympathetic ganglia in mammals by an antiserum to a nerve-growth protein. Proc Natl Acad Sci USA 46:384–391.
8. Shooter EM, Yankner BA, Landreth GE, Sutter A (1981) Biosynthesis and mechanism of action of nerve growth factor. Recent Prog Horm Res 37:417–446.
9. Haseltine FP, Ohno A (1981) Mechanisms of gonadal differentiation. Science 211:1272–1278. See also ibid., pp 1278–1324.
10. Dörner G (1980) Sexual differentiation of the brain. Vitam Horm 38:325–381.
11. Gurney ME, Heirich SP, Lee MR, Yin H-S (1986) Molecular cloning and expression of neuroleukin, a neurotrophic factor for spinal and sensory neurons. Science 234:566–574.
12. Pilar GR (1985) Functional interaction of developing neurons with their target tissues. A symposium. Fed Proc 44:2750–2779.
13. Audhya T, Kroon D, Heavner G, Viamontes G, Goldstein G (1986) Tripeptide structure of bursin, a selective B-cell-differentiating hormone of the bursa of Fabricius. Science 231:997–999.
14. Siegelman M, Bond MW, Gallatin WM, St John T, Smith HT, Fried VA,

Weissman IL (1986) Cell surface molecules associated with lymphocyte homing is a ubiquitinated branched-chain glycoprotein. Science 231:823–829.

15. Mosher DF (1984) Physiology of fibronectin. Annu Rev Med 35:561–565.
16. Orci L, Perrelet A (1977) Morphology of membrane systems in pancreatic islet. In Volk BW, Wellmann KF (eds): The Diabetic Pancreas. Plenum, New York, pp 171–210.
17. Andrew RD, MacVicar BA, Dudek FE, Hatton GI (1981) Dye transfer through gap junctions between neuroendocrine cells of rat hypothalamus. Science 211:1187–1189.
18. Loewenstein WR (1981) Junctional intercellular communication: The cell-to-cell membrane channel. Physiol Rev 61:829–913.
19. Revel J-P, Nicholson BJ, Yancey SB (1985) Chemistry of gap junctions. Annu Rev Physiol 47:263–279.
20. Veenstra RD, DeHaan RL (1986) Measurement of single channel currents from cardiac gap junctions. Science 233:972–974.
21. Verselis V, White RL, Spray DC, Bennett MVL (1986) Gap junctional conductance and permeability are linearly related. Science 234:461–464.
22. Spray DC, Bennett MVL (1985) Physiology and pharmacology of gap junctions. Annu Rev Physiol 47:282–303. See also ibid, pp 305–353.
23. Agrawal R, Daniel EE (1986) Control of gap junction formation in canine trachea by arachidonic acid metabolites. Am J Physiol 250:C495–C505.
24. Wade MH, Trosko JE, Schindler M (1986) A fluorescence photobleaching assay of gap junction–mediated communication between human cells. Science 232:525–528.
25. Saez JC, Spray DC, Nairn AC, Hertzberg E, Greengard P, Bennett VL (1986) cAMP increases junctional conductance and stimulates phosphorylation of the 27-kDa principal gap junction polypeptide. Proc Natl Acad Sci USA 83:2473–2477.
26. Dunbar JC, Walsh MF (1982) Glucagon and insulin secretion by dispersed islet cells: Possible paracrine relationships. Horm Res 16:257–267.
27. Pipeleers DG, in't Veld PA, Van de Winkel M, Maes E, Schuit FC, Gepts W (1986) A new in vitro model for the study of pancreatic A and B cells. Endocrinology 117:806–816. See also ibid., pp 817–823, 824–833.
28. Denef C, Andries M (1983) Evidence for paracrine interactions between gonadotrophs and lactotrophs in pituitary cell aggregates. Endocrinology 112:813–822.
29. Krieger DT, Perlow MJ, Gibson MJ, Davies TE, Zimmerman EA, Finn M, Charlton HM (1982) Brain grafts reverse hypogonadism of gonadotropin releasing hormone deficiency. Nature 298:468–471.
30. Bhasin S, Swerdloff RS (1985) Hypothalamic hypogonadism. In Cohen MP, Foà PP (eds): Special Topics in Endocrinology and Metabolism. Alan R. Liss, New York, Vol 7, pp 237–266.
31. Beamer WG, Eicher EM, Maltais LJ, Southard JL (1981) Inherited primary hypothyroidism in mice. Science 212:61–63.
32. Laron Z (1983) Deficiencies of growth hormone and somatomedin in man. In Cohen MP, Foà PP (eds): Special Topics in Endocrinology and Metabolism. Alan R. Liss, New York, Vol 5, pp 149–199.
33. Rogol AD, Blizzard RM, Foley TP Jr, Furlanetto R, Selden R, Mayo K, Thorner MO (1985) Growth hormone releasing hormone and growth hormone: Genetic studies in familial growth hormone deficiency. Pediatr Res 19:489–492.
34. Cheng TC, Beamer WG, Phillips JA III, Bartke A, Mallonee RL, Dowling C (1983) Etiology of growth hormone deficiency in Little, Ames and Snell dwarf mice. Endocrinology 113:1669–1678.

35. Eigenmann JE, Zanesco S, Arnold U, Froesch ER (1984) Growth hormone and insulin-like growth factor I in German shepherd dwarf dogs. Acta Endocrinol 105:289–293.
36. Carey DE (1985) Isolated ACTH deficiency in childhood, lack of response to corticotropin-releasing hormone alone or in combination with arginine vasopressin. J Pediatr 107:925–928.
37. Sandler R, Proudfoot GR (1985) Isolated ACTH deficiency contributing to frequent hypoglycemia in type 1 diabetes. Diabetes Care 8:302–304.
38. Rudman D, Kutner MH, Blackston RD, Cushman RA, Bain RP, Patterson JH (1981) Children with normal-variant short stature: Treatment with human growth hormone for six months. N Engl J Med 305:123–131.
39. Beck-Peccoz P, Amr S, Menezes-Ferreira MM, Faglia G, Weintraub BD (1985) Decreased receptor binding of biologically inactive thyrotropin in central hypothyroidism. Effect of treatment with thyrotropin-releasing hormone. N Engl J Med 312:1085–1090.
40. Tager HS (1987) Insulin gene mutations and abnormal products of the human insulin gene. In Cohen MP, Foà PP (eds): Hormone Resistance and Other Endocrine Paradoxes. Springer-Verlag, New York, pp 35–61.
41. Hammer RE, Palmiter RD, Brinster RL (1984) Partial correction of murine hereditary growth disorder by germ-line incorporation of a new gene. Nature 311:65–67.
42. Miller AD, Ong ES, Rosenfeld MG, Verma IM, Evans RM (1984) Infectious and selectable retrovirus containing an inducible rat growth hormone minigene. Science 225:993–998.
43. Nicolau G, Le Pape A, Soriano P, Fargette F, Jühel MF (1983) In vivo expression of rat insulin after intravenous administration of the liposome-entrapped gene for rat insulin I. Proc Natl Acad Sci USA 80:1068–1072.
44. Steiner DF, Oyer PE (1967) The biosynthesis of insulin and probable precursor of insulin by a human islet cell adenoma. Proc Natl Acad Sci USA 57:473–480.
45. Roberts JL, Chen C-LC, Eberwine JH, Evinger MJQ, Gee C, Herbert E, Schachter BS (1982) Glucocorticoid regulation of proopiomelanocortin gene expression in rodent pituitary. Recent Prog Horm Res 38:227–256.
46. Moore DD, Walker MD, Diamond DJ, Conkling MA, Goodman HM (1982) Structure, expression and evolution of growth hormone genes. Recent Prog Horm Res 38:197–225.
47. Schoelson S, Polonsky KS, Nakabayashi T, Jaspan JB, Tager HS (1986) Circulating forms of somatostatinlike immunoreactivity in human plasma. Am J Physiol 250:E428–E434.
48. Lechan RM, Wu P, Jackson IMD, Wolf H, Cooperman S, Mandel G, Goodman RH (1986) Thyrotropin-releasing hormone precursor: Characterization in rat brain. Science 231:159–161.
49. Heinrich G, Gros P, Lund PK, Bentley RC, Habener JF (1984) Pre-proglucagon messenger ribonucleic acid: Nucleotide and encoded amino acid sequences of the rat complementary deoxyribonucleic acid. Endocrinology 115:2176–2181.
50. Mojsov S, Heinrich G, Wilson IB, Ravazzola M, Orci L, Habener JF (1986) Preproglucagon gene expression in pancreas and intestine diversifies at the level of post-translational processing. J Biol Chem 261:11880–11889.
51. Kronenberg HM, Igarashi T, Freeman MW, Okazaki T, Brand SJ, Wiren KW, Potts JT Jr (1986) Structure and expression of the human parathyroid hormone gene. Recent Prog Horm Res 42:641–663. See also ibid., pp 665–703.
52. Sonnenberg H, Maack T (1986) Atrial natriuretic factors I and II. A symposium. Fed Proc 45:2079–2147.

53. Blaine EH (1986) Atrial natriuretic factor. A symposium. Fed Proc 45:2360–2391
54. Atlas SA (1986) Atrial natriuretic factor: A new hormone of cardiac origin. Recent Prog Horm Res 42:207–249.
55. Elbein SC, Gruppuso P, Schwartz R, Skolnick M, Permutt MA (1985) Hyperproinsulinemia in a family with a proposed defect in conversion is linked to the insulin gene. Diabetes 34:821–824.
56. Foà PP (1979) Clinical states associated with glucagon excess or glucagon deficiency. In Cohen MP, Foà PP (eds): Special Topics in Endocrinology and Metabolism. Alan R. Liss, New York, Vol. 1, pp 39–54.
57. Seino Y, Ishida H, Kitano N, Takeda J, Taminato S, Matsukura S, Imura H (1985) The heterogeneity of immunoreactive glucagon in a patient with glucagonoma. Biomed Res 6 (Suppl):45–50.
58. Uttenthal LO, Ghiglione M, George SK, Bishop AE, Polak JM, Bloom SR (1985) Molecular forms of glucagon-like peptide-1 in human pancreas and glucagonomas. J Clin Endocrinol Metab 61:472–479.
59. Baldissera FGA, Holst JJ (1986) Glycentin 1-61 probably represents a major fraction of glucagon-related peptides in plasma of anaesthetized uraemic dogs. Diabetologia 29:462–467.
60. Valenta LJ, Sigel MB, Lesniak MA, Elias AN, Lewis UJ, Friesen HG, Kershnar AK (1985) Pituitary dwarfism in a patient with circulating abnormal growth hormone polymers. N Engl J Med 312:214–217.
61. Spitz IM, LeRoith D, Hirsch H, Caryon P, Pekonen F, Liel Y, Sobel R, Chorer Z, Weintraub B (1981) Increased high-molecular-weight thyrotropin with impaired biologic activity in a euthyroid man. N Engl J Med 304:278–282.
62. Rosenfield RL, Miller WL (1983) Congenital adrenal hyperplasia. In Mahesh VB, Greenblatt RB (eds): Hisutism and Virilism: Pathogenesis, Diagnosis and Management. John Wright, Littleton, MA, pp 87–119.
63. Peterson RE, Imperato-McGinley J, Gautier T, Shackleton C (1985) Male pseudohermaphroditism due to multiple defects in steroid-bio-synthetic microsomal mixed-function oxydases: A new variant of congenital adrenal hyperplasia. N Engl J Med 313:1182–1191.
64. Imperato-McGinley J, Peterson RE, Gautier T, Sturla E (1979) Androgens and the evolution of male-gender identity among male pseudohermaphrodites with 5α-reductase deficiency. N Engl J Med 300;1233–1237.
65. Robertson D, Goldberg MR, Onrot J, Hollister AS, Wiley R, Thompson JG Jr, Robertson RM (1986) Isolated failure of autonomic noradrenergic neurotransmission: Evidence for impaired β-hydroxylation of dopamine. N Engl J Med 314:1494–1497.
66. Bartus RT, Dean RL, Beer B, Lippa AS (1982) The cholinergic hypothesis of geriatric memory dysfunction. Science 217:408–417.
67. Han VKM, Hynes MA, Jin C, Towle AC, Lauder JM, Lund PK (1986) Cellular localization of proglucagon/glucagon-like peptide I messenger RNAs in rat brain. J Neurosci Res 16:97–107.
68. Krieger DT (1984) Brain peptides. Vitam Horm 41:1–50.
69. Guillemin R, Brazeau P, Böhlen P, Esch F, Ling N, Wehrenberg WB, Bloch B, Mougin C, Zeytin F, Baird A (1984) Somatocrinin: The growth hormone releasing factor. Recent Prog Horm Res 40:233–299.
70. Howlett TA, Rees LH (1985) Ectopic hormones. In Cohen MP, Foà PP (eds): Special Topics in Endocrinology and Metabolism. Alan R. Liss, New York, Vol 7, pp 2–44.

71. Huff KK, Kaufman D, Gabbay KH, Spencer EM, Lippman ME, Dickson RN (1986) Secretion of an insulin-like growth factor-I-related protein by human breast cancer cells. Cancer Res 46:4613–4619.
72. Reiter RJ (ed) (1984) The Pineal Gland. Raven Press, New York.
73. Nestel PJ, Connor WE, Reardon MF, Connor S, Wong S, Boston R (1984) Suppression by diets rich in fish oil of very-low-density lipoprotein production in man. J Clin Invest 74:82–89.
74. Foà PP (1985) The many faces of glucagon. Biomed Res 6 (Suppl):3–13.
75. Subramanian MG, Gala RR (1986) Do prolactin levels measured by RIA reflect biologically active prolactin? J Clin Immunoass 9:42–51.
76. Pollard HB, Pazoles CJ, Creutz CE (1981) Mechanism of calcium action and release of vesicle-bound hormones during exocytosis. Recent Prog Horm Res 37:299–332.
77. Griffith G, Simons K (1986) The *trans* Golgi network: Sorting at the exit site of the Golgi complex. Science 234:438–443.
78. Polonsky K, Frank B, Pugh W, Addis A, Harrison T, Meier P, Tager H, Rubenstein A (1986) The limitation to and valid use of C-peptide as a marker of the secretion of insulin. Diabetes 35:379–386.
79. Schambelan M, Sebastian A (1987) States of aldosterone deficiency or pseudodeficiency. In Cohen MP, Foà PP (eds): Hormone Resistance and Other Endocrine Paradoxes. Springer-Verlag, New York, pp 204–230.
80. Okusa M, Bia MJ (1987) Bartter's syndrome. In Cohen MP, Foà PP (eds): Hormone Resistance and Other Endocrine Paradoxes. Springer-Verlag, New York, pp 231–263.
81. Yamada K, Hanafusa T, Fujno-Hurihara H, Miyazaki A, Nakajima J, Kono N, Nonaka K, Tarui S (1986) Demonstration of islet cell surface antibodies in sera of New Zealand black mice and inhibitory effect on insulin release. Diabetes 35:1262–1267.
82. Frawley LS, Clark CL (1986) Ovine prolactin (PRL) and dopamine preferentially inhibit PRL release from the same subpopulation of rat mammotropes. Endocrinology 119:1462–1466.
83. Pontiroli AE, Alberetto M, Pelicciotta G, De Castro e Silva E, De Pasqua A, Girari AM, Pozza G (1980) Interaction of dopaminergic and antiserotoninergic drugs in the control of prolactin and LH release in normal women. Acta Endocrinol 93:271–276.
84. Jarjour LT, Handelsman DJ, Raum WJ, Swerdloff RS (1986) Mechanism of action of dopamine on the in vitro release of gonadotropin-releasing hormone. Endocrinology 119:1726–1732.
85. Cronin MJ, Thorner MO, Hellmann P, Rogol AD (1984) Bromocriptine inhibits growth hormone release from rat pituitary cells in primary culture. Proc Soc Exp Biol Med 175:191–195.
86. Pontiroli AE, Micossi P, Foà PP (1980) Glucagon (IRG) and insulin (IRI) response to arginine (ARG) in the rat pancreas in vitro: Effect of histamine (HA) and serotonin (5HT). Horm Metab Res 12:703–704.
87. Wilczynski EA, Osmond DH (1986) Evidence for β-adrenergic regulation of renal and extrarenal plasma prorenin and renin in dogs. Proc Soc Exp Biol Med 182:208–214.
88. Carey RM, Sen S (1986) Recent progress in the control of aldosterone secretion. Recent Prog Horm Res 42:251–296.
89. Missale C, Liberini P, Memo M, Carruba MD, Spano P (1986) Characterization of dopamine receptors associated with aldosterone secretion in rat adrenal glomerulosa. Endocrinology 119:2227–2232.
90. Foà C (1912) Ipertrofia dei testicoli e della cresta dopo asportazione della ghiandola pineale nel gallo. Pathologica 90:1–29.

22   P. P. Foà

91. Aschoff J, Ceresa F, Halberg F (eds) (1974) Chronobiological aspects of endocrinology. Chronobiologia 1 (Suppl 1):1–509.
92. Lincoln DW, Fraser HM, Lincoln GA, Martin GB, McNeilly AS (1985) Hypothalamic pulse generators. Recent Prog Horm Res 41:369–419.
93. Vance ML, Kaiser DL, Evans WS, Furlanetto R, Vale W, Rivier J, Thorner MO (1985) Pulsatile growth hormone secretion in normal man during a continuous 24-hour infusion of human growth hormone releasing factor (1-40). J Clin Invest 75:1584–1590.
94. Jaspan JB, Lever E, Polonsky KS, Van Cauter E (1986) In vivo pulsatility of pancreatic islet peptides. Am J Physiol 251:E215–E226.
95. Sirek A, Vaitkus P, Norwich KH, Sirek OV, Unger RH, Harris V (1985) Secretory patterns of glucoregulatory hormones in prehepatic circulation of dogs. Am J Physiol 249:E34–E42.
96. Weigle DS, Goodner CJ (1986) Evidence that the physiological pulse frequency of glucagon secretion optimizes glucose production by perfused rat hepatocytes. Endocrinology 118:1606–1613.
97. Marsh BD, Marsh DJ, Bergman RN (1986) Oscillations enhance the efficiency and stability of glucose disposal. Am J Physiol 250:E576–E582.
98. Leyendecker G, Wildt L, Hausmann M (1980) Pregnancies following chronic intermittent (pulsatile) administration of Gn-RH by means of a portable pump ("Zyklomat"). A new approach to the treatment of infertility in hypothalamic amenorrhea. J Clin Endocrinol Metab 51:1214–1216.
99. Hoffman AR, Crowley WF Jr (1982) Induction of puberty in men by long-term pulsatile administration of low-dose gonadotropin-releasing hormone. N Engl J Med 307:1237–1241.
100. Creutzfeldt W, Ebert R (1985) New developments in the incretin concept. Diabetologia 28:565–573.
101. McDonald TJ, Dupre J, Greenberg GR, Tepperman F, Brooks B, Tatemoto K, Mutt V (1986) The effect of galanin on canine plasma glucose and gastroenteropancreatic hormone responses to oral nutrient and intravenous arginine. Endocrinology 119:2340–2345.
102. Jarrousse C, Niel H, Andousset-Puech MP, Martinez J, Bataille D (1986) Oxyntomodulin and its C-terminal octapeptide inhibit liquid meal-stimulated acid secretion. Peptides (Fayetteville NY) 7 (Suppl 1):253–256.
103. La Barre J (1927) Sur l'augmentation de la teneur en insuline du sang veineux pancréatique après excitation du nerf vague. C R Seances Soc Biol (Paris) 96:193–196.
104. Nauck M, Stöckmann F, Ebert R, Creutzfeldt W (1986) Reduced incretin effect in type 2 (non-insulin-dependent) diabetes. Diabetologia 29:46–52.
105. Means AR (1981) Calmodulin: Properties, intracellular localization and multiple roles in cell regulation. Recent Prog Horm Res 37:333–367.
106. Silinsky EM (1982) Recent approaches to secretion and stimulus-secretion coupling. A symposium. Fed Proc 41:2169–2192.
107. Schettini G, Cronin MJ, Hewlett EM, Thorner MO, MacLeod RM (1984) Human pancreatic tumor growth hormone–releasing factor stimulates anterior pituitary adenylate cyclase activity, adenosine 3',5'-monophosphate accumulation and growth hormone release in a calmodulin-dependent manner. Endocrinology 115:1308–1314.
108. Marone G, Columbo M, Poto S, Giugliano R, Condorelli M (1986) The possible role of calmodulin (CaM) in the control of histamine release from human basophil leukocytes. Life Sci 39:911–922.
109. Bridges CDB, Peters T Jr, Smith JE, Goodman, DeWS, Fong S-L, Griswold MD, Musto NA (1986) Biosynthesis and secretion of transport proteins. Interstitial and serum retinol-binding proteins, transhyretin, transferrin,

serum albumin and extracellular sex steroid-binding proteins. Fed Proc 45:2291–2303.

110. Süteri PK, Murai JT, Hammond GL, Nisker JA, Raymoure WJ, Kuhn RW (1982) The serum transport of steroid hormones. Recent Prog Horm Res 38:457–510.

111. Westphal U (1985) Steroid-Protein Interactions II. Springer-Verlag, New York.

112. Romanus JA, Terrell JE, Yang YW-H, Nissley SP, Rechler MM (1986) Insulin-like growth factor carrier proteins in neonatal and adult rat serum are immunologically different. Demonstration using a new radioimmunoassay for the carrier protein from BRL-3A rat liver cells. Endocrinology 118:1743–1758.

113. Grimaldi S, Bartalena L, Carlini F, Robbins J (1986) Purification and partial characterization of a novel thyroxine-binding protein (27K protein) from human plasma. Endocrinology 118:2362–2369.

114. Murata Y, Takamatsu J, Refetoff S (1986) Inherited abnormality of thyroxine-binding globulin with no demonstrable thyroxine-binding activity and high serum levels of denatured thyroxine-binding globulin. N Engl J Med 314:694–699.

115. Safran M, Braverman LE (1987) Hyperthyroxinemia. In Cohen MP, Foà PP (eds): Hormone Resistance and Other Endocrine Paradoxes. Springer-Verlag, New York, pp 62–91.

116. Glass AR, Young RA, Anderson J (1986) Decreased serum 3,5,3'-triiodothyronine (T3) and abnormal serum binding of T3 in calorie-deficient rats: Adaptation after chronic underfeeding. Endocrinology 118:2464–2469.

117. Chopra IJ, Solomon DH, Chua Teco GN, Eisenberg JB (1982) An inhibitor of the binding of thyroid hormones to serum proteins is present in extrathyroidal tissues. Science 215:407–409.

118. Hennemann G, Krenning EP, Polhuys M, Mol JA, Bernard BF, Visser TJ, Docter R (1986) Carrier-mediated transport of thyroid hormone into rat hepatocytes is rate-limiting in total cellular uptake and metabolism. Endocrinology 119:1870–1872.

119. Frank HJL, Pardridge WM, Morris WL, Rosenfeld RG, Choi TB (1986) Binding and internalization of insulin and insulin-like growth factors by isolated brain microvessels. Diabetes 35:654–661.

120. Banks WA, Kastin AJ, Fischman AJ, Coy DH, Strauss SL (1986) Carrier-mediated transport of enkephalins and N-Tyr-MIF-1 across blood-brain barrier. Am J Physiol 251:E477–E482.

121. Pardridge WM, Landaw EM (1985) Testosterone transport in brain: Primary role of plasma protein-bound hormone. Am J Physiol 249:E534–E542.

122. Spitzer JA (ed) (1983) The physiology and biochemistry of receptors: 1983 refresher course syllabus. Physiologist 26:185–205.

123. Roth J, Lesniak MA, Bar RS, Muggeo M, Megyesi K, Harrison LC, Flier JS, Wachslicht-Rodbard H, Gorden P (1979) An introduction to receptors and receptor disorders. Proc Soc Exp Biol Med 162:3–12. See also ibid., pp 13–21.

124. Jensen EV, Greene GL, Closs LE, DeSombre ER, Nadji M (1982) Receptors reconsidered: A 20-year perspective. Recent Prog Horm Res 38:1–40.

125. Schmidt TJ, Litwack G (1982) Activation of the glucocorticoid-receptor complex. Physiol Rev 62:1131–1192.

126. Lefkowitz RJ, Caron MC, Stiles GL (1984) Mechanisms of membrane-receptor regulation. Biochemical, physiological, and clinical insights derived from studies of the adrenergic receptors. N Engl J Med 310:1570–1579.

127. Exton JH (1985) Mechanisms involved in α-adrenergic phenomena. Am J Physiol 248:E633–E647.

128. Posner BJ, Bergeron JJM, Josefberg Z, Kahn MN, Kahn RJ, Patel BA, Sikstrom RA, Verma AK (1981) Polypeptide hormones: Intracellular receptors and internalization. Recent Prog Horm Res 37:539–582.
129. Forsayeth J, Maddux B, Goldfine ID (1986) Biosynthesis and processing of the human insulin receptor. Diabetes 35:837–846.
130. Roth J, Taylor SI (1982) Receptors for peptide hormones, alterations in diseases of humans. Annu Rev Physiol 44:639–651.
131. Brickman AS, Carlson HE (1987) Pseudohypoparathyroidism: Target organ resistance to parathyroid hormone and other metabolic defects. In Cohen MP, Foà PP (eds): Hormone Resistance and Other Endocrine Paradoxes. Springer-Verlag, New York, pp 92–119.
132. Grunfeld C (1984) Insulin resistance: Pathophysiology, diagnosis, and therapeutic implications. In Cohen MP, Foà PP (eds): Special Topics in Endocrinology and Metabolism. Alan R. Liss, New York, Vol 6, pp 194–240.
133. Hedo JA, McElduff A, Taylor SI (1984) Defects in receptor biosynthesis in patients with genetic forms of extreme insulin resistance. Trans Assoc Am Physicians 97:151–160.
134. Craig JW, Larner J, Locker EF, Widom B, Elders MJ (1984) Mechanisms of insulin resistance in cultured fibroblasts from a patient with leprechaunism: Impaired post-binding actions of insulin and multiplication-stimulating activity. Metab Clin Exp 33:1084–1086.
135. Jansson J-O, Downs TR, Beamer WG, Frohman LA (1986) Receptor-associated resistance to growth hormone–releasing factor in dwarf "little" mice. Science 232:511–512.
136. McLeod JF, Haddad JG Jr (1987) Syndromes of vitamin D resistance. In Cohen MP, Foà PP (eds): Hormone Resistance and Other Endocrine Paradoxes. Springer-Verlag, New York, pp 120–156.
137. Lipsett MB, Chrousos GP, Tomita M, Brandon DD, Loriaux DL (1985) The defective glucocorticoid receptor in man and nonhuman primates. Recent Prog Horm Res 41:199–247.
138. Yamauchi K, Hashizume K (1986) Glucagon alters insulin binding to isolated rat epididymal adipocytes: Possible role of adenosine 3',5'-monophosphate in modification of insulin action. Endocrinology 119:218–223.
139. Walsh MF, Dunbar JC (1984) Glucagon binding to liver membranes of Mt-T-W15 tumor-bearing and hypophysectomized rats. A possible role for insulin and growth hormone. Diabetes 33:978–983.
140. Singh P, Muldoon TG (1986) Baseline and leuteinizing hormone–releasing hormone. Perturbed patterns of estrogen receptor distribution in anterior pituitary cell nuclei. Endocrinology 118:2355–2361.
141. Chung CS, Etherton TD (1986) Characterization of porcine growth hormone (pGH) binding to porcine liver microsomes: Chronic administration of pGH induces pGH binding. Endocrinology 119:780–786.
142. Soloff MS, Alexandrova M, Fernstrom MJ (1979) Oxytocin receptors: Triggers for parturition and lactation? Science 104:1313–1315.
143. Horwitz KB, Wei LL, Sedlacek SM, D'Arville CN (1985) Progestin action and progesterone receptor structure in human breast cancer: A review. Recent Prog Horm Res 41:249–316.
144. Douglas JG (1986) Regulation of angiotensin receptors. News Physiol Sci 1:67–69.
145. Israel A, Saavedra JM, Plunkett L (1985) Water deprivation upregulates angiotensin II receptors in rat anterior pituitary. Am J Physiol 248:E264–E267.
146. DeGroot LJ, Rue PA (1983) Pathophysiologic control of nuclear triiodothyronine receptor capacity. Acta Endocrinol (Copenh) 104:57–63.

147. Khan SG, Boyle PC, Lachance PA (1986) Decreased triiodothyronine binding to isolated nuclei from livers of preobese and obese (ob/ob) mice. Proc Soc Exp Biol Med 182:84–87.
148. Begum N, Tepperman HW, Tepperman J (1985) Insulin-induced internalization and replacement of insulin receptors in adipocytes of rats adapted to fat feeding. Diabetes 34:1272–1277.
149. Tanti J-F, Grémeaux T, Brandenburg D, Van Obberghen E, Le Marchand-Brustel Y (1986) Brown adipose tissue in lean and obese mice. Insulin-receptor binding and tyrosine kinase activity. Diabetes 35:1243–1248.
150. Mash DC, Flynn DD, Potter LT (1985) Loss of M2 muscarine receptors in the cerebral cortex in Alzheimer disease and experimental cholinergic denervation. Science 228:1115–1117.
151. Moncada VY, Hedo JA, Serrano-Rios M, Taylor SI (1986) Insulin-receptor biosynthesis in cultured lymphocytes from an insulin-resistant patient (Rabson-Mendenhall syndrome). Evidence for defect before insertion of receptor into plasma membrane. Diabetes 35:802–807.
152. Grünberger G, Taylor SI, Dons RF, Gorden P (1983) Insulin receptors in normal and disease states. Clin Endocrinol Metab 12:191–219.
153. Pou M-A, Bismuth J, Gharbi-Chihi J, Torresani J, Ghiringhelli O, Savary J (1986) Triiodothyronine (T3)-induced down-regulation of the nuclear T3 receptor in mouse preadipocyte cell liver. Endocrinology 119:2360–2367.
154. Snavely MD, Mahan LC, O'Connor DT, Insel PA (1983) Selective down-regulation of adrenergic receptor subtypes in tissues from rats with pheochromocytoma. Endocrinology 113:354–361.
155. Aarons RD, Nies AS, Gerber JG, Molinoff PB (1983) Decreased beta adrenergic receptor density on human lymphocytes after chronic treatment with agonists. J Pharmacol Exp Ther 224:1–6.
156. Brodde O-E, Kretsch R, Ikezono K, Zerkowski H-R, Reidemeister JC (1986) Human β-adrenoreceptors: Relation of myocardial and lymphocyte β-adrenoreceptor density. Science 231:1584–1585.
157. Bar RS, Levis WR, Rechler MM, Harrison LC, Siebert C, Podskalny J, Roth J, Muggeo M (1978) Extreme insulin resistance in ataxia telangiectasia. Defect in affinity of insulin receptor. N Engl J Med 298:1164–1171.
158. Vincent A (1980) Immunology of acetylcholine receptors in relation to myasthenia gravis. Physiol Rev 60:756–824.
159. Peper K, Bradley RJ, Dreyer F (1982) The acetylcholine receptor at the neuromuscular junction. Physiol Rev 62:1271–1340.
160. Kahn CR (1985) The molecular mechanism of insulin action. Annu Rev Med 36:429–451.
161. Inomata H, Sasaki N, Tamaru K, Ushiku H, Nimi H, Nakajima H (1986) Relationship between potency of blocking type thyrotropin-binding inhibitor immunoglobulin in three women with primary myxedema and thyroid function of their neonates. Endocrinol Jpn 33:353–359.
162. Ealey PA, Valente MA, Ekins RP, Kohn LD, Marshall NJ (1984) Characterization of monoclonal antibodies raised against solubilized thyrotropin receptors in a cytochemical bioassay for thyroid stimulators. Endocrinology 116:124–131.
163. Adams DD (1980) Thyroid-stimulating autoantibodies. Vitam Horm 38:120–203.
164. De Pirro R, Roth RA, Rossetti L, Godfine ID (1984) Characterization of the serum from a patient with insulin resistance and hypoglycemia. Evidence for multiple populations of insulin receptor antibodies with different receptor binding and insulin mimicking activities. Diabetes 33:301–304.
165. Morgan DO, Roth RA (1986) Mapping surface structures of the human

insulin receptor with monoclonal antibodies: Localization of main immunogenic regions to the receptor kinase domain. Biochemistry 25:1364–1371.
166. Boland D, Lee PWK, Goren HJ (1986) A study of insulin receptor structure function with monoclonal antibodies: Preparation of anti-insulin receptor antibodies. Proc West Pharmacol Soc 29:451–453.
167. Kobayashi M, Haneda M, Maegawa H, Watanabe N, Takada Y, Shigeta Y, Inouye K (1984) Receptor binding and biological activity of [Ser B^{24}]-insulin, an abnormal mutant insulin. Biochem Biophys Res Commun 119:49–57.
168. McKee RL, Pelton JT, Trivedi D, Hruby VJ, Sueira-Diaz J, Coy DH (1986) Receptor binding and adenylate cyclase activities of glucagon analogs modified in the N-terminal region. Biochemistry 25:1650–1656.
169. Wood SP, Tickle IJ, Treharne AM, Pitts JE, Marcarenhas Y, Li JY, Husain J, Cooper S, Blundell TL, Hruby VJ, Buku A, Fischman AJ, Wyssbrod HR (1986) Crystal structure analysis of deamino-oxytocin: Conformational flexibility and receptor binding. Science 232:633–636.
170. Maletti M, Carlquist M, Portha B, Kergoat M, Mutt V, Rosselin G (1986) Structural requirements for gastric inhibitory polypeptide (GIP) receptor binding and stimulation of insulin release. Peptides (Fayetteville NY) 7 (Suppl 1):75–78.
171. Robberecht P, Waelbroeck M, Camus J-C, DeNeef P, Coy DH, Christophe J (1986) Comparative efficacy of seven synthetic glucagon analogs, modified in position 1, 2, and/or 12, on liver and heart adenylate cyclase from rat. Peptides (Fayetteville NY) 7 (Suppl 1):109–112.
172. Luborsky JL, Slater WT, Behrman HR (1984) Luteinizing hormone (LH) receptor aggregation: Modification of ferritin-LH binding and aggregation by prostaglandin F$_{2\alpha}$ and ferritin-LH. Endocrinology 115:2217–2226.
173. Fletcher JM, Basset JM (1986) Increased placental growth and raised plasma glucocorticoid concentrations in fetal rabbits injected with insulin in utero. Horm Metab Res 18:441–445.
174. Freinkel N (1980) The Banting lecture 1980: Of pregnancy and progeny. Diabetes 29:1023–1035.
175. King GL, Goodman AD, Buzney S, Mones A, Kahn CR (1985) Receptors and growth-promoting effects of insulin and insulin-like growth factors on cells from bovine retinal capillaries and aorta. J Clin Invest 75:1028–1036.
176. Capron L, Jarnet J, Kazandjian S, Housset E (1986) Growth-promoting effects of diabetes and insulin on arteries. An in vivo study of rat aorta. Diabetes 35:973–978.
177. Silardi S, Cacciari E, Ballardini D, Righetti F, Capelli M, Cicognani A, Zucchini S, Natali G, Tassinari D (1986) Relationships between growth factors (somatomedin-C and growth hormone) and body development, metabolic control and retinal changes in children and adolescents with IDDM. Diabetes 35:832–836.
178. Tomita M, Hirata Y, Uchihashi M, Fujita T (1986) Characterization of epidermal growth factor receptors in cultured vascular smooth muscle cells of rat aorta. Endocrinol Jpn 33:177–184.
179. Greene GL, Gilna P, Waterfield M, Baker A, Hort Y, Shine J (1986) Sequence and expression of human estrogen receptor complementary DNA. Science 231:1150–1154.
180. Perrot-Applanat M, Logeat F, Groyer-Picard MT, Milgrom E (1985) Immunocytochemical study of mammalian progesterone receptor using monoclonal antibodies. Endocrinology 116:1473–1484.
181. Weinberger C, Hollenberg SM, Ong ES, Harmon JM, Brower ST, Cidlowski J, Thompson EB, Rosenfeld MG, Evans RM (1985) Identification of human

glucocorticoid receptor complementary DNA clones by epitope selection. Science 228:740–742.

182. Raftery MA, Hunkapiller MW, Strader CD, Hood LE (1980) Acetylcholine receptor: Complex of homologous subunits. Science 208: 1454–1457.

183. Conti-Tronconi BM, Raftery MA (1982) Cholinergic receptor: Correlation of molecular structure with functional properties. Annu Rev Biochem 51:491–530.

184. Iyengar R, Herberg JT (1984) Structural analysis of the hepatic glucagon receptor. Identification of a guanine nucleotide–sensitive hormone-binding region. J Biol Chem 259:5222–5229.

185. Ulrich A, Bell JR, Chen EY, Herrera R, Petruzzelli LM, Dull TJ, Gray A, Coussens L, Liao YC, Tsubokawa M, Mason A, Seeburg PH, Grunfeld C, Rosen OM, Ramachandran J (1985) Human insulin receptor and its relationship to the tyrosine kinase family of oncogenes. Nature 313:756–761.

186. Czech MP, Massague J, Yu K, Oppenheimer CL, Mottola C (1984) Subunit structures and actions of the receptors for insulin and the insulin-like growth factors. In Federlin K, Scholtholt J (eds): The Importance of Islets of Langerhans for Modern Endocrinology. Raven Press, New York, pp 41–53.

187. Rotwein P, Pollock KM, Didier DK, Krivi GG (1986) Organization and sequence of the human insulin-like growth factor I gene. Alternative RNA processing produces two insulin-like growth factor I precursor peptides. J Biol Chem 261:4828–4832.

188. Abraham JA, Mergia A, Whang JL, Tumolo A, Friedman J, Hjerild KA, Gospodarowicz D, Fiddes JC (1986) Nucleotide sequence of a bovine clone encoding the angiogenic protein, basic fibroblast growth factor. Science 233:545–548.

189. Jaye M, Howk R, Burgess W, Ricca GA, Chiu I-M, Ravera MW, O'Brien SJ, Modi WS, Maciag T, Drohan WN (1986) Human endothelial cell growth factor: Cloning, nucleotide sequence and chromosome localization. Science 233:541–545.

190. Betsholtz C, Johnsson A, Heldin C-H, Westermark B, Lind P, Urdea MS, Eddy R, Shows TB, Philpott K, Mellor AL, Knott TJ, Scott J (1986) cDNA sequence and chromosomal localization of human platelet derived growth factor A-chain and its expression in tumor cell lines. Nature 320:695–699.

191. Downward J, Yarden Y, Mayes E, Scrace G, Totty N, Stockwell P, Ullrich A, Schlessinger J, Waterfield MD (1984) Close similarity of epidermal growth factor receptor and v-erb B oncogene protein sequences. Nature 307:521–527.

192. Coussens L, Yang-Feng TL, Liao Y-C, Chen E, Gray A, McGrath J, Seeburg PH, Libermann TA, Schlessinger J, Francke U, Levinson A, Ullrich A (1985) Tyrosine kinase receptor with extensive homology to EGF receptor shares chromosomal location with neu oncogene. Science 230:1132–1139.

193. Akiyama T, Sudo C, Ozawara H, Toyoshima K, Yamamoto T (1986) The product of the human c-erb B-2 gene: A 185-kilodalton glycoprotein with tyrosine kinase activity. Science 232:1644–1646.

194. Yang-Feng TL, Francke U, Ullrich A (1985) Gene for human insulin receptor: Localization to site on chromosome 19 involved in pre-B-cell leukemia. Science 228:728–731.

195. Roth RA, Cassell DJ (1983) Insulin receptor: Evidence that it is a protein kinase. Science 219:299–301.

196. Graves CB, Goewert RR, McDonald JM (1985) The insulin receptor contains a calmodulin-binding domain. Science 230:827–829.

197. Rees Jones RW, Quarum ML, Taylor SI (1985) An endogenous substrate for

the insulin receptor-associated tyrosine kinase. J Biol Chem 260:4461–4467.

198. Carrascosa JM, Wieland OH (1986) Evidence that (a) serine specific protein kinase(s) different from protein kinase C is responsible for the insulin-stimulated actin phosphorylation by placental membranes. FEBS Lett 201:81–86.
199. Olefsky JM, Kolterman OG (1981) Mechanisms of insulin resistance in obesity and non-insulin-dependent (type II) diabetes. Am J Med 70:151–168.
200. Okamoto M, White MF, Maron R, Kahn CR (1986) Autophosphorylation and kinase activity of insulin receptor in diabetic rats. Am J Physiol 251:E542–E550.
201. Birnbaumer L, Codina J, Mattera R, Cerione RA, Hildebrandt JD, Sunyer T, Rojas FJ, Caron MG, Lefkowitz RJ, Iyengar R (1985) Regulation of hormone receptors and adenylyl cyclase by guanine nucleotide binding N proteins. Recent Prog Horm Res 41:41–99.
202. Roberts J, Steinberg GM (1986) Effects of aging on adrenergic receptors. A symposium. Fed Proc 45:40–64.
203. Kadata T, Ui M (1982) Direct modification of the membrane adenylate cyclase system by islet-activating protein due to ADP-ribosylation of a membrane protein. Proc Natl Acad Sci USA 79:3129–3133.
204. Spiegel AM, Gierschik P, Levine MA, Down RW Jr (1985) Clinical implication of guanine nucleotide–binding proteins as receptor-effector couplers. N Engl J Med 312:26–33.
205. Farfel Z, Cohen Z (1984) Adenylate cyclase in the maturing human reticulocyte: Selective loss of the catalytic unit, but not of the receptor-cyclase coupling protein. Eur J Clin Invest 14:79–82.
206. Robberecht P, Gillard M, Waelbroeck M, Camus J-C, DeNeef P, Christophe J (1986) Alterations of rat cardiac adenylate cyclase activity with age. Eur J Pharmacol 126:91–95.
207. Rodan SB, Rodan GA (1986) Dexamethasone effects on β-adrenergic receptors and adenylate cyclase regulatory proteins Gs and Gi in ROS 17/2.8 cells. Endocrinology 118:2510–2518.
208. Faye JC, Fargin A, Bayard F (1986) Dissimilarities between the uterine estrogen receptor in cytosol of castrated and estradiol-treated rats. Endocrinology 118:2276–2283.
209. Mariash CN, Oppenheimer JH (1982) Thyroid hormone–carbohydrate interaction at the hepatic nuclear level. Fed Proc 41:2671–2676.
210. Marver D (1980) Aldosterone action in target epithelia. Vitam Horm 38:57–117.
211. Eisen LP, Harmon JM (1986) Activation of the rat kidney mineralocorticoid receptor. Endocrinology 119:1419–1426.
212. Dickson RB, McManaway ME, Lippman ME (1986) Estrogen-induced factors of breast cancer cells partially replace estrogen to promote tumor growth. Science 232:1540–1543.
213. Bigsby RM, Cunha GR (1986) Estrogen stimulation of deoxyribonucleic acid synthesis in uterine epithelial cells which lack estrogen receptors. Endocrinology 119:390–396.
214. Yu KT, Peters MA, Czech MP (1986) Similar control mechanisms regulate the insulin and type I insulin-like receptor kinases. Affinity-purified insulin-like growth factor I receptor kinase is activated by tyrosine phosphorylation of its β subunit. J Biol Chem 261:11341–11349.
215. Gorden P, Carpentier J-L, Freychet P, Orci L (1980) Internalization of polypeptide hormones. Mechanism, intracellular localization and significance. Diabetologia 18:263–274.
216. Berridge MJ (1985) Calcium: A universal second messenger. Triangle 24:79–90.
217. Denton RM, McCormack JG (1985) Ca^{++} transport by mammalian mitochondria and its role in hormone action. Am J Physiol 249: E543–E554.
218. Pearson RB, Woodgett JR, Cohen P, Kemp BE (1985) Substrate specificity of a

multifunctional calmodulin-dependent protein kinase. J Biol Chem 260:14471–14476.

219. Soderling TR (1982) Role of hormones and protein phosphorylation in metabolic regulation. A symposium. Fed Proc 41:2615–2638.
220. Schramm M, Selinger Z (1984) Message transmission: Receptor controlled adenylate cyclase system. Science 225:1350–1356.
221. Nishizuka Y (1986) Studies and perspectives of protein kinase C. Science 233:305–312.
222. Williamson JR (1986) Inositol lipid metabolism and intracellular signaling mechanisms. News Physiol Sci 1:72–76.
223. Parker PJ, Coussens L, Totty N, Rhee L, Young S, Chen E, Stabel S, Waterfield MD, Ullrich A (1986) The complete primary structure of protein kinase C, the major phorbol ester receptor. Science 233:853–866.
224. Yap WH, Teo TS, Tan YH (1986) An early event in the interferon-induced transmembrane signaling process. Science 234:355–368.
225. Agranoff BW (1986) Inositol triphosphate and related metabolism. A symposium. Fed Proc 45:2627–2652.
226. Bollag GE, Roth RA, Beaudoin J, Mochly-Rosen D, Koshland DE Jr (1986) Protein kinase C directly phosphorylates the receptor in vitro and reduces its protein-tyrosine kinase activity. Proc Natl Acad Sci USA 83:5822–5824.
227. DiSalvo J (1984) Protein phosphatase targets for cellular regulation. A symposium. Proc Soc Exp Biol Med 177:1–41.
228. Purrello F, Burnham DB, Goldfine ID (1983) Insulin regulation of protein phosphorylation in isolated rat liver nuclear envelopes: Potential relationship to mRNA metabolism. Proc Natl Acad Sci USA 80:1189–1193.
229. Hecht LB, Straus DS (1986) Insulin-sensitive, serum-sensitive protein kinase activity that phosphorylates ribosomal protein S6 in cultured fibroblast-melanoma hybrid cells. Endocrinology 119:470–480.
230. Goodridge AG (1986) Regulation of the gene for fatty acid synthase. Fed Proc 45:2399–2405.
231. Makino H, Kanatsuka A, Suzuki T, Kuribayashi S, Hashimoto N, Yoshida S, Nishimura M (1985) Insulin resistance of fat cells from spontaneously diabetic KK mice. Analysis of insulin-sensitive phosphodiesterase. Diabetes 34:844–849.
232. Saltiel AR, Steigerwalt RW (1986) Purification of putative insulin-sensitive cAMP phosphodiesterase or its catalytic domain from rat adipocytes. Diabetes 35:698–704.
233. Sugimoto T, Fukase M, Tsutsumi M, Nakada M, Hishikawa R, Tsunenari T, Yoshimoto Y, Fujita T (1986) Impaired parathyroid hormone-stimulated adenosine 3',5'-monophosphate release by isolated perfused bones obtained from vitamin D–deficient rats. Endocrinology 118:1808–1813.
234. Solomon SS, Deaton J, Shankar TP, Palazzolo M (1986) Cyclic AMP phosphodiesterase in diabetes. Effect of glyburide. Diabetes 35:1233–1236.
235. Field JB (1980) Diseases of glycogen metabolism. In Isselbacher KJ, Adams RD, Braunwald E, Petersdorf RG, Wilson JD (eds): Harrison's Principles of Internal Medicine, 9th Ed. McGraw-Hill, New York, pp 500–507.
236. Jarett L, Kiechle FL (1984) Intracellular mediator of insulin action. Vitam Horm 41:51–78.
237. Begum N, Terjung RL, Tepperman HM, Tepperman J (1986) Effect of acute exercise on insulin generation of pyruvate dehydrogenase activator by rat liver and adipocyte plasma membranes. Diabetes 35:785–790.
238. Begum N, Tepperman HM, Tepperman J (1984) Effect of dexamethasone on adipose tissue and liver pyruvate dehydrogenase and its stimulation by insulin-generated chemical mediator. Endocrinology 114:99–107.

239. Schubart UK (1986) Regulation of gene expression in the rat hepatocytes and hepatoma cells by insulin: Quantitation of messenger ribonucleic acid's coding for tyrosine amino transferase, typtophan oxygenase, and phophoenolpyruvate carboxykinase. Endocrinology 119:1741–1749.
240. Mueckler M, Caruso C, Baldwin SA, Panico M, Blench I, Morris HR, Allard WJ, Lienhard GE, Lodish HF (1985) Sequence and structure of a human glucose transporter. Science 229:941–945.
241. Wheeler TJ, Hinkle PC (1985) The glucose transporter of mammalian cells. Annu Rev Physiol 47:503–517.
242. Vinten J, Noergaard Petersen L, Sonne B, Galbo H (1985) Effect of physical training on glucose transporters in fat cell fractions. Biochim Biophys Acta 841:223–227.
243. Matthaei S, Horuk R, Olefsky JM (1986) Blood-brain glucose transfer in diabetes mellitus. Decreased number of glucose transporters at blood-brain barrier. Diabetes 35:1181–1184.
244. Yki-Järvinen H, Koivisto VA (1986) Natural course of insulin resistance in type I diabetes. N Engl J Med 315:224–230.
245. Arsenis G, Livingstone JN (1986) Alterations in the tyrosine kinase activity of the insulin receptor produced by in vitro hyperinsulinism. J. Biol Chem 261:147–153.
246. Hjøllund E, Pedersen O, Espersen T, Klebe JG (1986) Impaired insulin receptor binding and postbinding defects of adipocytes from normal and diabetic pregnant women. Diabetes 35:598–603.
247. Van Putten JPM, Wieringa T, Krans HMJ (1985) Low pH and ketoacids induce insulin receptor binding and postbinding alterations in cultured 3T3 adipocytes. Diabetes 34:744–750.
248. Hansen I, Tsalikian E, Beaufrere B, Gerich J, Haymond M, Rizza R (1986) Insulin resistance in acromegaly: Defects in both hepatic and extrahepatic insulin action. Am J Physiol 250:E269–E273.
249. Schwenk WF, Rizza RA, Mandarino LJ, Gerich JE, Hayles AB, Haymond MW (1986) Familial insulin resistance and acanthosis nigricans. Presence of a postbinding defect. Diabetes 35:33–37.
250. Grunfeld C (1984) Insulin resistance: Pathophysiology, diagnosis and therapeutic implications. In Cohen MP, Foà PP (eds): Special Topics in Endocrinology and Metabolism, Vol 6. Alan R. Liss, New York, pp 194–240.
251. Tabor CW, Tabor H (1984) Polyamines. Annu Rev Biochem 53:749–790.
252. Jänne J, Höltta E, Kallio A, Käpyaho K (1983) Role of polyamines and their antimetabolites in clinical medicine. In Cohen MP, Foà PP (eds): Special Topics in Endocrinology and Metabolism, Vol 5. Alan R. Liss, New York, pp 227–293.
253. Peavy DE, Edmonson JW, Duckworth WC (1984) Selective effects of inhibitors of hormone processing on insulin action in isolated adipocytes. Endocrinology 114:753–760.
254. Phillis JW, Barraco RA (1985) Adenosine, adenylate cyclase and transmitter release. Adv Cyclic Nucleotide Phosphoryl Res 19:243–257.
255. Snyder SH (1985) Adenosine as a neuromodulator. Annu Rev Neurosci 8:103–124.
256. Berne RM (1986) Adenosine: An important physiological regulator. News Physiol Sci 1:163–167.
257. Gilman AG, Goodman LS, Rall TW, Murad F (eds) (1985) Goodman and Gilman's The Pharmacological Basis of Therapeutics, 7th Ed. Macmillan, New York.
258. Ceda GP, Hoffman AR (1985) Growth hormone–releasing factor desensitization in rat anterior pituitary cells in vitro. Endocrinology 116:1334–1340.

259. Bilezikjian LM, Seifert H, Vale W (1986) Desensitization to growth hormone–releasing factor (GRF) is associated with down-regulation of GRF-binding sites. Endocrinology 118:2045–2052.
260. Komatsu M, Iwahana H, Mitsuhashi S (1986) Effects of an LHRH agonist on endocrine function, LHRH receptor and LH/hCG receptor in the pituitary-gonadal axis of male rats. Endocrinol Jpn 33:185–195.
261. Crowley WF, Comite F, Vale W, Rivier J, Loriaux DL, Cutler GB Jr (1981) Therapeutic use of pituitary desensitization with long-acting LHRH agonist: A potential new treatment for idiopathic precocious puberty. J Clin Endocrinol Metab 52:370–372.
262. Sibley DR, Lefkowitz RJ (1985) Molecular mechanisms of receptor desensitization using the β-adrenergic receptor-coupled adenylate cyclase system as a model. Nature 317:124–129.
263. Ercolani L, Lin HL, Ginsberg BH (1985) Insulin-induced desensitization at the receptor and postreceptor level in mitogen-activated human T-lymphocytes. Diabetes 34:931–937.
264. De Feo P, Perriello G, De Cosmo S, Ventura MM, Campbell PJ, Brunetti P, Gerich JE, Bolli GB (1986) Comparison of glucose counterregulation during short-term and prolonged hypoglycemia in normal humans. Diabetes 35:563–569.
265. Francis AJ, Hanning I, Alberti KGMM (1985) The influence of insulin antibody levels on the plasma profiles and action of subcutaneously injected human and bovine short-acting insulins. Diabetologia 28:330–334.
266. Ooi GT, Herington AC (1986) Covalent cross-linking of insulin-like growth factor-I to a specific inhibitor from human serum. Biochem Biophys Res Commun 137:411–417.
267. Bratusch-Marrain PR (1983) Insulin counteracting hormones; their impact on glucose metabolism: Diabetologia 24:74–79.
268. Gerich J, Rizza R, Haymond M, Cryer P (1981) Hormonal mechanisms in acute glucose counterregulation: The relative roles of glucagon, epinephrine, norepinephrine, growth hormone and cortisol. Metabolism 29 (Suppl 2):1164–1175.
269. Hoelzer DR, Dalsky GP, Clutter WE, Shah DS, Holloszy JO, Cryer PE (1986) Glucoregulation during exercise: Hypoglycemia is prevented by redundant glucoregulatory systems, sympathochromaffin activation, and changes in islet hormone secretion. J Clin Invest 77:212–221.
270. Bolli GB, Dimitriadis G, Pehling G, Baker B, Haymond M, Cryer P, Gerich J (1984) Abnormal glucose counterregulation after subcutaneous insulin in insulin-dependent diabetes mellitus. Engl J Med 310:1706–1711.
271. Koivisto VA, Yki-Järvinen H, Helve E, Karonen S-L, Pelkonen R (1986) Pathogenesis and prevention of the dawn phenomenon in diabetic patients treated with CSII. Diabetes 35:78–82.
272. Foà PP, Dunbar JC Jr, Klein SP, Levy SH, Malik MA, Campbell BB, Foà NL (1980) Reactive hypoglycemia and A-cell ("pancreatic") glucagon deficiency in the adult. JAMA 244:2281–2285.
273. Bolli GB, Gottesman IS, Campbell PJ, Haymond MW, Cryer PE, Gerich JE (1984) Glucose counterregulation and waning of insulin in the Somogyi phenomenon (posthypoglycemic hyperglycemia). N Engl J Med 311:1214–1219.
274. King GL, Johnson SM (1985) Receptor-mediated transport of insulin across endothelial cells. Science 227:1583–1586.
275. Bar RS, Boes M, Yorek M (1986) Processing of insulin-like growth factors I and II by capillary and large vessel endothelial cells. Endocrinology 118:1072–1080.
276. Hachiya HL, Carpentier J-L, King GL (1986) Comparative studies on

insulin-like growth factor II and insulin processing by vascular endothelial cells. Diabetes 35:1065–1072.

277. Levy JR, Olefsky JM (1986) Retroendocytosis of insulin in rat adipocytes. Endocrinology 119:572–579.

278. Duckworth WC, Kitabchi AE (1981) Insulin metabolism and degradation. Endocr Rev 2:210–283.

279. Banskota NK, Carpentier J-L, King GL (1986) Processing and release of insulin and insulin-like growth factor I by macro- and microvascular endothelial cells. Endocrinology 119:1904–1923.

280. Sessler FM, Jacquez JA, Malvin RL (1986) Different production and decay rates of six renin forms isolated from rat plasma. Am J Physiol 250:E551–E557.

281. Boden G, Shimoyama R, Ray TK, Savage RC Jr (1985) Effect of anti-insulin receptor antibodies on in vivo insulin metabolism. Diabetes 34:342–346.

282. Tokuyama K, Himms-Hagen J, (1986) Brown adipose tissue thermogenesis, torpor, and obesity of glutamate-treated mice. Am J Physiol 251:E407–E415.

283. Bottazzo GF, Pozzilli P, Mirakian R, Dean BM, Doniach D (1984) Early immunological events in diabetes. In Andreani D, Di Mario U, Federlin KF, Heding LG (eds): Immunology in Diabetes. Kimpton, London, pp 95–104.

284. Barkan AL, Kelch RP, Marshall JC (1985) Isolated gonadotrope failure in the polyglandular autoimmune syndrome. N Engl J Med 312:1535–1540.

285. Robertson RP (1981) Prostaglandins, thromboxanes, and eicosanoids: Arachidonic acid metabolites relevant to medicine. In Isselbacher KJ, Adams RD, Braunwald E, Martin JB, Petersdorf RG, Wilson JD (eds): Harrison's Principles of Internal Medicine, Update 1. McGraw-Hill, New York, pp 191–207.

286. Hall PF (1985) Role of cytochromes P-450 in the biosynthesis of steroid hormones. Vitam Horm 42:315–368.

287. Harris AL (1985) Aminoglutethimide. A new endocrine therapy in breast cancer. A cancer research review. Exp Cell Biol 53:1–8.

288. Imperato-McGinley J, Binienda Z, Arthur A, Miniberg DT, Vaughan ED Jr, Quimby FW (1985) The development of a male pseudohermaphroditic rat using an inhibitor of the enzyme 5α-reductase. Endocrinology 116:807–812.

289. Tseng L, Mazella J, Sun B (1986) Modulation of aromatase activity in human endometrial stromal cells by steroids, tamoxifen and RU 486. Endocrinology 118:1312–1318.

290. Osawa Y, Yarborough C, Osawa Y (1982) Norethisterone, a major ingredient of contraceptive pills, is a suicide inhibitor of estrogen biosynthesis. Science 215:1249–1251.

291. Feuillan PP, Foster CM, Pescovitz OH, Hench KD, Shawker T, Dwyer A, Malley JD, Barnes K, Loriaux DL, Cutler GB Jr (1986) Treatment of precocious puberty in the McCune-Albright syndrome with the aromatase inhibitor testolactone. N Engl J Med 315:1115–1119.

292. Halban PA, Amherdt M, Orci L, Renold AE (1986) Tris (hydroxymethyl) aminomethane inhibits the synthesis and processing of proinsulin in isolated rat pancreatic islets without affecting release of insulin stores. Diabetes 35:433–439.

293. Gold G, Pou J, Gishizky ML, Landall MD, Grodsky GM (1986) Effects of tolbutamide pretreatment on the rate of conversion of newly synthesized proinsulin to insulin and the compartmental characteristics of insulin storage in isolated rat islets. Diabetes 35:6–12.

294. Bazan MC, Domene H, Heinrich JJ, Barontini M, Bergada C (1984) Comparison of single and combined tests for the evaluation of plasma growth hormone secretion in normal short children. J Endocrinol Invest 7:295–298.

295. Arita J, Kimura F (1986) Characterization of in vitro dopamine synthesis in the

median eminence of rats with haloperidol-induced hyperprolactinemia and bromocriptine-induced hypoprolactinemia. Endocrinology 119:1666–1672.

296. Gross MD, Grekin RJ, Gniadek TC, Villareal JZ (1981) Suppression of aldosterone by cyproheptadine in idiopathic aldosteronism. N Engl J Med 305:181–185.

297. Coiro V, Chiodera P, Volpi R, D'Amato L, Camellini L, Rossi G, Pignatti D, Butturini U (1986) Muscarinic cholinergic modulation of insulin response to an intravenous glucose tolerance test in normal man. J Endocrinol Invest 9:27–30.

298. Kashiwagi A, Harano Y, Suzuki M, Kojima H, Harada M, Nishio Y, Shigeta Y (1986) New α_2-adrenergic blocker (DG-5128) improves insulin secretion and in vivo glucose disposal in NIDDM patients. Diabetes 35:1085–1089.

299. Ribeiro AM, Gomez MV (1986) The effect of calmodulin inhibitors on the release of acetylcholine and protein phosphorylation induced by tityustoxin, potassium and ouabain. Brain Res Bull 16:673–680.

300. Kabuto M, Namura I, Saitoh Y (1986) Nocturnal enhancement of plasma melatonin could be suppressed by benzodiazepines in humans. Endocrinol Jpn 33:405–414.

301. Nakai A, Nagasaka A, Hidaka H, Tanaka T, Ohyama T, Iwase K, Ohtani S, Shinoda S, Aono T, Masunaga R, Nakagawa H, Kataoka K (1986) Effect of calmodulin inhibitors on thyroid hormone secretion. Endocrinology 119:2279–2283.

302. Lamberts SWJ, Uitterlinden P, Verschoor L, Van Dongen KJ, Del Pozo E (1985) Long-term treatment of acromegaly with the somatostatin analogue SMS 201-995. N Engl J Med 313:1576–1580.

303. Boden G, Ryan IG, Eisenschmid BL, Shelmet JJ, Owen OE (1986) Treatment of inoperable glucagonoma with the long-acting somatostatin analogue SMS 201-995. N Engl J Med 314:1686–1689.

304. Horiuchi N, Holick MF, Potts JT Jr, Rosenblatt M (1983) A parathyroid hormone inhibitor in vivo: Design and biological evaluation of a hormone analog. Science 220:1053–1055.

305. Rosenblatt M (1986) Peptide hormone antagonists that are effective in vivo. Lessons from parathyroid hormone. N Engl J Med 315:1004–1013.

306. Sawyer WH, Pang PKT, Seto J, McEnroe M, Lammek B, Manning M (1981) Vasopressin analogs that antagonize antidiuretic responses by rats to the antidiuretic hormone. Science 212:49–51.

307. Nekola MV, Horvath A, Ge L-J, Coy DH, Schally AV (1982) Suppression of ovulation in the rat by an orally active antagonist of leuteinizing hormone–releasing hormone. Science 218:160–161.

308. Rivier C, Rivier J, Vale W (1980) Antireproductive effects of a potent gonadotropin releasing hormone antagonist in the male rat. Science 210:93–95.

309. Felicetta JV, Czanko R, Huber-Smith MJ, McCann DS (1986) Cholecystographic agents and sulfobromophthalein inhibit the binding of L-thyroxine to plasma membranes of rat hepatocytes. Endocrinology 118:2500–2504.

310. Auf'Mkolk M, Ingbar JC, Kubota K, Amir SM, Ingbar SH (1985) Extracts and auto-oxidized constituents of certain plants inhibit the receptor-binding and the biological activity of Graves' immunoglobulins. Endocrinology 116:1687–1693.

311. Seamon KB (1984) Forskolin and adenylate cyclase; new opportunities in drug design. Annu Rep Med Chem 19:293–302.

312. West WL (1982) Calmodulin-regulated enzymes: Modifications by drugs and disease. A symposium. Fed Proc 41:2251–2299.

313. Hof RP (1985) Pharmacological characterization of calcium antagonists: What they do and where they act. Triangle 24:143–156.

314. Login IS, Judd AM, Cronin MJ, Koike K, Schettini G, Yasumoto T,

MacLeod RM (1985) The effects of maitotoxin on ⁴⁵Ca²⁺ flux and hormone release in GH₃ rat pituitary cells. Endocrinology 116:622–627.

315. Tutwiler GF, Brentzel HJ, Kiorpes TC (1985) Inhibition of mitochondrial carnitine palmitoyl transferase A in vivo with methyl 2-tetradecylglycidate (methyl palmoxirate) and its relationship to ketonemia and glycemia. Proc Soc Exp Biol Med 178:288–296.
316. Katsuhiro H, Kohei K, Masafumi M, Masako T, Toshio K (1985) Sulfonyl-urea stimulates liver fructose 2,6-bisphosphate formation in proportion to its hypoglycemic action. Diabetes Res Clin Pract 1:49–53.
317. George FW, Wilson JD (1986) Hormonal control of sexual development. Vitam Horm 43:145–196.
318. Pilar GR (1985) Functional interaction of developing neurons with their target tissue. A symposium. Fed Proc 44:2750–2779.
319. Vartanian T, Szuchet S, Dawson G, Campagnoni AT (1986) Oligodendrocyte adhesion activates protein kinase C–mediated phosphorylation of myelin basic protein. Science 234:1393–1398.
320. Mason AJ, Pitts SL, Nikolics K, Szonyi E, Wilcox JN, Seeburg PH, Stewart TA (1986) The hypogonadal mouse: Reproductive functions restored by gene therapy. Science 234:1372–1378.
321. Shull JD, Gorski J (1986) The hormonal regulation of prolactin gene expression: An examination of mechanisms controlling prolactin synthesis and the possible relationship of estrogen to these mechanisms. Vitam Horm 43:197–249.
322. Loeffler JP, Kley N, Pittius CW, Höllt V (1986) Calcium ion and cyclic adenosine 3′,5′-monophosphate regulate proopiomelanocortin messenger ribonucleic acid levels in rat intermediate and anterior pituitary lobes. Endocrinology 119:2840–2847.
323. Philippe J, Mojsov S, Drucker DJ, Habener JF (1986) Proglucagon processing in a rat islet cell line resembles phenotype of intestine rather than pancreas. Endocrinology 119:2833–2839.
324. Negro-Vilar A, Conte D, Valenca M (1986) Transmembrane signals mediating neural peptide secretion: Role of protein kinase C activators and arachidonic acid metabolites in luteinizing hormone–releasing hormone secretion. Endocrinology 119:2796–2802.
325. Grunberger G, Iacopetta B, Carpentier J-L, Gorden P (1986) Diacylglycerol modulation of insulin receptor from cultured human mononuclear cells. Effects of binding and internalization. Diabetes 35:1364–1370.
326. Imagawa W, Spencer EM, Larson L, Nandi S (1986) Somatomedin-C substitutes for insulin for the growth of mammary epithelial cells from normal virgin mice in serum-free collagen gel cell culture. Endocrinology 119:2695–2699.
327. Murphy WA, Coy DH, Lance VA (1986) Superactive amidated carboxy-terminal glucagon analogs with no methionine or tryptophan. Peptides (Fayetteville NY) 7 (Suppl 1):69–74.
328. Constant RB, Weintraub BD (1986) Differences in the metabolic clearance of pituitary and serum thyrotropin (TSH) derived from euthyroid and hypothyroid rats: Effects of chemical deglycosylation of pituitary TSH. Endocrinology 119:2720–2727.
329. Segal J, Rehder M-C, Ingbar SH (1986) Calmodulin mediates the stimulatory effect of 3,5,3′-triiodothyronine on adenylate cyclase activity in rat thymocyte plasma membranes. Endocrinology 119:2629–2634.
330. Marangou AG, Weber KM, Boston RC, Aitken PM, Heggie JCP, Kirsner RLG, Best JD, Alford FP (1986) Metabolic consequences of prolonged hyperinsulinemia in humans. Evidence for induction of insulin insensitivity. Diabetes 35:1383–1389.

2
Insulin Gene Mutations and Abnormal Products of the Human Insulin Gene

HOWARD S. TAGER

In many ways, the study of insulin, the pancreatic B cell, and diabetes has provided a paradigm for the analysis of various aspects of endocrine function, ranging from clinical diagnosis and management to biochemistry and cell and molecular biology. Diabetes is, of course, a complex disease with both multiple causes and multiple consequences. Although the terms juvenile onset and maturity onset, type I and type II, and most recently insulin-dependent and non-insulin-dependent have been applied to the disease, none of these terms are completely satisfactory, and none specifically address the cause of glucose intolerance. In this regard, one must deal at least with matters such as the potential loss of (a) B cell mass (arising most probably from autoimmune B cell death in insulin-dependent diabetes), (b) B cell secretetory responses to glucose and other secretogogues (notwithstanding adequate B cell numbers and associated insulin stores), and (c) peripheral insulin sensitivity (due to receptor or postreceptor defects at insulin target cells).

The subject of this review concerns yet another potential cause of diabetes: insulin gene mutations and defects in the biosynthesis of insulin. Although the primary structure of insulin had been known for nearly 15 years, Elliott et al. (1) and Kimmel and Pollack (2) postulated only in 1967 that structural alterations in insulin might account for the abnormal glucose regulation in some diabetic patients. It turns out that abnormal products of the human insulin gene can indeed contribute to diabetes in some rather complex and as yet not fully understood ways and that the human insulin gene is probably no less subject to detrimental mutation than, for example, the human globin genes. The framework for understanding abnormal insulins and their relationship to diabetes must include consideration of insulin biosynthesis, insulin action and structure-function relationships, and insulin secretion and metabolic clearance.

I. Insulin Biosynthesis

A. Proinsulin and Its Convention to Insulin

Insulin has a rather unusual structure in that it contains two separate polypeptide chains (the A chain containing 21 amino acid residues and the B chain containing 30 residues) which are linked by a pair of disulfide bonds. Although the mechanisms for the production of insulin in the pancreatic B cell long remained an enigma, Steiner and his colleagues conclusively demonstrated by studies first published 20 years ago that the two-chained hormone is synthesized by the limited processing of a single-chain polypeptide precursor called proinsulin (3,4). Fundamental aspects of proinsulin structure and its conversion to insulin are illustrated by the scheme of Fig. 2.1. The precursor, presumably initially existing in a random conformation, folds to yield the basic insulin structure in which (a) two interchain disulfide bonds are formed between the A and B chains, (b) an intrachain disulfide bond is formed within the A chain, and (c) the COOH terminus of the B chain is linked via a connecting peptide (called C peptide) to the NH_2 terminus of the A chain. Although the precursor at this stage presumably has gained insulin-like secondary and tertiary structure, it possesses only a small amount of the biological activity of insulin itself (5). Subsequent processing events involve the participation of two enzyme activities, one of which cleaves the precursor at the paired dibasic amino acid sites linking the insulin chains to the C peptide (an enzyme with trypsinlike specificity) and the other of which removes COOH-terminal basic amino acids that do not appear in the final products (a carboxypeptidase B-like enzyme) (4,6–8).

Although details of the participating converting enzymes remain unknown, the course of processing of proinsulin to insulin by limited proteolysis is clearly understood. As illustrated in Fig. 2.1, (a) the trypsinlike endopeptidase clips proinsulin first at either the Lys-Arg junction between the C peptide and A chain or the Arg-Arg junction between the B chain and C peptide; (b) the carboxypeptidaselike exopeptidase removes exposed COOH-terminal basic residues; (c) the endopeptidase clips the resulting two-chained proinsulin conversion intermediates on the contralateral side; and (d) again, the exopeptidase removes exposed COOH-terminal basic residues (4,7,9). The result is the production of insulin and C peptide. Conversion of proinsulin to insulin within the pancreatic B cell normally proceeds with about 95% efficiency. Figure 2.1 omits consideration of yet another peptide precursor, one with a hydrophobic NH_2-terminal extension on the B chain, which is called preproinsulin and which plays a role in guiding the translocation of the polypeptide along its intracellular course (6,10). A more complete description of the process of insulin biosynthesis includes consideration of (a) translation of the preproinsulin mRNA on the cytosolic side of the

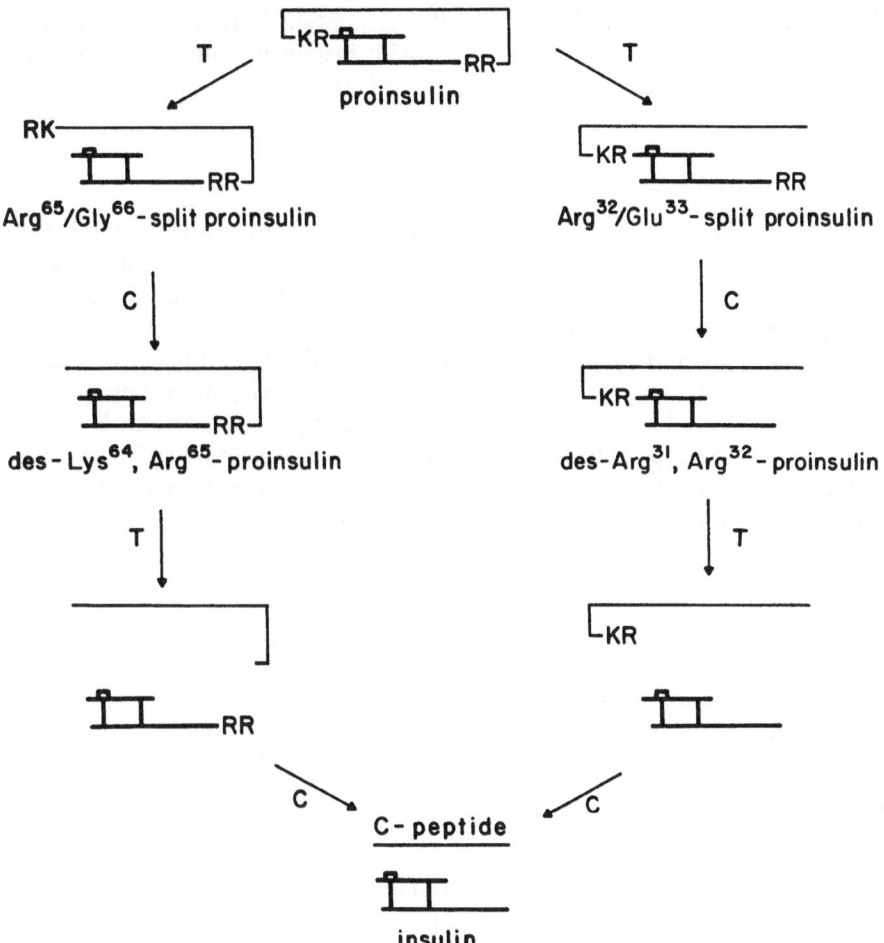

FIGURE 2.1. Scheme for the processing of proinsulin to major conversion intermediates and to insulin. The A chains and B chains of insulin are illustrated by heavy lines, whereas the C peptide is illustrated by a light line. (The length of the C peptide is exaggerated in the projection used). Paired dibasic amino acids at precursor conversion sites are shown explicitly and are identified by the one-letter amino acid code (K, lysine; R, arginine). The numbering of amino acid residues assumes that the peptides are derived from humans—i.e., that the C peptide is 31 residues in length. The letters T and C identify reactions thought to be catalyzed by trypsinlike and carboxypeptidase B-like enzymes, respectively.

rough endoplasmic reticulum, (b) translocation of the synthesized peptide to the cisternum of the endoplasmic reticulum concomitant with the removal of the NH_2-terminal signal sequence, (c) movement of proinsulin to the Golgi apparatus, (d) folding of the product, (e) packaging of the precursor (along with the converting enzymes) into secretion granules, and (f) conversion of the single-chained precursor into two-chained insulin plus the C peptide (4,6,11). Importantly, the C peptide is retained within the secretion granule and is released along with insulin in equimolar amounts during appropriate stimulation of the B cell (12).

B. The Insulin Gene

Our knowledge of the mechanisms for insulin biosynthesis has extended in recent years to the structure of the gene that encodes the human insulin sequence (13,14) and to the movement of relevant information from the nucleus to the cytoplasm. Figure 2.2 illustrates diagrammatically the structure of the human insulin gene and identifies (a) the 5' untranslated region, which is split by an intervening sequence; (b) the region coding for the hydrophobic signal sequence; (c) the region coding for the insulin B chain; (d) the region coding for the C peptide, which is again split by an intervening sequence; (e) the region coding for the insulin A chain; and (f) the 3' untranslated region. Although the first intervening sequence is rather

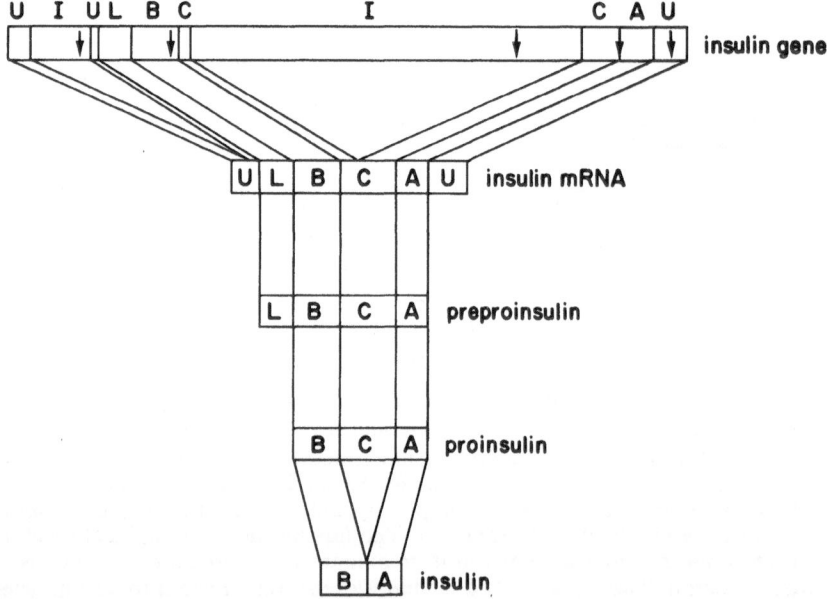

FIGURE 2.2. Diagrammatic illustration of the structure of the human insulin gene (top) and the processing of information and structure in the biosynthesis of insulin. The topmost structure also corresponds to that of the mRNA precursor which arises by direct transcription of the gene. U, untranslated region; I, intervening sequence; L, leader or signal peptide sequence; B, insulin B chain; C, C peptide; A, insulin A chain. Structures in the figure illustrate those of the insulin gene, the mature mRNA (resulting from excision of the intervening sequences), preproinsulin (the first peptide product), proinsulin, and insulin, from top to bottom, respectively.

conserved in its length (ranging from 119 to 151 base pairs in different species), the second varies considerably from one species to another (ranging from 0 to about 3,500 base pairs) (15); the second intervening sequence is 786 base pairs in the human insulin gene (13,14). As we now know, and as illustrated in Fig. 2.2, the insulin gene is initially transcribed to yield an mRNA precursor that contains information corresponding to the intervening sequences, the intervening sequences are subsequently removed within the nucleus (with ligation of the mRNA), and the mature mRNA is translocated to the cytoplasm, where it directs the biosynthesis of insulin. Together, Figs. 2.1 and 2.2 illustrate both the multiplicity of processing events that occur from the initial transcription of the insulin gene through the final production of the hormone and the rather considerable amount of information (encoded by the structure of the gene) that is used and lost as insulin biosynthesis proceeds: although a gene containing 153 base pairs would in theory be sufficient to code for the structure of insulin, the human insulin gene actually contains about 10 times that number. The remainder of this chapter will deal with mutations that we know to occur in the human insulin gene and with the consequences of those mutations for human physiology, insulin biosynthesis, insulin action, and diabetes.

II. Abnormal Insulin Gene Products

A. Patients Expressing Abnormal Insulin Genes

Although we will consider later in greater depth the clinical significance and physiology of abnormal insulin gene expression, it is important at this stage to present, as already described by Rubenstein and his colleagues, an account of just how patients with insulin gene mutations have been identified (16–18). First, all patients present with fasting hyperinsulinemia, typically about 80–120 μU of immmunreactive insulin per milliliter as determined by direct serum assay. Second, the patients exhibit either fasting hyperglycemia or euglycemia. Third, the patients show no evidence of insulin resistance as demonstrated by normal levels of counterregulatory hormones, the absence of insulin antibodies or insulin receptor antibodies, and normal numbers of monocyte insulin receptors. Fourth, the serum insulin from these patients shows reduced biologic activity when evaluated during a glucose clamp study with hyperglycemia, or by in vitro assay. Fifth, the patients exhibit a normal response to exogenously administered insulin, as reflected by a normal insulin tolerance test or normal response to an insulin clamp study with euglycemia.

From the above information alone, one can see that patients expressing abnormal insulin genes do not respond as expected (i.e., with hypoglycemia) to their existing serum levels of endogenous insulin but respond normally to fully active insulin. The conclusion is inescapable: the insulin from such patients, although it is immunreactive, has greatly diminished biologic activity. Therefore, it must be a hormone that has been detrimentally altered by structural change. Interestingly, diabetes and glucose intolerance may or may not be present. Another important characteristic of these patients is a reduced C peptide:insulin molar ratio in the plasma (17,18). Although carefully executed clinical studies are the key to identifying patients expressing abnormal insulin genes, the final proof of an insulin gene mutation of course requires the identification of the abnormal insulin gene and gene product. The following sections describe several cases we and others have studied. Details of the structural alterations in the insulin gene products in these patients are illustrated in Fig. 2.3.

B. Abnormal Insulins

1. INSULIN CHICAGO

Insulin Chicago was the first abnormal human insulin to be fully characterized. It was discovered in a 51-year-old nonobese man who exhibited fasting hyperinsulinemia (70–120 μU/ml), fasting hyperglycemia (140–170 mg/ml), and no sign of insulin resistance (as evaluated by measurement of contrainsulin hormones, insulin tolerance, and monocyte insulin receptor numbers) (16). Small amounts of insulin were purified from the patient's serum by immunoaffinity chromatography and, although the insulin showed normal mobility on polyacrylamide gel electrophoresis at pH 8.9, it exhibited only about 15% of normal biologic activity in vitro (relative to immunometrically equivalent amounts of insulin purified from the serum of normal subjects) when evaluated by assays involving competition for binding of ^{125}I-labeled insulin to IM-9 cultured human lymphocytes or rat adipocytes, or stimulation of glucose oxidation or 2-deoxy-D-glucose uptake into rat adipocytes (16).

Although the above studies suggested the existence of an abnormal insulin, other possible explanations (e.g., partial degradation of serum insulin by a serum protease) could not be excluded by the data obtained. We were fortunate, however, to receive a 3-g biopsy of the patient's pancreas (obtained during laparotomy for a pancreatic cyst) and to be able to isolate from the specimen small but chemically significant amounts of insulin (19). Although the purified insulin was immunometrically equivalent to normal insulin when normalized for total protein, the patient's pancreatic insulin exhibited significantly decreased receptor binding and biologic potency. Importantly, amino acid analysis showed that the insulin was deficient in phenylalanine and contained excess leucine, a result suggesting the occurrence of a Leu-for-Phe replacement. Analysis of the tryptic heptapeptide B$^{23-29}$ confirmed the existence of a Leu-for-Phe

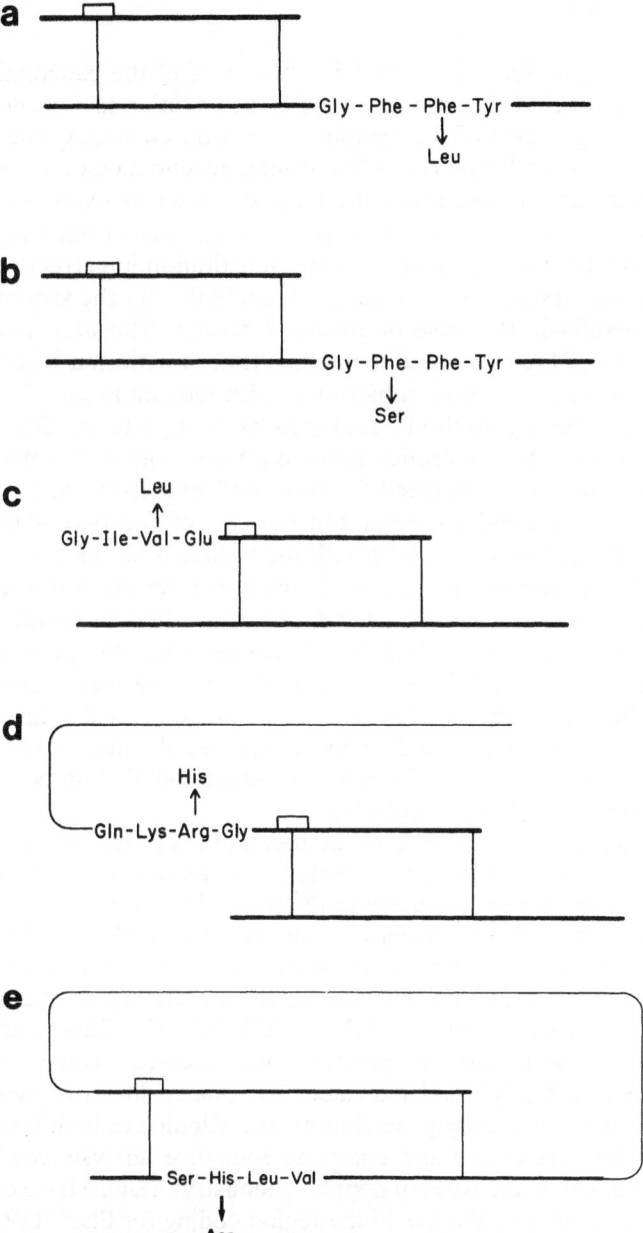

FIGURE 2.3. Diagrammatic structures of abnormal products of the human insulin gene. In each case, the simple diagram illustrates the site of amino acid substitution, the amino acid residue that is replaced, and the amino acid residue that appears in the abnormal insulin. The insulin chains are shown in heavy lines, and the C peptide (where applicable) is shown in lighter lines. **a**, Insulin Chicago ([Leu^{B25}]insulin); **b**, Insulin Los Angeles ([Ser^{B24}]insulin); **c**, Insulin Wakayama ([Leu^{A3}]insulin); **d**, Proinsulin Tokyo ([His^{65}]-des-Arg^{31},Arg^{32}-proinsulin); **e**, Pro-insulin Providence ([Asp^{B10}]proinsulin). In the last example, the form of the final secreted product remains unknown.

substitution at position B24 or B25 in about half of the pancreatic insulin. Although we concluded that a Leu-for-Phe substitution had occurred in the receptor-binding region of the hormone (a region containing the sequence Gly^{23}-Phe^{24}-Phe^{25}-Tyr^{26}), that it resulted from a genetic mutation, and that the isolated pancreatic insulin arose from the codominant expression of both normal and abnormal insulin alleles (19), the amount of material available was too small to allow delineation of the substitution in correct sequence.

Two approaches have been used to evaluate further the structure of the abnormal insulin in the case of Insulin Chicago. The first involved the semisynthesis of the two possible insulins (one substituted with leucine at position B24, and the other substituted with leucine at position B25) by use of the pioneering method developed by Inouye et al. (20). Development of a system for isocratic, reverse-phase, high-performance liquid chromatography that separated insulins differing from one another by single amino acid residues permitted the use of the two semisynthetic insulin standards for comparison with the patient's serum and pancreatic insulins (21). As shown in Fig. 2.4, the patient's serum insulin chromatographed at a position slightly ahead of normal human insulin. Further studies showed that the abnormal hormone chromatographed at the position of human [Leu^{B25}]insulin was well separated from normal insulin and from human [Leu^{B24}]insulin and cochromatographed with [Leu^{B25}]insulin when the two were studied in admixture. We thus concluded that Insulin Chicago was [Leu^{B25}]insulin and suggested that it be designated human insulin B25 (Phe → Leu) (21).

The second approach, applied by Steiner and his colleagues, involved the use of the techniques of molecular biology and recombinant DNA analysis. Treatment of the patient's leucocyte DNA by the restriction endonuclease MboII, followed by Southern analysis, indeed confirmed that only one of the patient's insulin alleles appeared to be normal; that is, the abnormal allele was not cleaved by the restriction endonuclease in the region of the gene that codes for the sequence Phe^{B24}-Phe^{B25} (TTCTTC) (22). This result demonstrated that a nucleotide replacement had occurred within the MboII restriction site (TCTTC) and indicated that one of the two phenylalanine codons had been altered by genetic mutation. Cloning of both insulin alleles in a phage lambda vector and complete sequence analysis confirmed the assignment made on the basis of peptide chemistry: a single base change had occurred in the abnormal allele in the region coding for Phe^{B25} (TTC → TTG) (23), accounting for the replacement of Phe^{B25} by Leu in Insulin Chicago [(human insulin B25 (Phe → Leu)] (see Fig. 2.3).

2. INSULIN LOS ANGELES

The propositus in the example of Insulin Los Angeles was a 28-year-old nonobese woman with mild diabetes, fasting hyperinsulinemia, and reduced fasting C peptide:insulin molar ratio (1.1–1.5, where normal is

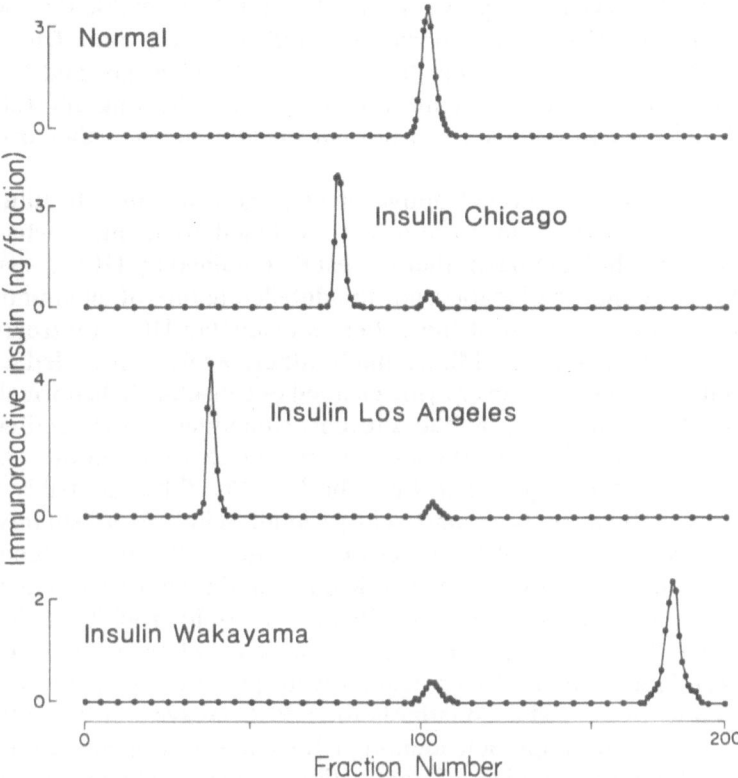

FIGURE 2.4. Analysis of human serum insulins by reverse-phase HPLC. The four panels show the isocratic, HPLC elution profiles of serum insulin from a normal individual, a subject expressing Insulin Chicago ([Leu^{B25}]insulin), a subject expressing Insulin Los Angeles ([Ser^{B24}]insulin), and a subject expressing Insulin Wakayama ([Leu^{A3}]insulin). In each case, insulin was purified from serum by immunoaffinity methods, the sample was applied to a C-18 column, the column was eluted with a mixture of acetonitrile and NaClO$_4$-containing triethylammonium phosphate buffer, fractions were collected, and aliquots of each fraction were subjected to a radioimmunometric assay for immunoreactive insulin. (Reprinted by permission from *Nature*, Vol. 302, 540–543, Copyright © 1983 Macmillan Journals Limited.)

>4) (17). She exhibited a markedly abnormal tolerance to oral glucose, no sign of abnormal levels of contrainsulin hormones, normal response to exogenous insulin in a hypoglycemia test and in an insulin clamp study with euglycemia, and substantially reduced biologic activity of endogenously secreted insulin as assessed by a glucose clamp study with hyperglycemia (17). As was the case with the patient expressing Insulin Chicago, the immunoaffinity-purified serum insulin from this woman showed decreased receptor binding and biologic activity in vitro. The HPLC analysis of serum insulin obtained from the patient expressing

Insulin Los Angeles is presented in Fig. 2.4 (21). It is notable that (a) the patient's insulin eluted at a position significantly different from that observed for normal human insulin or for Insulin Chicago, and (b) this insulin eluted from the column rather early, indicating its relative hydrophilicity when compared to normal insulin and the other insulins shown.

Although results described above clearly indicated that Insulin Los Angeles was abnormal in structure, they failed to identify what the abnormality might have been; that is, results obtained by HPLC analysis can indicate neither the location nor the detailed nature of an amino acid substitution. Investigation of the patient's leucocyte DNA by treatment with the restriction enzyme MboII and Southern analysis revealed a very important clue: whereas the enzyme cleaved one insulin allele normally, it failed to cleave the other at the site corresponding to that coding for Phe^{B24}-Phe^{B25} (TTCTTC) (21). It could thus be said that a nucleotide change had occurred in this region and that Phe^{B24} or Phe^{B25} had probably been replaced by a hydrophilic amino acid residue. With (a) the supposition that the genetic change had resulted from a single nucleotide replacement, (b) information provided by the genetic code, and (c) our knowledge that the amino acid replacement had resulted in a more hydrophilic molecule, we guessed that Phe^{B24} or Phe^{B25} had been replaced by serine. Human [Ser^{B24}]insulin and human [Ser^{B25}]insulin were prepared by semisynthetic methods and were used as standards for HPLC analysis of the patient's serum insulin. The approach identified Insulin Los Angeles as human [Ser^{B24}]insulin (24). Cloning of the patient's insulin alleles and DNA sequence analysis by Haneda et al. occurred simultaneously with peptide analysis in this case and identified both the Ser-for-Phe replacement at position B24 and the single nucleotide change (TTC → TCC) that had resulted in the expression of human insulin B24 (Phe → Ser) (25) (see Fig. 2.3).

3. Insulin Wakayama

The propositus expressing Insulin Wakayama was a 56-year-old nonobese Japanese woman with fasting hyperglycemia and hyperinsulinemia. She showed no sign of insulin resistance, her plasma insulin exhibited markedly decreased biologic and receptor-binding activity, and her plasma C peptide:insulin molar ratio was significantly lower than normal (18). HPLC analysis of serum insulin purified by immunoaffinity chromatography demonstrated an abnormal circulating insulin having properties distinct from any insulin previously examined (18,21); as shown in Fig. 2.4, the insulin was notably more hydrophobic than Insulin Chicago, Insulin Los Angeles, or normal human insulin. Southern analysis, however, revealed no change in restriction site cleavage with MboII. Thus, in contrast to the previous examples, no clues were available to suggest the location of the essential amino acid substitution, and structural analysis had to await the sequence determination of the abnormal insulin allele. The insulin alleles from the patient's leucocyte DNA were cloned by Nanjo et al. (18), by use of the πVX system, with the result that a single nucleotide change was demonstrated in one of the two alleles, and the substitution of Val[A3] by leucine (GTG → TTG) was identified. Insulin Wakayama, human insulin A3 (Val → Leu), represents the first example of an abnormal insulin involving amino acid replacement within the insulin A chain (see Fig. 2.3). Most important, two additional examples of Insulin Wakayama have now been identified in Japanese individuals unrelated to the propositus and unrelated to each other. In one of these, the abnormal insulin structure was identified by gene cloning and DNA sequencing (26), whereas in the other, the structure was identified by protein sequence determination of pancreatic insulin purified from a tissue biopsy (27,28). In all cases, Val[A3] was found to be replaced by leucine.

4. OTHER EXAMPLES AND NEGATIVE FINDINGS

Although the search for abnormal insulins has been conducted for only a few years, another abnormal insulin that elutes from reverse-phase HPLC columns at a position different from that of normal human insulin has recently been demonstrated by Seino et al. (29) in a patient exhibiting hyperinsulinemia of unknown etiology; although the insulin exhibited hydrophobicity relatively greater than that of the normal hormone, the causative structural change has not yet been identified. Other methods have also been applied to the detection of abnormal insulins: a sensitive radioreceptor assay for plasma insulin (involving the use of guinea pig kidney membranes and the radioimmunometric standardization of hormone concentrations) has been described and has been applied to the detection of the third example of Insulin Wakayama (28). Although the method does not provide structural or physicochemical information, it is readily suited to patient screening and is considerably less cumbersome and less expensive than HPLC analysis. Another approach, undertaken by Bell and his colleagues, has involved the screening of more than 200 unselected patients with non-insulin-dependent diabetes by use of Southern analysis of leucocyte DNA and application of the restriction enzymes MboII, RsaI, and PvuII to probe for nucleotide changes in the corresponding restriction endonuclease cleavage sites (30). Although it was found that one insulin allele in a single individual (a 45-year-old black male) was not cleaved by MboII at the genomic sequence TCTTC (corresponding to the region encoding Phe[B24]-Phe[B25]), the patient's serum insulin appeared to be normal on HPLC analysis. This situation can arise from a so-called silent mutation in which one phenylalanine codon has been replaced by another. The result is that the final gene product has undergone no structural change.

During several years of studies involving HPLC analysis of serum insulin, we have not obtained evidence for abnormal circulating forms of the hormone in any of more than a dozen normal subjects examined or in a number of samples from patients with insulin-dependent diabetes who had been treated with beef or pork insulin. A study of 8 patients with benign insulinoma has also revealed normal insulin in each case (9,29). Analysis of an additional 8 samples of patient serum sent to us over the years by colleagues in the United States, Europe, and South America has yielded negative results. Each of these patients had clinical characteristics consistent with the expression of an abnormal insulin gene, although a few also showed some signs of peripheral insulin resistance (occasionally in association with acanthosis nigricans). Many of these patients presented with a milder form of fasting hyperinsulinemia (40–50 μU/ml), however, and exhibited greatly exaggerated, but sometimes unsustained, insulin responses to oral glucose; in a few of these individuals, serum insulin levels as high as 2,500 μU/ml were reached after a 75-g oral glucose load.

C. Abnormal Proinsulins and Proinsulin Conversion Intermediates

1. PROINSULIN TOKYO/BOSTON

The syndrome of familial hyperproinsulinemia was first described 10 years ago by Gabbay and colleagues as a condition associated with increased serum levels of 9,000 molecular weight insulin-immunoreactive material (31). Glucose intolerance is generally absent, and levels of total serum insulinlike immunoreactivity are about 100 μU/ml. Although the characteristic of the syndrome (the accumulation of circulating proinsulinlike material) could arise from a variety of circumstances, it was accurately attributed to the failure of proinsulin to be completely converted to insulin within the pancreatic B cell and to the secretion of a higher-molecular-weight, insulin-containing form. Even so, the physiologic abnormality could have arisen from a defect in a required proinsulin-converting enzyme, or indeed from any of several possible defects in the cell biology of proinsulin conversion (see above). Structural studies were of course hindered by the small amounts of material obtainable from blood by immunoaffinity chromatography. Nevertheless, in the case of Proinsulin Boston (a molecule probably identical to Proinsulin Tokyo), preliminary studies indicated that the circulating proinsulinlike material in patient serum was really a two-chained intermediate of proinsulin conversion rather than intact human proinsulin (32).

Further structural analysis of proinsulin-related material in patients with familial hyperproinsulinemia (initially applied to material from patients expressing Proinsulin Tokyo, and later to that from patients expressing Proinsulin Boston) relied on (a) immunochemical methods involving separate radioimmunoassays for insulin, C peptide, and bis-S-sulfo-insulin B chain; (b) separation methods involving both gel filtration and native polyacrylamide gel electrophoresis at pH 8.9; and (c) chemical modifications and biochemical manipulations involving oxidative sulfi-tolysis of disulfide bonds, acetylation of amino groups with acetic anhydride, and digestion of native and derivatized peptides by trypsin (33). All studies were accomplished with picomole amounts of material. Results for material from each of the two separate cohorts exhibiting familial hyperproinsulinemia showed that the serum proinsulinlike material was indeed a two-chained intermediate of proinsulin conversion in which the C peptide remained attached to the insulin A chain (33,34); that is, the peptide was an intermediate similar to des-Arg^{31},Arg^{32}-proinsulin (see Fig. 2.1).

Chemical manipulations examined the structure of the dibasic amino acid conversion site and demonstrated the presence of the lysine residue in the Lys^{64}-Arg^{65} sequence connecting the C peptide to the A chain but showed the absence of the arginine residue. The proinsulinlike material from each of the two cohorts was indistinguishable on HPLC analysis by use of two different solvent systems and migrated to the position taken by a des-Arg^{31},Arg^{32} conversion intermediate (34). It was presumed that Arg^{65} had been replaced by another amino acid residue in both examples and that the replacement prevented the further processing of the biosynthetic intermediate by the B cell proinsulin converting enzymes. The abnormal allele from the Japanese cohort has now been cloned and sequenced by Shibasaki et al. (35). Results confirmed the work described above and identified a single nucleotide change (CGC → CAC) that results in the replacement of Arg^{65} by histidine. Proinsulin Tokyo thus gives rise to [His^{65}]des-Arg^{31};Arg^{32}-proinsulin (see Fig. 2.3) and can be identified as human proinsulin 65 (Arg → His).

2. PROINSULIN PROVIDENCE

The third example of familial hyperproinsulinemia involves as the propositus a 12-year-old girl with fasting hyperinsulinemia and borderline glucose intolerance. Findings by Gruppuso et al. (36) were somewhat different from those obtained for Proinsulin Tokyo or Proinsulin Boston, however. First, the 9,000-molecular-weight insulin immunoreactive material obtained from serum represented about 60% (rather than 90%) of total serum insulin immunoreactivity. Second, the high-molecular-weight material (in contrast to related material from the other kindreds) was readily digested by trypsin to yield an insulinlike peptide. Third, HPLC analysis of the material revealed what appeared to be intact proinsulin (rather than an intermediate of proinsulin conversion). Taken together, these results suggested that the causative abnormality might have arisen from a defect in an enzyme necessary for proinsulin conversion rather than from a structural defect in the product of the insulin gene (36).

Very recent studies (involving the cloning and sequencing of the insulin alleles from two members of the kindred), however, have revealed a most interesting result. In each case, one allele was normal whereas the other was not (37). Although the gene sequence of the abnormal allele identified the presence of the usual dibasic amino acid pairs (Arg^{31}-Arg^{32} and Lys^{64}-Arg^{65}) connecting the insulin chains to the C peptide, it also identified a single nucleotide change (CAC → GAC) that predicts the replacement of His^{B10} by aspartic acid (37). Thus, Proinsulin Providence can now be identified as [Asp^{B10}]proinsulin or as human proinsulin B10 (His → Asp) (see Fig. 2.3). A number of important questions remain, and it is not yet clear why the abnormal proinsulin apparently escapes normal precursor processing, whether the presence of the abnormal precursor also interferes with the conversion of the normal insulin gene product, and whether the abnormal gene product might not actually have escaped detection by HPLC analysis and by radioimmunometric methods. Future studies will undoubtedly provide much needed further information.

III. Structure-Function Relationships and Biologic Activity

Although a detailed analysis of the structure-function relationships of the abnormal human insulins (see Fig. 2.3) and their biologic activities both in vitro and in vivo is beyond the scope of this chapter, the subject is one of considerable importance both to the physiology of abnormal insulin secretion in man and to our understanding of how insulin exerts its biologic effects by specific interactions with its plasma membrane receptor. Given that the amounts of abnormal insulins that can be isolated from plasma or even from small biopsies of whole pancreas are low, the success of many of these studies has relied on new methods for the semisynthesis of insulin analogs. Suffice it to say that [Leu^{B25}]insulin, [Ser^{B24}]insulin, and [Leu^{A3}]insulin (Insulins Chicago, Los Angeles, and Wakayama, respectively) have all been prepared by semisynthetic methods (most often in several laboratories) and that their receptor binding and biologic potencies have been found to be greatly decreased relative to normal human insulin (24,38–45); most have only about 1% of the receptor binding affinity of the normal hormone.

Notwithstanding preliminary (and incorrect) suggestions that Insulin Chicago might behave as an insulin antagonist, all of the abnormal insulins show equivalently decreased receptor binding and biologic potency. Although the processing intermediate corresponding to Pro-insulin Tokyo/Boston has yet to be synthesized, the low biologic and receptor-binding potencies of proinsulin conversion intermediates in general, and of des-Arg^{31},Arg^{32}-proinsulin in particular (46), suggest that it too would have no more than about 15% of the biologic activity of native insulin. Thus, the abnormal products of mutant insulin genes are very poorly active in modulating normal target cell responses to insulin. Given the low concentrations of the abnormal forms that circulate in the plasma (even under the hyperinsulinemic conditions usually associated with expression of an abnormal insulin gene), it is doubtful that the abnormal gene products contribute in any meaningful way to glucose homeostasis. Under most circumstances, patients secreting abnormal insulins ex-pressed from a mutant insulin allele must rely for glucose regulation on the product of a normal insulin allele that coexists within the first in the heterozygous state.

Almost all of the amino acid substitutions in the abnormal insulin products of mutant alleles occur at sites in the insulin molecule that have remained invariant during animal evolution. It is noteworthy that the sequences Gly^{B23}-Phe^{B24}-Phe^{B25}-Tyr^{B26} and Gly^{A1}-Ile^{A2}-Val^{A3}-Glu^{A4} or Asp^{A4} (the sites of amino acid replacements in Insulins Chicago, Los Angeles, and Wakayama), in fact, represent the two longest invariant sequences in the insulin molecule. The Lys^{64}-Arg^{65} sequence in proinsulin is also invariant, and replacement of Arg^{65} by histidine in Proinsulin Tokyo is certain to prevent normal cleavage of the precursor to the product. Although the evolutionary invariance of a particular amino acid residue within a protein does not always presage its importance in determining appropriate biologic activity, in the case of the abnormal insulins the dictum holds true; that is, in each case of the abnormal insulin gene products, replacement of the invariant residue has yielded a hormone-related peptide with low biologic potency. In many ways the identification of insulin gene mutations after clinical presentation for glucose intoler-ance represents a bias that has selected for mutations that would cause physiologic perturbations and clinical manifestations. Although someone in the world has probably undergone an insulin gene mutation resulting in replacement of Thr^{B30} by alanine (ACC → GCC), the mutation would go undetected. The result of such a mutation would be that the subject would secrete porcine rather than human insulin, but this would permit totally normal insulin physiology and glucose homeostasis given the equivalent activities of the two hormones.

Insulins Chicago, Los Angeles, and Wakayama each result from an amino acid substitution that occurs in or near the region of the molecule that has previously been included in what is called the active site of the hormone (47,48). As illustrated in Fig. 2.5, this region includes (a) the COOH-terminal region of the B chain, (b) the NH$_2$-terminal region of the A chain, and (c) the very COOH-terminal region of the A chain. Given the

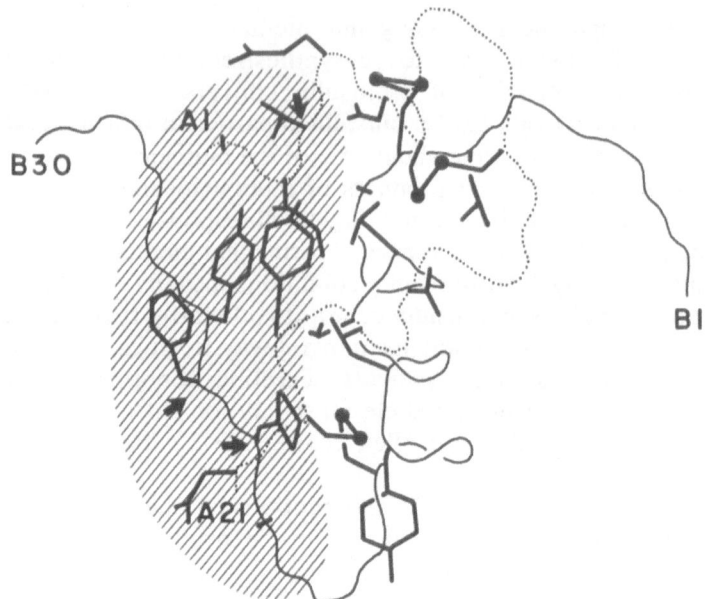

FIGURE 2.5. Representation of the structure of insulin as seen in the 2-zinc insulin hexamer and as determined by X-ray crystallography. The NH$_2$ termini and COOH termini of the A and B chains are indicated. The main peptide chain is shown in all cases, but side chains are shown only for invariant residues. The sites of known amino acid replacements in abnormal insulins are indicated by arrows: residue A3, top; residue B25, upper left; residue B24, lower left. The portion of the molecule thought to be most important in conferring receptor binding affinity and biologic activity of the hormone is indicated by hatching. (Reprinted by permission from *Nature*, Vol. 273, 504–509. Copyright © 1978 Macmillan Journals Limited. From Dodson EJ, Dodson GG, Hubbard RE, Reynolds CD (1983) Insulin's structural behavior and its relationship to activity. Biopolymers 22:281–291. Copyright © 1983 by John Wiley & Sons, Inc. Reprinted by permission of John Wiley & Sons, Inc.)

importance of the first of these regions, it is not surprising that two of the three known abnormal insulins have resulted from amino acid substitutions at positions B24 and B25, both normally filled by phenylalanine. In fact, the identification of these abnormal insulins specifically prompted insulin chemists to study the significance of amino acid replacements in this region. The results have broad importance. First, it has been found that replacements at position B24 have somewhat less of an impact on insulin activity than do replacements at position B25 (24,38–43). Second, though replacements at position B24 alter the overall secondary and tertiary structure of the hormone, those at position B25 do not (42,49). Third, though both residues have the potential for interaction with the receptor, the side chain at position B25 has an especially favorable disposition for direct receptor contact (47,48). Fourth, amino acid replacements at position B25 have been shown to affect markedly the potential for insulin-receptor interactions that seem to involve negative homotropic interactions (i.e., negative cooperativity) (39–41,50–52). Fifth, changes in the phenylalanine side chain at position B25, in particular, seem more dependent on the loss of the natural residue per se than they do on the nature of the amino acid replacement: substitution of phenylalanine by leucine (a branched, hydrophobic amino acid) or by serine (a small, hydrophilic amino acid) results in insulin analogues with equivalently perturbed biologic activity (about 1% of normal) (24,38–43).

Further studies based on the replacement of Phe[B25] by natural and unnatural amino acids have been most useful in identifying both the nature of the negative effect attending amino acid substitution at position B25 and possible conformational changes that normally occur when insulin interacts with its receptor (52). These findings, while being a direct outcome of investigations of abnormal human insulins resulting from insulin gene mutations, are of considerable importance in understanding fundamental aspects of insulin-receptor interactions and of the mechanism by which the insulin signal is transduced across the target cell plasma membrane.

IV. Clinical Relevance

A. Inheritance and Physiology

In all examples studied so far, individuals having insulin gene mutations resulting in the secretion of an abnormal insulin, an abnormal intermediate of proinsulin conversion, or an abnormal proinsulin have been shown or are suspected to represent the heterozygous situation—that is, a situation where one insulin allele has undergone a nucleotide replacement and where the other insulin allele is completely normal (17–19,21–28,53). In all kindreds as well, the insulin gene abnormality appears to be inherited in an autosomal-dominant pattern (whether assessed by HPLC analysis, Southern analysis, or radioreceptor assays) (17,18,21,26–28,31,35,36), and it is likely that all cases reflect the inheritance of nonlethal mutant insulin genes rather than the appearance of insulin gene mutations de novo. For both Insulin Los Angeles (17,54) and Insulin Wakayama (26,28), the abnormal allele has been traced through three generations, involving the mother or father of the propositus, the siblings of the propositus, and either the children of the propositus or the children of affecting siblings. Remarkably, in the example of Insulin Los Angeles, all three siblings of the propositus have inherited the abnormal insulin allele from their affected father (17,54). The pedigree for inheritance of Proinsulin Boston (as assessed by direct measurements of serum *insulin* immunoreactivity) is the most detailed and complete of any studied so far (31).

The heterozygous state does not necessarily imply codominant expression of normal and abnormal alleles, however, and here our knowledge is rather incomplete. Concrete evidence for codominant expression of normal and abnormal alleles had been obtained in only two examples of insulin gene mutations—i.e., in those examples where assessment of the pancreatic insulin stores was possible. Both in the patient with Insulin Chicago (19) and in one patient with Insulin Wakayama (27), insulin extracted from pancreatic biopsies was shown to contain approximately equimolar amounts of the respective abnormal insulin and the normal hormone. In both of these examples, histologic examination of the biopsies showed hyalinization of pancreatic B cells and either fibrosis or an increase in intracellular vacuoles (16,27). All matters considered, it is probable that codominant expression of normal and abnormal alleles is the rule rather than the exception.

Although the finding was not emphasized earlier, small amounts of normal insulin have always been identified by HPLC analysis in the serum of individuals who secrete abnormal insulin gene products (18,21,24,28,34,54). The normal insulin, in general, comprises only about 5–10% of the total, but the amount is small only in comparison to the much greater amount of the abnormal hormone: in a typical patient exhibiting a fasting insulin level of 100 μU/ml, about 10 μU/ml arises from expression of the normal insulin allele, and about 90 μU/ml arises from expression of the abnormal allele (21,54). To put the matter another way, levels of normal insulin (at least in subjects without severe glucose intolerance) are often not far from the normal range; the hyperinsulinemia associated with abnormal insulin gene expression therefore arises from the measurement of the mutant form of the hormone. The important matters of insulin degradation and insulin clearance from the plasma identify (a) the source of fasting hyperinsulinemia per se, (b) the explanation for the abnormal hormone predominating in the peripheral blood (under the circumstance that both normal and abnormal forms are cosecreted from the pancreatic B cell in approximately equimolar amounts), (c) the reason for the sustained elevated insulin levels that are achieved by most affected subjects upon B cell stimulation by oral glucose, and (d) the source of the decreased C-peptide:insulin molar ratio that is typical of subjects secreting the product of an abnormal insulin gene.

It has long been recognized that most insulin degradation at cells occurs by means of receptor-mediated processes that require the prior and specific binding of hormone to plasma membrane receptors (55,56). Given that the receptor binding and biologic potencies of abnormal human insulin gene products are greatly diminished compared to those of normal insulin (see above), it is not surprising that abnormal insulins would be degraded and cleared from the circulation more slowly than normal insulin and would thus appear in the serum in concentrations greater than those found for the normal hormone. Although a few studies of insulin degradation in vitro have suggested the slow metabolism of the abnormal human insulins (44,49,57), the most pertinent data have been accumulated from experiments using whole animals. Thus, Insulin Los Angeles (human [Ser^{B24}]insulin) (54,58) and Insulin Wakayama (human [Leu^{A3}]insulin) (26) have been shown to have greatly increased biologic half-lives in the circulation of dogs as assessed both by insulin levels achieved during constant peripheral infusions at steady state and by die-away curves obtained after termination of the hormone infusions. An even more exacting test involves the graded intraportal infusion of an equimolar mixture of abnormal insulin and normal insulin in pancreatectonized dogs, as shown in Fig. 2.6 (54). Results obtained both by measurement of total serum insulin levels and by HPLC analysis of serum insulin demonstrated that (a) the hyperinsulinemia of abnormal insulin secretion is easily mimicked by the equimolar infusion of normal and abnormal insulins, (b) the hyperinsulinemia is sustained relative to that which would be observed from the infusion of normal insulin alone, and (c) the fraction of total insulin in the peripheral circulation that arises from the normal hormone during the coinfusion amounts to only about 0.05 during a low basal rate of infusion and to only 0.2 during the infusion of the insulin mixture at four times the basal rate. The magnitude of total and abnormal insulin levels that are achieved and the prolonged elevations of those levels both arise from the slow clearance of the abnormal form from the plasma. Since C peptide and insulin are cosecreted from the pancreatic B cell in equimolar amounts, a decreased overall metabolic clearance of immunoreactive insulin (under circumstances where C peptide is cleared at the normal rate) could easily give rise to the decreased C peptide:insulin molar ratio that is typically observed in patients expressing abnormal insulin genes.

B. Diabetes

Although many of the individuals identified as propositi in kindreds expressing abnormal insulin genes exhibit fasting hyperglycemia, it is often the case that affected family members are asymptomatic (17,18,25,26,28,31,36,53,54); that is, within any kindred, many or even most individuals expressing an abnormal insulin gene exhibit only a mild degree of intolerance to oral glucose. Figure 2.7 shows the responses of the propositus of the kindred expressing [Ser^{B24}]insulin (Insulin Los Angeles) and those of two of her siblings (who also express the abnormal insulin) to oral glucose (54). First, it should be noted that the propositus exhibits a diabetic glucose tolerance curve whereas her siblings exhibit only a degree of glucose intolerance. Second, it should be noted that the insulin and C peptide responses of the propositus to oral glucose are markedly blunted when compared to the related responses of her siblings.

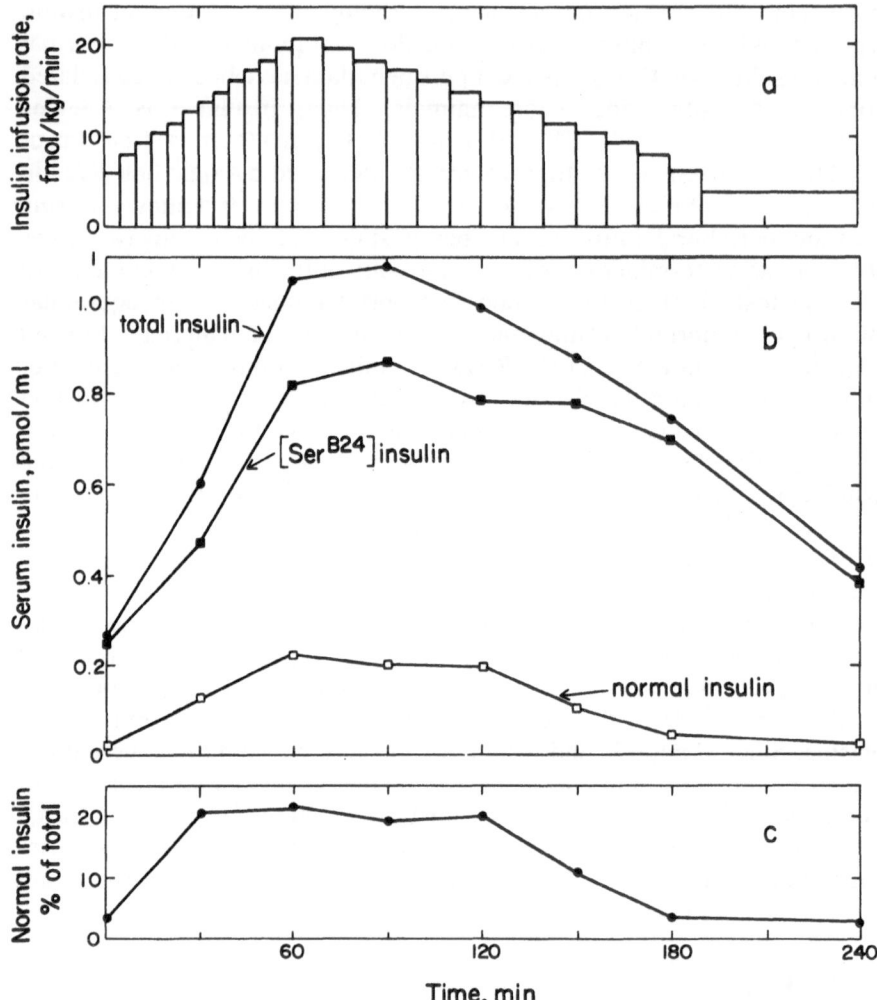

FIGURE 2.6. HPLC analysis of serum insulin derived from the graded intraportal infusion of an equimolar mixture of human [Ser^B24]insulin (Insulin Los Angeles) and normal human insulin in a pancreatectomized dog. Details of the animal preparation are described in Shoelson et al. (54). (a) Rate of infusion of total insulin for each time segment after a 60-min period in which the rate of infusion was 3.3 pmol/kg/min. (b) Results of HPLC analysis applied to individual serum samples to separate and quantitate human [Ser^B24]insulin and normal human insulin. (c) Percentage of total insulin identifiable as normal human insulin at each time point. Data at time zero show values at the close of the constant basal infusion. The abcissa records the time after initiation of the graded intraportal infusion. Similar results were found for two other dogs infused with the insulin mixture in an equivalent way. (Reproduced from *The Journal of Clinical Investigation*, 1984, 73:1351–1358 by copyright permission of The American Society for Clinical Investigation.)

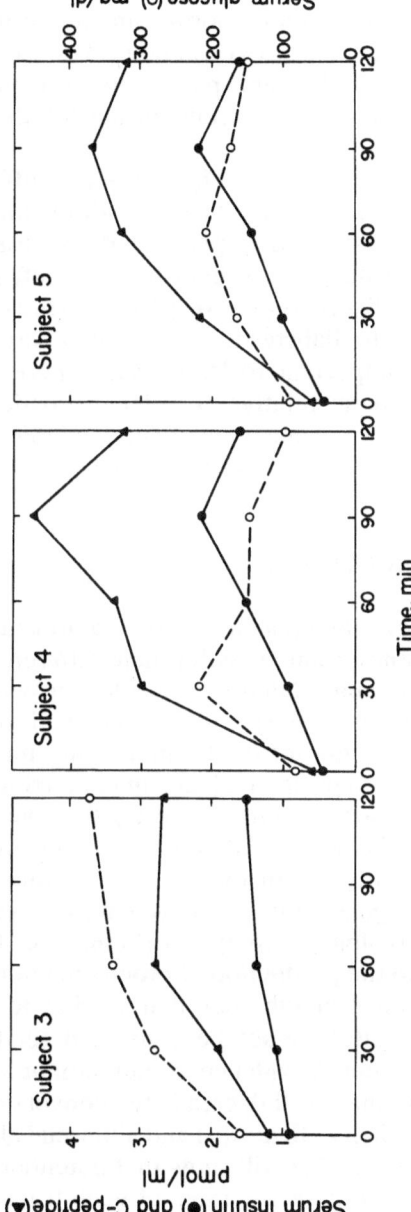

FIGURE 2.7. Insulin, C peptide, and glucose responses of three subjects expressing Insulin Los Angeles ([Ser^{B24}]insulin) during an oral glucose tolerance test. Subject 3 is the diabetic propositus; subjects 4 and 5 are two of her nondiabetic siblings. Details of the measurements and of the oral glucose tolerance test are provided in Shoelson et al. (54). Insulin (●) and C peptide (▲) concentrations are plotted using the left-hand ordinate scale; glucose concentrations (○), using the right-hand ordinate scale. The abscissa shows the time after the administration of oral glucose that blood was drawn for analysis. (Reproduced from The Journal of Clinical Investigation, 1984, 73:1351–1358 by copyright permission of The American Society for Clinical Investigation.)

Thus, it appears that nondiabetic individuals expressing abnormal insulins can compensate, at least in part, for the fact that only 50% of pancreatic insulin is secreted in biologically active form (material expressed from the normal allele) by the more vigorous and perhaps more sustained release of total insulin (material expressed from both normal and abnormal alleles). The diabetic propositus of Fig. 2.7, notwithstanding her elevated total serum insulin level, responds only very poorly to oral glucose and apparently secretes insufficient insulin to compensate for the glycemic load.

The conclusion arising from the data of Fig. 2.7 is not different from that made in other examples of insulin secretion within kindreds expressing an abnormal insulin allele. It appears that expression of an abnormal insulin allele cannot itself cause diabetes but that, perhaps in concert with a defect in insulin secretory response, expression of an abnormal allele can contribute to diabetes in an important way. Needless to say, it will be extremely important to learn over a period of years whether or not the asymptomatic relatives of diabetic patients (all of whom express abnormal insulin genes) eventually exhibit increasing intolerance to oral glucose and perhaps overt diabetes.

V. Summary and Perspectives

Studies reported in this review have resulted in the identification of five different abnormal insulin gene products within nine different kindreds. Analysis of the gene products and related peptides has provided important information on (a) the roles of specific amino acid residues in directing the high-affinity interaction of the hormone with its target cell receptor, (b) the participation of specific residues in conferring to insulin its normal secondary and tertiary structure, and (c) the role of dibasic amino acid pairs in directing the appropriate conversion of proinsulin to insulin during hormone biosynthesis. In the cases of Insulins Chicago, Los Angeles, and Wakayama, gene mutations have led to the synthesis of abnormal insulins with low biologic activity. In Proinsulins Tokyo and Boston, mutations have led to the production of processing intermediates that contain the usual insulin structure but that (owing to structural alterations in the processing site) cannot be converted to the normal product. In the case of Proinsulin Providence, a mutation at a distal site within the structure of insulin may well decrease the conversion of even normal precursor to product when both normal and abnormal allelic genes are coexpressed. Reference to Fig. 2.2 will show that potential mutations in the insulin gene could as well alter the rates of gene transcription at promoter, initiator, and enhancer sites; change the efficiency or pattern of excision of the intervening sequences; alter mRNA-ribosome interactions in the 5′ untranslated region; decrease the efficiency of preproinsulin translocation by perturbations of the signal sequence; or alter the termination site of protein biosynthesis (resulting in the formation of either truncated or elongated proinsulin-related molecules). Each would have its own unique and deleterious effect on the synthesis of biologically active insulin.

As noted earlier, each example of an abnormal insulin gene being expressed within a human kindred has involved an autosomal-dominant pattern of inheritance, heterozygosity of normal and abnormal insulin alleles, and the presumed codominant expression of both alleles. In some ways, these individuals may be considered to bear a physiologic burden resulting from the fact that only about half of their stored insulin is structurally normal and has full biologic activity. The severity of diabetes in affected individuals is extremely variable, however, and depends on other factors in addition to the expression of an abnormal insulin allele per se. Such factors certainly include the competency of the total pancreatic B cell mass to produce the hormone and, probably most important, the magnitude of secretory responses of the pancreatic B cell arising from stimulation by glucose and other secretogogues. Varying degrees of insulin resistance may also play a role in specific patients expressing abnormal insulins. In all cases, it is probably fair to say that most non-insulin-resistant, heterozygous individuals would be euglycemic and asymptomatic if only the B cell response to stimulation were of a great enough magnitude. It remains to be seen, of course, whether the secretion of an abnormal insulin has additional physiologic or pathophysiologic consequences. Although no homozygous patient with abnormal insulin alleles has yet been identified, it is probable that the occurrence of homozygosity would under most circumstances be extremely rare and would result in overt and severe diabetes.

Additional examples of peptide hormone variants have been reported. However, in several cases (including a variant of human growth hormone arising from alternative splicing of an mRNA precursor (59,60), a variant of human insulinlike growth factor II arising from replacement of Ser[33] by the sequence of Cys-Gly-Asp (61), and a variant of insulinlike growth factor I involving deletion of the NH$_2$-terminal three residues of the hormone (62)), the variant peptides most probably arise from expression of normal hormone genes, from expression of hormone-related genes, or from posttranslational changes, rather than from gene mutations. In a few cases (involving both human luteinizing hormone (63) and human nerve growth factor (64)), the production of an abnormal hormone has been proposed to be linked to a specific pathophysiologic state. Although relatively little information is available, it is likely that mutant forms of all of the peptide hormones will eventually be identified and that their physiologic and clinical correlates will eventually be documented. The understanding of the consequences of such mutations, and of the consequences of mutations within the human insulin gene, will require many years of collaborative investigation and the combined efforts of investigators in the disciplines of genetics, biochemistry, cell biology, histology, physiology, and the clinical sciences.

Acknowledgments. I thank my many colleagues and co-workers who have contributed to work from this laboratory on the subject of abnormal human insulins. These individuals include Drs. R. Assoian, S. Chan, M. Fickova, B. Given, M. Haneda, K. Inouye, E.T. Kaiser, Y. Kanazawa, S. Kwok, J. Markese, K. Nanjo, S. Nakagawa, J. Olefsky, K. Polonsky, R. Poucher, D. Robbins, S. Shoelson, W. Wishner, and, most of all, A. Rubenstein and D. Steiner. Work performed in this laboratory was supported by grants DK 18347 and DK 20595 from the National Institutes of Health.

References

1. Elliott RB, O'Brien D, Roy CC (1966) An abnormal insulin in juvenile diabetes mellitus. Diabetes 14:780–787.
2. Kimmel JR, Pollack HG, (1967) Studies of human insulin from nondiabetic and diabetic pancreas. Diabetes 16:687–694.
3. Steiner DF, Cunningham DD, Spigelman S, Aten B (1967) Insulin biosynthesis: Evidence for a precursor. Science 157:697–700.
4. Steiner DF, Clark JL, Nolan C, Rubenstein AH, Margoliash E, Aten B, Oyer PE (1969) Proinsulin and the biosynthesis of insulin. Recent Prog Horm Res 25:207–282.
5. Peavy DE, Abram JD, Frank BH, Duckworth WC (1984) In vitro activity of biosynthetic human proinsulin. Diabetes 33:1062–1067.
6. Steiner DF (1977) Insulin today. Diabetes 26:322–340.
7. Tager HS, Patzelt C, Assoian RK, Chan SJ, Duguid JR, Steiner DF (1980) Biosynthesis of islet cell hormones. Ann NY Acad Sci 343:133–147.
8. Steiner DF, Quinn PS, Chan SJ, Marsh J, Tager HS (1980) Processing mechanisms in the biosynthesis of proteins. Ann NY Acad Sci 343:1–16.
9. Given BD, Cohen RM, Shoelson SE, Frank BH, Rubenstein AH, Tager HS (1985) Biochemical and clinical implications of proinsulin conversion intermediates. J. Clin Invest 76:1398–1405.
10. Chan SJ, Keim P, Steiner DF (1976) Cell-free synthesis of rat preproinsulins: Characterization and partial amino acid sequence determination. Proc Natl Acad Sci USA 73:1964–1968.
11. Orci L, Ravazzola M, Amherdt M, Madsen O, Vassalli J-D, Perrelet A (1985) Direct identification of prohormone conversion site in insulin-secreting cells. Cell 42:671–681.
12. Rubenstein AH, Steiner DF, Horwitz DL, Mako ME, Block MB (1977) Clinical significance of circulating proinsulin and C-peptide. Recent Prog Horm Res 33:435–475.
13. Bell GI, Pictet RL, Rutter WJ, Cordell B, Tischer E, Goodman HM (1980) Sequence of the human insulin gene. Nature 284:26–32.
14. Ullrich A, Dull TJ, Gray A, Brosius J, Sives I (1980) Genetic variation in the human insulin gene. Science 209:612–615.
15. Chan SJ, Nanjo K, Miyano M, Lu Y-J, Welsh M, Nielsen D, Haneda M,

Kwok SCM, Tager HS, Rubenstein AH, Steiner DF (1986) Abnormalities of insulin gene structure and expression. In Serrano-Rios, M & Lefebre PJ (eds): Diabetes 1985., pp 486–494. New York, Elsevier.
16. Given BD, Mako ME, Tager HS, Baldwin D, Markese J, Rubenstein AH, Olefsky J, Kobayashi M, Kolterman O, Poucher R (1980) Diabetes due to secretion of an abnormal insulin. N Engl J Med 302:129–135.
17. Haneda M, Polonsky KS, Bergenstal RM, Jaspan JB, Shoelson SE, Blix PM, Chan SJ, Kwok SCM, Wishner WB, Zeidler A, Olefsky JM, Friedenberg G, Tager HS, Steiner DF, Rubenstein AH (1984) Familial hyperinsulinemia due to a structurally abnormal insulin: Definition of an emerging new clinical syndrome. N Engl J Med 310:1288–1294.
18. Nanjo K, Sanke T, Miyano M, Okai K, Sowa R, Kondo M, Nishimura S, Iwo K, Miyamura K, Given BD, Chan SJ, Tager HS, Steiner DF, Rubenstein AH (1986) Diabetes due to secretion of a structurally abnormal insulin (Insulin Wakayama). J Clin Invest 77:514–519.
19. Tager H, Given B, Baldwin D, Mako M, Markese J, Rubenstein AH, Olefsky J, Kobayashi M, Kolterman O, Poucher P (1979) A structurally abnormal insulin causing human diabetes. Nature 281:122–125.
20. Inouye K, Watanabe K, Morihara K, Tochino Y, Kanaya T, Emura J, Sakakibara S (1979) Enzyme-assisted semisynthesis of human insulin. J Am Chem Soc 101:751–752.
21. Shoelson S, Haneda M, Blix P, Nanjo K, Sanke T, Inouye K, Steiner D, Rubenstein A, Tager HS (1983) Three mutant insulins in man. Nature 302:540–543.
22. Kwok SCM, Chan SJ, Rubenstein AH, Poucher R, Steiner DF (1981) Loss of restriction endonuclease cleavage site in the gene of a structurally abnormal insulin. Biochem Biophys Res Commun 98:844–849.
23. Kwok SCM, Steiner DF, Rubenstein AH, Tager HS (1983) Identification of the mutation giving rise to Insulin Chicago. Diabetes 32:872–875.
24. Shoelson S, Fickova M, Haneda M, Nahum A, Musso G, Kaiser ET, Rubenstein AH, Tager HS (1983) Identification of a mutant insulin predicted to contain a serine-for-phenylalanine substitution. Proc Natl Acad Sci USA 80:7390–7394.
25. Haneda M, Chan SJ, Kwok SCM, Rubenstein AH, Steiner DF (1983) Studies on mutant insulin genes: Identification and sequence analysis of a gene encoding [Ser^{B24}]insulin. Proc Natl Acad Sci USA 80:6366–6370.
26. Nanjo K, Given B, Sanke T, Kondo M, Miyano M, Okai K, Miyama K, Chan S, Tager H, Steiner D, Polonsky K, Rubenstein A (1986) Pancreatic function in the mutant insulin syndrome. Diabetes 35 (Suppl 1):77A.
27. Iwamoto Y, Sakura H, Yui R, Fujita T, Sakamoto Y, Matsuda A, Kuzuya T (1986) Identification and characterization of a mutant insulin isolated from the pancreas of a patient with abnormal insulinemia. Diabetes 35 (Suppl 1):77A.
28. Iwamoto Y, Sakura H, Ishii Y, Yamamoto R, Kumakura S, Sakamoto Y, Masuda A, Kuzuya T (1986) Radioreceptor assay for serum insulin as a useful method for detection of abnormal insulin with a description of a new family of abnormal insulinemia. Diabetes 35:1237–1242.
29. Seino S, Funakoshi A, Fu ZZ, Vinik A (1985) Identification of insulin variants in patients with hyperinsulinemia by reversed-phase, high performance liquid chromatography. Diabetes 34:1–7.
30. Sanz N, Karam JH, Horita S, Bell GI (1985) DNA screening for insulin gene mutations in non-insulin-dependent diabetes mallitus (NIDDM). Diabetes 34 (Suppl 1):85A.
31. Gabbay KH, DeLuca K, Fisher NJ Jr, Mako ME, Rubenstein AH (1976) Familial hyperproinsulinemia: An autosomal dominant defect. N Engl J Med 249:911–915.

32. Gabbay KH, Bergenstal RM, Wolff J, Mako ME, Rubenstein AH (1979) Familial hyperproinsulinemia: Partial characterization of circulating pro-insulin-like material. Proc Natl Acad Sci USA 76:2882–2885.
33. Robbins DC, Blix PM, Rubenstein AH, Kanazawa Y, Kosaka K, Tager HS (1981) A human proinsulin variant at arginine 65. Nature 291:679–681.
34. Robbins DC, Shoelson SE, Rubenstein AH, Tager HS (1984) Familial hyperproinsulinemia: Two cohorts indistinguishable type II intermediates of proinsulin conversion. J Clin Invest 73:714–719.
35. Shibasaki Y, Kawakami T, Kanazawa Y, Akamura Y, Takaku T (1985) Posttranslational cleavage of proinsulin is blocked by a point mutation in familial hyperproinsulinemia. J Clin Invest 76:378–380.
36. Gruppuso PA, Gorden P, Kahn RC, Cornblath M, Zeller WP, Schwartz R (1984) Familial hyperproinsulinemia due to a proposed defect in conversion of proinsulin to insulin. N Engl J Med 311:629–634.
37. Chan SJ, Seino S, McCormick MB, Gruppuso PA, Schwartz R, Steiner DF (1986) An unusual mutation in the insulin gene associated with familial hyperproinsulinemia. Diabetes 35 (Suppl 1):97A.
38. Tager H, Thomas N, Assoian R, Rubenstein A, Saekow M, Olefsky J, Kaiser ET (1980) Semisynthesis and biological activity of porcine ([Leu^B24]insulin and [Leu^B25]insulin. Proc Natl Acad Sci USA 77:3181–3185.
39. Inouye K, Watanabe K, Tochino Y, Kobayashi M, Shigeta Y (1981) Semi-synthesis and properties of some insulin analogs. Biopolymers 20:1845–1858.
40. Kobayashi M, Ohgaku S, Iwasaki M, Maegawa H, Shigeta Y, Inouye K (1982) Characterization of [Leu^B24]- and [Leu^B25]-insulin analogs. Biochem J 206:597–603.
41. Kobayashi M, Haneda M, Maegawa H, Watanabe N, Takato Y, Shigeta Y, Inouye K (1984) Receptor binding and biological activity of [Ser^B24]-insulin, an abnormal mutant insulin. Biochem Biophys Res Commun 119:49–57.
42. Wolmer A, Strassburger W, Glatler V, Dodson GG, McCall M, Danho W, Brandenburg D, Gattner H-G, Rittel W (1981) Two mutant forms of human insulin: Structural consequences of the substitution of invariant B24 or B25 by leucine. Hoppe Seylers Z Physiol Chem 362:581–592.
43. Gattner H-G, Danho W, Ben C, Zahn H (1980) Hoppe Seylers Z Physiol Chem 361:1135–1138.
44. Kobayashi M, Takata Y, Ishibashi O, Sasoka T, Iwasaki M, Shigeta Y, Inouye K (1986) Receptor binding and negative cooperativity of a mutant insulin [Leu^A3]-insulin. Biochem Biophys Res Commun 137:250–257.
45. Haneda M, Kobayashi M, Maegawa H, Watanabe N, Takata Y, Ishibashi O, Shigeta Y, Inouye K (1985) Decreased biological activity and degradation of human [Ser^B24]-insulin, a second mutant insulin. Diabetes 34:568–573.
46. Peavy DE, Brunner MR, Duckworth WC, Hooker CS, Frank BH (1985) Receptor binding and biological potency of several split forms (conversion intermediates) of human proinsulin. J Biol Chem 26:13989–13994.
47. DeMeyts P, Van Obberghen E, Roth J, Wollmer A, Brandeburg D (1978) Mapping of the residues responsible for the negative cooperativity of the receptor-binding region of insulin. Nature 273:504–509.
48. Dodson EJ, Dodson GG, Hubbard RE, Reynolds CD (1983) Insulin's structural behavior and its relationship to activity. Biopolymers 22:281–291.
49. Assoian RK, Thomas NE, Kaiser ET, Tager HS (1982) [Leu^B24]insulin and [Ala^B24]insulin: Altered structures and cellular processing of B24-substituted insulin analogs. Proc Natl Acad Sci USA 79:5147–5151.
50. Olefsky JM, Green A, Ciaraldi TP, Saekow M, Rubenstein AH, Tager HS (1981) Relationship between negative cooperativity and insulin action. Biochemistry 20:4488–4492.

51. Keefer LM, Piron M-A, DeMeyts P, Gattner HG, Diaconescue C, Saunders D, Brandenburg D (1981) Impaired negative cooperativity of the semisynthetic analogs human [Leu$_{B24}$]- and [Leu$_{B25}$]-insulins. Biochem Biophys Res Commun 100:1229–1236.
52. Nakagawa SH, Tager HS (1986) Role of the phenylalanine B25 side chain in directing insulin interaction with its receptor. J Biol Chem 261:7332–7341.
53. Robbins DC, Tager HS Rubenstein AH (1984) Biologic and clinical importance of proinsulin. N Engl J Med 310:1165–1175.
54. Shoelson SE, Polonsky KS, Zeidler A, Rubenstein AH, Tager HS (1984) Human insulin (Phe → Ser): Secretion and metabolic clearance of the abnormal insulin in man and in a dog model. J Clin Invest 73:1351–1358.
55. Terris S, Steiner DF (1975) Binding and degradation of ^{125}I-insulin by rat hepatocytes. J Biol Chem 250:8389–8398.
56. Terris S, Steiner DF (1976) Retention and degradation of ^{125}I-insulin by perfused livers from diabetic rats. J Clin Invest 57:885–896.
57. Green A, Tager H, Rubenstein A, Frank B, Olefsky J (1982) Internalization and degradation of a low affinity insulin analog ([Leu^{B24}]insulin) by rat adipocytes. FEBS Lett 144:255–258.
58. Kobayashi M, Haneda M, Ishibashi O, Takata Y, Maegawa H, Watanabe N, Shigeta Y (1984) Prolonged disappearance rate of a structurally abnormal mutant insulin from the blood. Diabetes (Suppl 1) 33:17A.
59. Chapman GE, Rogers KM, Brittain T, Bradshaw RA, Bates OH, Turner C, Cary PD, Crane-Robinson C (1981) The 20,000 molecular weight variant of human growth hormone. J Biol Chem 256:2395–2401.
60. Wallis M (1980) Growth hormone: Deletions in the protein and introns in the gene. Nature 284:512.
61. Zumstein PP, Luthi C, Humbel RE (1985) Amino acid sequence of a variant proform of insulin-like growth factor II. Proc Natl Acad Sci USA 82:3169–3172.
62. Sara VR, Carlsson-Skwirut C, Andersson C, Hall E, Sjogren B, Holmgren A, Jornvall H (1986) Characterization of somatomedins from human fetal brain: Identification of a variant form of insulin-like growth factor I. Proc Natl Acad Sci USA 83:4904–4907.
63. Beitins IZ, Axelrod L, Ostrea T, Little R, Badger TM (1981) Hypogonadism in a male with an immunologically active, biologically inactive luteinizing hormone: Characterization of the abnormal hormone. J Clin Endocrinol Metab 52:1143–1149.
64. Schwartz JP, Breakefield XO (1980) Altered nerve growth factor in fibroblasts from patients with familial dysautonomia. Proc Natl Acad Sci USA 77:1154–1158.

3
Euthyroid Hyperthyroxinemia

MARJORIE SAFRAN and LEWIS E. BRAVERMAN

The circulating concentration of thyroxine (T_4) usually correlates with the clinical status of the patient. Serum T_4 concentrations are almost always increased in thyrotoxicosis and usually decreased in hypothyroidism. However, since T_4 is tightly bound to plasma proteins in the circulation, alterations in the serum-binding proteins (particularly thyroxine-binding globulin-TBG) may cause corresponding alterations in the total T_4 concentration (Fig. 3.1). In these situations, the free T_4 (FT_4) concentration remains normal, and clinical status does not correlate with total T_4 levels, since only the free hormone is believed to be biologically active (1–3). Tests such as the free thyroxine index (FT_4I) and direct measurement of the serum FT_4 concentration have been developed to unmask this problem. The FT_4I, an indirect measure of the FT_4 concentration, is calculated as the product of the serum T_4 concentration and the resin triiodothyronine (T_3) uptake (RT_3U), an in vitro test used to estimate unoccupied binding sites on serum TBG. FT_4 concentrations can be determined by equilibrium dialysis, a laborious procedure, or by radioimmunoassay (RIA) using either ^{125}I-labeled T_4 (2-step) or ^{125}I-labeled T_4 analogs (1-step) methods that are widely available. Therefore, measurement of the serum FT_4 concentration or calculation of the FT_4I have been suggested as the best screening methods to evaluate thyroid function (4).

Despite this refinement, there remain a number of clinical situations in which the FT_4I or FT_4 concentration are increased in the absence of thyroid disease. In such situations, patients are clinically euthyroid despite elevations in these values, and the basal serum thyrotropin (thyroid-stimulating hormone; TSH) level and the serum TSH response to thyrotropin-releasing hormone (TRH) are almost always normal. To avoid unnecessary treatment of these patients, it is important to recognize those situations in which an increase in serum T_4 or FT_4 concentrations or the FT_4I does not represent hyperthyroidism.

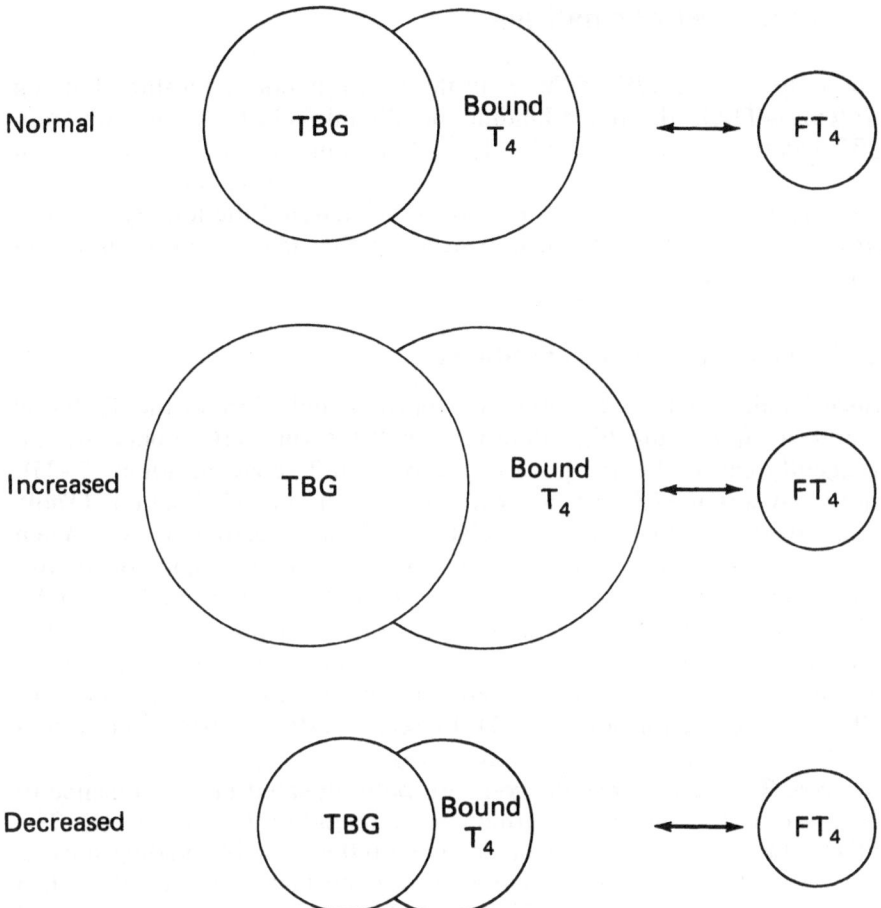

FIGURE 3.1. Effect of changes in serum TBG concentration on serum total T_4 and FT_4 concentrations.

TABLE 3.1. Affinity constants and binding capacities of thyroxine-binding proteins.

	Affinity constant $K_a(M^{-1})$	Binding capacity (μg T_4/100 ml serum)
TBG	2×10^{10}	20
TBPA	1.5×10^8	300
Albumin	1.5×10^6	[a]

Source: Refetoff (5).
[a] Greater than 800 μg T_4/100 ml serum.

I. Binding Abnormalities

Approximately 99.97% of the T_4 in the serum is bound by three binding proteins—TBG, thyroxine-binding prealbumin (TBPA), and albumin. TBG has the highest affinity for T_4, and albumin the least (Table 3.1). In contrast, albumin has the greatest T_4-binding capacity, and TBG has the least (1,2). Increases in both the concentration and the affinity of these proteins for T_4 have been described, and either may result in euthyroid hyperthroxinemia.

A. Elevated TBG Concentrations

Since binding to TBG accounts for approximately 75% of the T_4 that is bound at equilibrium (6), alterations in the serum TBG concentration frequently cause abnormalities in serum total T_4 concentrations (7–25). Increased serum TBG can occur in a variety of clinical situations (Table 3.2). Several of these are associated with hyperestrogenemia. When circulating estrogen levels are increased, as during pregnancy or the use of oral contraceptives, hepatic synthesis of TBG is increased (26). On the other hand, it is likely that hepatocellular damage allows leakage of TBG into the circulation in hepatic disorders such as acute viral hepatitis or chronic active liver disease (21), although increased hepatic synthesis of TBG has also been postulated (27). Increased TBG synthesis alco occurs in patients with familial TBG excess (19).

Serum T_4 and T_3 concentrations are both elevated in the presence of increased TBG, since both hormones are bound by the protein (1). The RT_3U is low because it varies inversely with the available binding sites on TBG. Therefore, the FT_4I is almost always normal. Occasionally, when TBG concentrations are very high, the RT_3U is not sufficiently decreased in proportion to the TBG increase, probably owing to methodologic problems, and the FT_4I is slightly increased. However, the presence of a low RT_3U is an important clue that the patient is not hyperthyroid, since the RT_3U is usually high normal or elevated in patients with thyrotoxicosis. The serum TBG concentration can be directly measured by RIA to confirm the presence of an elevated TBG. The serum FT_4 concentration is normal in these patients (28–30).

B. Familial Dysalbuminemic Hyperthyroxinemia (FDH)

Albumin ordinarily has a low affinity for T_4 and T_3 and normally accounts for only 5–10% of the T_4 and T_3 binding in serum (6). As first described, patients with FDH were believed to have an abnormal circulating albumin which has a greater affinity for T_4 than for T_3 (31–37). Thus, serum T_4 concentrations were elevated, and serum T_3 concentrations were in the normal range or only slightly elevated in these patients (31,34–37). The

TABLE 3.2. Euthyroid hyperthyroxinemic states associated with increased serum TBG concentration.

Hyperestrogenemic states
  Pregnancy
  Newborn
  Estrogen therapy
  Oral contraceptive agents
  Estrogen-producing tumor
  Hydatidiform mole[a]
  Choriocarcinoma[a]
Drugs
  Heroin
  Methadone
  Perphenazine
  Tamoxifen
  5-Fluorouracil
Diseases
  Acute intermittent porphyria
  Acute viral hepatitis
  Chronic active liver disease
Genetic
  X-linked familial increase in serum TBG
  X-linked hypogammaglobulinemia

[a] Some patients are hyperthyroid.

abnormal albumin bound T_3 to a small extent, but the binding affinity for T_3 relative to that of T_4 was much less than that of TBG (37–39). Since the abnormal albumin binds relatively little T_3, the RT_3U, which is performed using ^{125}I-labeled T_3, is normal, and the FT_4I is elevated. In contrast, the FT_4 by equilibrium dialysis is normal. Since the latter test uses ^{125}I-labeled T_4 as the tracer and the abnormal albumin avidly binds T_4, the percent FT_4 is very low. Direct measurement of the serum FT_4 concentration by those RIA methods that employ ^{125}I-labeled T_4 yields normal serum FT_4 values in patients with FDH (33,40–42). However, the one-step RIA methods that employ ^{125}I-labeled T_4 analogs, which do not bind appreciably to normal serum proteins, result in spuriously elevated FT_4 values, because these analogs are bound by the abnormal albumin (40–42).

FDH is inherited in an autosomal-dominant pattern. Patients are clinically euthyroid, but serum thyroid hormone concentrations are sometimes measured because of complaints or symptoms suggestive of thyrotoxocosis. Patients rarely have goiters, and the 24-h radioactive iodine uptake (RAIU) and the TSH response to TRH are both normal. The definitive diagnosis of FDH is made by the demonstration of increased T_4 binding by the abnormal serum albumin using paper or gel electrophoresis of serum enriched with tracer concentrations of ^{125}I-

PAGE OF NORMAL AND FDH SERA

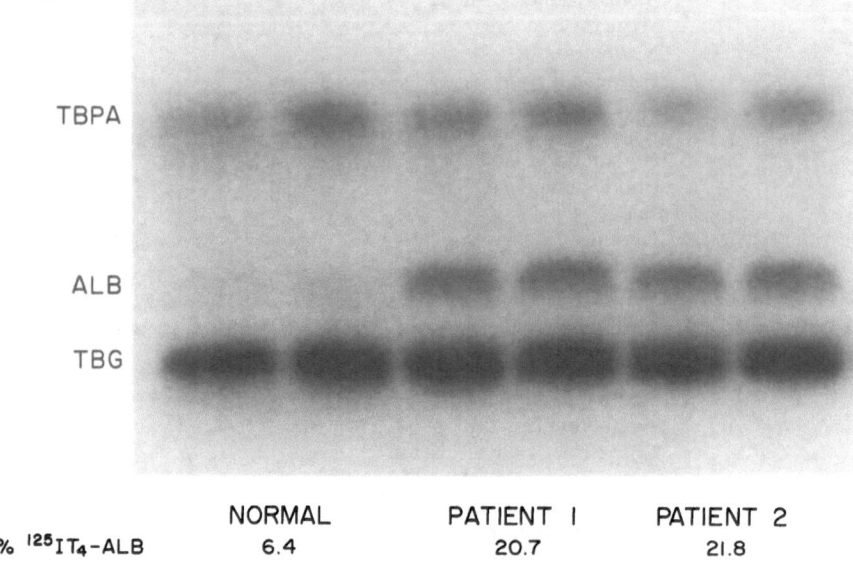

	NORMAL	PATIENT 1	PATIENT 2
% $^{125}IT_4$–ALB	6.4	20.7	21.8

FIGURE 3.2. Polyacrylamide gel electrophoresis of ^{125}I-T$_4$ binding to serum proteins from 2 patients with FDH compared to normal.

labeled T$_4$ (31,32,34,35,37) (Fig. 3.2). The results of recent studies of the serum albumin obtained from patients with FDH suggest that the abnormality in T$_4$ binding in this syndrome is due largely or entirely to an overabundance of albumins that have a particularly high affinity for T$_4$; these albumins are present in normal serum but at a lower concentration (43). The affinity of one of these albumins for T$_4$ is in the intermediate range, and disulfide linkage is essential for its association with T$_4$ (38).

Additional variant albumins with increased affinity for T$_3$, or for T$_3$ and reverse T$_3$ (rT$_3$), as well as T$_4$ have been described (44,45). Thus, FDH may represent a heterogeneous disorder. Most patients will have only elevated serum T$_4$ concentrations, and a few will have increased T$_4$ and rT$_3$ or an increase in all three iodothyronines. Since the FT$_4$ and FT$_3$ values will be elevated if assessed by methods employing ^{125}I-labeled T$_4$ and T$_3$ analogs, other tests of thyroid function should be carried out in these patients. Such tests include the thyroid radioactive iodine uptake, measurement of free T$_4$ and T$_3$ by equilibrium dialysis, measurement of basal and TRH-stimulated TSH values, and, finally, paper or gel electrophoresis of serum enriched with ^{125}I-labeled T$_4$ or T$_3$ to confirm the presence of abnormal serum albumin binding of T$_4$ and/or T$_3$.

Occasionally, patients with FDH will also have underlying thyroid disease such as hyperthyroid Graves' disease or hypothyroidism. When FDH and hyperthyroidism coexist, the serum T_4 concentration is markedly and disproportionately elevated, and the T_3/T_4 ratio is lower than in hyperthyroid patients with normal serum binding (46). The serum T_4 may be normal in patients with FDH and mild to moderate hypothyroidism, but the serum TSH concentration will be elevated. Replacement T_4 therapy results in strikingly elevated serum T_4 concentrations.

C. Increased T_4 Binding to TBPA

Normally, TBPA binds approximately 15% of the circulating T_4 and practically none of the circulating T_3 (6). T_4 binding to TBPA can be increased secondary to an increase in the circulating TBPA concentrations or an increase in the affinity of TBPA for T_4 (47–50).

Two patients with glucagon-secreting islet cell carcinomas and elevated serum T_4 concentrations have been reported (48,50). In both patients, serum T_3 and FT_4 concentrations and RT_3U were normal. ^{125}I-labeled T_4 binding to TBPA was markedly increased, whereas ^{125}I-labeled T_4 binding to TBG and albumin was decreased (Fig. 3.3). TBPA from one of the patients was shown to be similar to normal TBPA by electrophoretic mobility, immunoreactivity, and binding inhibition studies (50), but serum TBPA concentrations in both individuals were found to be 3–4 times normal (48,50). Accordingly, an increase in TBPA synthesis by some islet cell carcinomas has been postulated.

Moses and associates described 3 members of a family who were euthyroid and in whom thyroid function studies gave results similar to those observed in patients with FDH (49). However, paper electrophoresis of serum enriched with ^{125}I-labeled T_4 revealed an abnormally high percentage of ^{125}I-labeled T_4 tracer associated with TBPA, a low percentage associated with TBG, and a normal percentage associated with albumin. Immunoabsorption also disclosed an unusually high percent of serum T_4 associated with immunoreactive TBPA. The mild increase in the serum TBPA concentration was not great enough to account for the increase in the serum T_4 concentration, suggesting an increased binding affinity of the variant TBPA for T_4. This has been recently confirmed by other investigators in some patients with this abnormality (47,51).

The normal serum T_3 concentration and normal or slightly elevated RT_3U value can be explained by the fact that little or no T_3 is bound by TBPA (1). Although serum FT_4I values are markedly elevated in patients with increased T_4 binding to TBPA, serum FT_4 concentrations measured by RIA methods that employ ^{125}I-labeled T_4 or T_4 analogs are either normal or slightly increased (52). Because of the differences in the total T_4-bind-

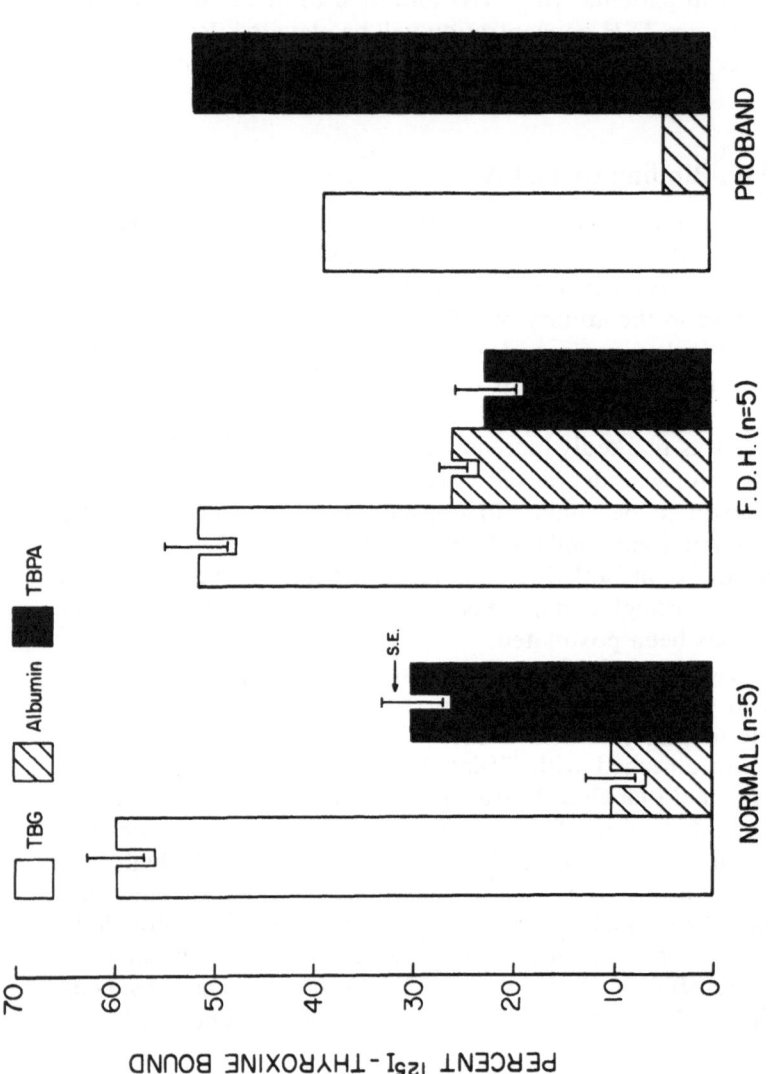

FIGURE 3.3. ^{125}I-T$_4$ binding to TBG, albumin, and TBPA in serum from 5 normal subjects, 5 patients with FDH (familial dysalbuminemic hyperthyroxinemia), and 1 patient with an islet cell carcinoma with elevated serum TBPA concentrations. Sera were fractionated by polyacrylamide gel electrophoresis. (Reprinted with permission from Rajatanavin R, Liberman C, Lawrence GD, D'Arcangues CM, Young RA, Emerson CH. Euthyroid hyperthyroxinemia and thyroxine-binding prealbumin excess in islet cell carcinoma. J Clin Endocrinol Metab 61:17–21, © by The Endocrine Society (1985).)

ing capacities in sera of normal subjects and patients with FDH or TBPA-associated hyperthyroxinemia, a nonbarbital buffer T_4 resin uptake test in serum enriched with an optimum concentration of T_4 may help differentiate these two conditions from one another and from normal variants (53,54).

D. Autoantibodies Against T_4

Robbins and associates first reported the occurrence of T_4 binding to gamma globulin in a patient with papillary carcinoma of the thyroid gland (55). Premachandra and Blumenthal (56) and Ikekubo et al. (57) subsequently found that several patients with Hashimoto's thyroiditis had autoantibodies to T_4 and/or T_3. Similar antibodies have also been observed in patients with primary and secondary hypothyroidism (58), hyperthyroidism (58), and Waldenstrom's macroglobulinemia (59) and during therapy with desiccated thyroid (60). However, the majority of patients with such autoantibodies have Hashimoto's thyroiditis (61). Most autoantibodies against thyroid hormones are of the IgG class, although in a few patients IgA, IgM, and IgE antibodies have been reported (61).

Disturbances in thyroid function tests are primarily due to interference by the antibody to T_4 in the RIA methods for measuring serum T_4. The serum T_4 concentration may be high or low, depending on the RIA methods employed. Methods based on the absorption of free hormone or precipitation of all hormone-antibody complexes by polyethylene glycol, dextran-coated charcoal, or ammonium sulfate will result in spuriously low values for the serum T_4 concentration (58,60,62). On the other hand, RIA methods based on double-antibody or solid-phase techniques will result in spuriously high values. The serum T_4 concentration measured by RIA in ethanol extracts of serum or by competitive protein-binding methods will be either normal or slightly elevated, depending on the avidity of T_4 binding to autoantibody (57). Therefore, the serum FT_4I values will depend on the method employed for measuring T_4. Similarly, the serum FT_4 values measured by equilibrium dialysis will depend on the method used for measuring the serum total T_4, although the percent FT_4 is always decreased when antibodies against T_4 are present. Most RIA methods give normal FT_4 concentrations in these patients (63). However, the serum FT_4 concentration determined by RIA methods that employ ^{125}I-labeled T_4 analogs will be spuriously elevated, suggesting that the T_4 analog is also bound by the T_4 autoantibody (63–65).

The presence of autoantibodies to T_4 or T_3 can be confirmed by the demonstration of radioactivity associated with gamma globulin upon electrophoresis of the patient's serum enriched with ^{125}I-labeled T_4 or ^{125}I-labeled T_3 tracer (56). Immunoprecipitation techniques or column chromatography of the labeled hormone-enriched sera can also be used to confirm the diagnosis (57). Removal of the endogenous thyroid hormones by treatment with acid dextran-coated charcoal before adding labeled thyroid hormones has been reported to enhance detection of thyroid hormone autoantibodies (61). The presence of T_4 antibody in the serum can be inferred by a gross discrepancy between the serum T_4 concentration measured by RIA and the value obtained by competitive protein-binding assays.

Patients with thyroid hormone autoantibodies are usually euthyroid unless, as noted above, autoimmune thyroid disease is present. However, a few such patients with hypothyroidism in the absence of Hashimoto's thyroiditis have been reported, perhaps secondary to the very intense binding of the thyroid hormones by the autoantibodies (59,66).

II. Hormone Resistance Syndromes

Resistance to thyroid hormones has been described as a familial syndrome (67) and in sporadic cases (68). In the familial form, the pattern of genetic transmission is either autosomal-dominant (69) or autosomal-recessive (67). The peripheral tissues and the anterior pituitary can be equally resistant, or resistance can occur in varying degrees in different tissues (Fig. 3.4). Theoretically, if the degree of resistance in the anterior pituitary is much greater than that in the peripheral tissues, the patient will present with hyperthyroxinemia and hyperthyroidism (70). If the anterior pituitary is not resistant or is much less resistant than the peripheral tissues, the patient will present with hypothyroidism (71). When both the anterior pituitary and peripheral tissues are nearly equally resistant, the patient will be euthyroid with elevated serum thyroid hormone concentrations (68,71a). In practice, most patients fall into this last group and will present with euthyroid hyperthyroxinemia.

The basal serum TSH concentration and the TSH increment after TRH stimulation may be normal or elevated; either is inappropriate in the presence of high serum thyroid hormone concentrations (71a,72). In patients with generalized resistance, serum TSH concentrations are usually normal if the patient has not received therapy for his or her hyperthyroxinemia (72,73). However, many patients have been previously treated inappropriately with radioactive iodine and/or thyroidectomy and have elevated serum TSH concentrations (72–75). Patients with isolated pituitary resistance tend to have higher serum TSH concentrations (70).

Since both serum TSH and T_3 concentrations increase after TRH administration, endogenous TSH is biologically active (68,71a). However, the secretion of TSH is not completely autonomous, because there is a decrease in the serum TSH concentration after the administration of supraphysiologic doses of thyroid hormone (68,76). The eumet-

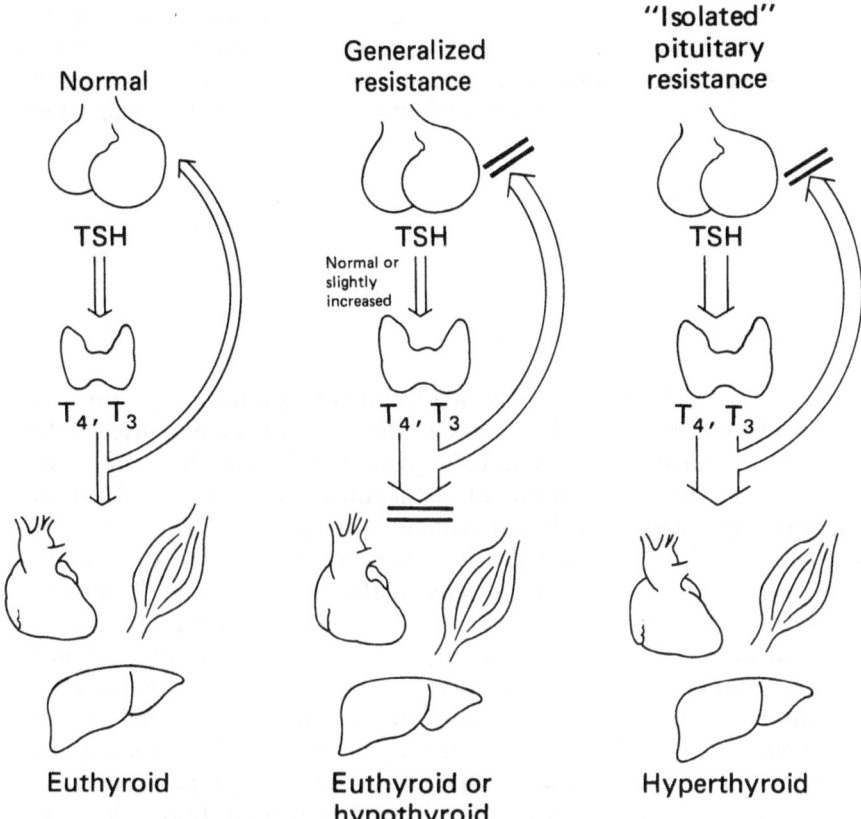

FIGURE 3.4. Changes observed in patients with thyroid hormone resistance who have not been previously treated for hyperthyroxinemia.

abolic state in these patients is maintained at the expense of increased serum thyroid hormone concentrations. Nevertheless, the clinical picture is complicated by the fact that there are variable degrees of resistance to thyroid hormone action in different tissues (67). The patients usually have a goiter and an elevated thyroid radioactive iodine uptake, findings that further contribute to the false diagnosis of thyrotoxicosis (70,73,75).

Studies to determine the pathogenesis of this syndrome have resulted in conflicting data. The T_3 nuclear binding affinities in peripheral lymphocytes from patients with thyroid hormone resistance are reported to be either low (77) or normal (78,79). There are also discrepant findings in the magnitude of T_3 nuclear binding capacities in peripheral lymphocytes (77–79). Abnormal T_3 nuclear binding sites in fibroblasts from these patients have been observed (78,80). A study of fibroblasts cultured from members of a family with this syndrome demonstrated a normal binding capacity and affinity for T_3 but a failure of T_3 to stimulate low-density lipoprotein receptor activity, suggesting that the abnormality resides after receptor binding (81). Both decreased red cell plasma membrane uptake of T_4 (82) and decreased fibroblast nuclear uptake of T_3 (83) have also been described in this syndrome.

III. Nonthyroidal Illness

A. Sick Euthyroid Syndrome

Abnormal thyroid function tests are frequently found in patients with acute or chronic nonthyroidal illnesses who are otherwise euthyroid (84). It is well recognized that hospitalized patients will often have decreased serum T_3 and increased serum rT_3 concentrations owing to decreased outer ring deiodination of iodothyronines (T_4 to T_3 and rT_3 to 3,3'-diio-dothyronine) (Fig. 3.5). The serum T_4 concentration and FT_4I are frequently normal but may be low, especially in the more severely ill patients. The serum FT_4 concentration measured by equilibrium dialysis or by some RIA methods is often increased in sick patients in whom the serum T_4 concentration is normal and normal in patients with a low serum T_4 concentration (85). In contrast, values for serum FT_4 measured by other RIA methods, including those employing ^{125}I-labeled T_4 analogs, are usually normal in sick patients with a normal serum T_4 concentration and low in those patients in whom the serum T_4 concentration is low. The mechanism(s) responsible for these abnormalities in serum T_4 and FT_4 concentrations and FT_4I are unclear. They may be due to the combination of an abnormal thyroid hormone clearance rate and alterations in T_4 binding to serum proteins. Recently, an inhibitor of T_4 binding to serum proteins has been identified in serum (86,87) and in extrathyroidal tissues (88) from sick patients. The inhibiting substances may be unsaturated free fatty acids, which are often elevated in severely ill patients (89). The normal or low serum TSH (90) and elevated serum rT_3 (91) concentrations are evidence against primary hypothyroidism, although the serum TSH may rise above the normal range as some of these patients recover from their illness (92,93).

Less frequent is the occurrence of increased serum T_4 and FT_4 concentrations and FT_4I values that can occur in some sick patients who are nevertheless euthyroid (94–96). Gooch and associates (97) reported that serum T_4 and FT_4 concentrations and the FT_4I were elevated in 3.9%, 5.4%, and 11.7%, respectively, of patients without manifestations of thyroid disease who were admitted to a medical service. This may be confused with the clinical state termed T_4 toxicosis (98,98a). Indeed, some of these patients have hyperthyroidism, but an intercurrent systemic

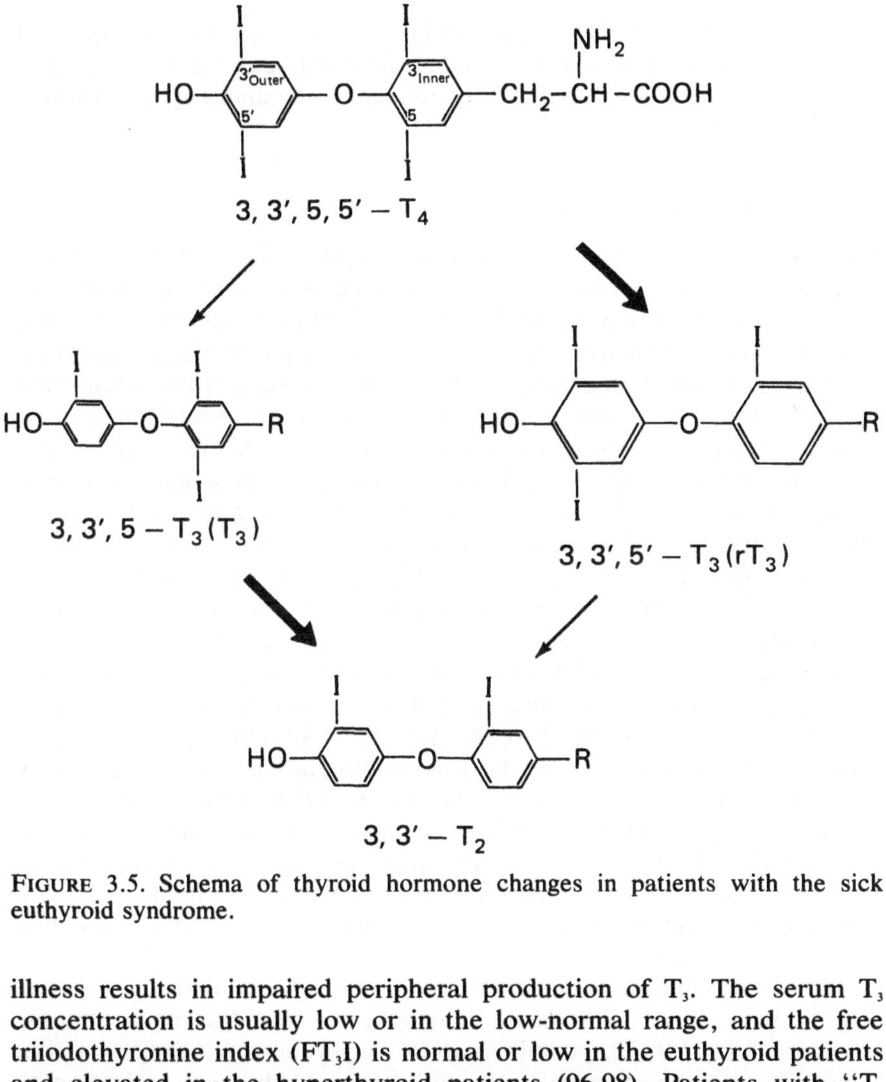

FIGURE 3.5. Schema of thyroid hormone changes in patients with the sick euthyroid syndrome.

illness results in impaired peripheral production of T_3. The serum T_3 concentration is usually low or in the low-normal range, and the free triiodothyronine index (FT_3I) is normal or low in the euthyroid patients and elevated in the hyperthyroid patients (96,98). Patients with "T_4 toxicosis" may eventually develop classic hyperthyroidism with elevations of both serum T_4 and T_3 concentrations after the intercurrent systemic illness has remitted (96,98–100). These patients usually have a goiter and many have a history of recent iodine ingestion (iodine-induced hyperthyroidism; Jod-Basedow's disease). Occasionally, true "T_4 toxicosis" may occur in the absence of intercurrent illness (98). Serum TSH measured by supersensitive assays and the TRH stimulation test are most helpful in differentiating hyperthyroidism from euthyroidism in patients with other illnesses. A normal basal serum TSH concentration and/or a normal TSH response to TRH rules out hyperthyroidism, whereas a low basal serum TSH and/or a flat TSH response suggests hyperthyroidism.

However, an absent TSH response to TRH and/or a low basal serum TSH may not always be diagnostic, since euthyroid, extremely ill patients occasionally have a low serum TSH concentration and may not respond to TRH (90,101).

B. Acute Psychiatric Illness

It is well known that patients with thyroid dysfunction may present with psychiatric symptoms (102,103). It has also been reported that elevations of serum T_4 or FT_4I may occur during the initial phase of depression and revert to normal after treatment (104,105). Serum T_3 in these patients is either normal (106) or decreased (105). Responsiveness of the serum TSH to TRH stimulation in depressed patients may be decreased (107).

Recent studies have documented that elevations of the serum T_4 and FT_4 concentrations and the FT_4I occur in euthyroid patients with a wide variety of acute psychiatric illnesses (108–110). Spratt and associates observed that 33% of patients with acute psychiatric illnesses have an elevated serum T_4 concentration and 18% have an elevated FT_4I (110), although this was not confirmed in a more recent report (111). The serum T_3 concentration in the majority of these patients is in the normal range but may occasionally be elevated (109). These changes in thyroid function are transient, since serum thyroid hormone concentrations decline to normal in all patients within 2 weeks (108–110). An important observation in this group of patients is that the TRH stimulation test does not reliably rule out hyperthyroidism. There may be no TSH response to TRH in patients with psychiatric disorders, regardless of their initial serum T_4 concentration (110). When the TRH stimulation tests are repeated after the serum T_4 concentration and FT_4I revert to normal, some of those with an initially flat response return to normal. However, an abnormal test may persist in others (110).

These acute changes in thyroid function were observed in all categories of psychiatric illness, but acute schizophrenia was the most common diagnosis (109,110). The administration of psychotropic drugs was not responsible for the changes in thyroid function, although patients with amphetamine abuse may present with an elevated serum T_4 concentration (150). The mechanism(s) underlying these changes in thyroid function tests that accompany acute psychiatric disease are unknown.

C. Hyperemesis Gravidarum

Compared to values in normal pregnancy, the serum protein-bound iodine (PBI) concentration is increased in some patients with hyperemesis gravidarum (112). Bouillon et al. found a very high incidence of increased serum T_4 concentrations and elevated FT_4I values in patients with hyperemesis gravidarum compared to levels in normal pregnant women (113). Elevation of the FT_4I was observed in 75% of a group of 33 patients with this disorder. The mean serum T_3 concentration was not different from that found in normal pregnancy, but the FT_3I was increased in 4 of 11 patients. These abnormalities were transient, and the serum T_4 concentration and FT_4I returned to normal within 3 weeks after resolution of the hyperemesis, whether or not the patients received treatment with antithyroid drugs. In 5 hyperthyroxinemic patients, there was no TSH response to TRH, suggesting the possibility of hyperthyroidism. However, signs and symptoms of hyperthyroidism were absent, and the results of thyroid function tests in these patients returned to normal within a few weeks.

The mechanism(s) underlying these changes are unknown. Although human chorionic gonadotropin (hCG) has weak thyroid-stimulating properties (114), the concentration of hCG in the serum of the hyperthyroxinemic patients was within the normal range for pregnant women (113). The changes in thyroid function may represent a subset of patients with nonthyroidal illness with an unusually high incidence of hyperthyroxinemia. Given that the flat TSH response to TRH in some of these patients resembled that observed in patients with acute psychiatric illness (110), it may represent concomitant psychiatric disturbance. It has been suggested that hyperemesis gravidarum may have a psychiatric component (115), and, indeed, psychiatric examination performed in 11 patients with hyperthyroxinemia and hyperemesis gravidarum revealed psychologic conflict in all (113).

D. Hyponatremia

Severe hyponatremia has recently been associated with hyperthyroxinemia (116). Cogan and Abramow reported elevated serum T_4 concentrations or FT_4I values in 7 of 12 patients with a mean plasma sodium of 117 mEq/L and neurologic manifestation of hyponatremia. All values returned to normal when patients were evaluated 2 weeks after correction of their low serum sodium. The mean serum T_3 concentration was higher initially than upon recovery, and 2 of the hyperthyroxinemic patients had frankly elevated FT_3I values. None of the 10 patients with hyponatremia of a similar degree without neurologic complications were found to have hyperthyroxinemia. The authors suggested that the rate of change of the serum sodium was of importance in causing the hyperthyroxinemia; this postulate is analogous to that proposed to explain the finding that the presence of neurologic complications of hyponatremia are closely related to the rate of change of the serum sodium and not to its ultimate value.

IV. Drugs

A number of drugs affect thyroid function test in different ways. Since a complete review is available elsewhere (117), only those drugs that cause hyperthyroxinemia will be discussed here.

A. Radiologic Contrast Agents

Iopanoic acid (Telepaque) was the first oral cholecystographic agent demonstrated to induce changes in thyroid function (118). The drug is a potent inhibitor of outer ring monodeiodination of T_4 and rT_3 in the peripheral tissues and will cause decreased serum T_3 and elevated serum rT_3 concentrations within 24 h after administration (118,119). Iopanoic acid also inhibits the intrapituitary conversion of T_4 to T_3 (120,121). The decreased intrapituitary conversion of T_4 to T_3 and the decreased serum T_3 concentration probably explain the marginally increased serum TSH concentration and the augmented TSH response to TRH that frequently occur following iopanoic acid administration (119). These iodinated dyes also decrease the hepatic uptake and binding of T_4 (122). The latter findings, coupled with the slight increase in serum TSH, contribute to the elevated serum T_4 and FT_4I often observed after cholecystography (123). Other oral cholecystographic agents such as sodium ipodate (Oragrafin), iodobenzamic acid (Osbil), and sodium tyropanoate (Tyropaque) also decrease out-ring deiodination of T_4 to T_3, resulting in similar abnormalities in thyroid function. These changes may persist for a few weeks.

In contrast to the abnormalities in thyroid function observed after administration of iodinated dyes for gallbladder visualization, the contrast agent diatrizoate, most frequently employed in cardiac catheterization studies, does not influence serum thyroid hormone or TSH concentrations when samples are obtained 5–7 days after catheterization (124). However, since all contrast agents are rich in iodine, there is a possibility that iodine-induced hyperthyroidism might occur (125).

B. Amiodarone

Amiodarone is a benzofuranic derivative that is widely used throughout Europe as an antianginal and antiarrhythmic agent (126–131) and has recently been released in the United States. Amiodarone has been shown to be a potent inhibitor of hepatic outer-ring (5') deiodination of T_4 and reverse T_3 (rT_3) (132–137). In man, this inhibition is associated with elevated serum T_4 and rT_3 concentrations and decreased serum T_3 concentrations during amiodarone therapy (133–139). The production rate of T_3 is decreased (140), whereas that of T_4 is either normal or increased with short-term treatment (3 weeks) (138,140). The increase in serum T_4 concentration after amiodarone administration is mainly due to a decrease in the T_4 metabolic clearance rate (138). These changes in serum hormone concentrations are accompanied by an increase in basal and TRH-stimulated serum TSH concentrations during the first 3 months of therapy (133,136). However, after more prolonged treatment the serum TSH usually returns to normal (136). In contrast, a number of patients on chronic therapy will have suppressed serum TSH concentrations and an absent TSH response to TRH. However, such patients are clinically euthyroid and have normal or low-normal serum T_3 concentrations (134,139,141).

Amiodarone contains 37.5% iodine (75 mg of iodine per 200-mg tablet), and its use in man is associated with iodine-induced hyper- and hypothyroidism (126–128,139,142–149). The incidence of thyroid dysfunction during amiodarone therapy appears to vary from area to area depending on the ambient iodine intake (139,145), and the release of iodine from the drug probably contributes to the development of abnormalities in thyroid function. Thyrotoxicosis occurs more frequently in areas of iodine deficiency (126,139,145,146), and hypothyroidism appears to be a more frequent complication in areas with sufficient iodine intake (127,139,143–145,149). Patients with amiodarone-induced hyperthyroidism differ from patients who remain euthyroid with amiodarone therapy but have elevated serum T_4 concentrations with or without flat or blunted TSH responses to TRH in that the former almost always have serum T_3 concentrations above the normal range (139). The euthyroid patients have low or normal serum T_3 concentrations (139), as would be expected from the effects of amiodarone on T_4 deiodination. The explanation for an absent TSH response to TRH in a number of patients receiving chronic amiodarone therapy who remain euthyroid is unknown. However, the results of a recent study in which amiodarone decreased TSH secretion in hypothyroid rats suggested that amiodarone may function as a thyroid hormone agonist at the level of the pituitary (150).

C. Amphetamines

Elevation of the serum T_4 and FT_4 concentrations and the FT_4I has been reported during heavy amphetamine abuse (109,151). The serum T_3 concentration in these patients is normal or in the upper limit of the normal range. In spite of the hyperthyroxinemia, there is an exaggerated TSH response to TRH. All abnormalities revert to normal after withdrawal of the drug. Although some patients appeared to have symptoms resembling hyperthyroidism, such symptoms could be ascribed to excess amphetamine ingestion without invoking abnormalities in thyroid function. These patients were probably euthyroid, since the serum T_4 concentration was elevated only during drug ingestion, and the serum T_3 concentration was normal.

Studies using the vervet monkey as an experimental model of substance abuse suggest that a rise in serum TSH is responsible for the elevated serum T_4 concentration observed during amphetamine administration, prompting the postulate that amphetamines exert their effect at the hypothalamic or pituitary level (151). However, in these experiments the rise in serum TSH was small and the increase in serum T_4 large, and other mechanisms may be operative.

Hyperthyroxinemia induced by amphetamines may pose a diagnostic problem when patients deny a history of amphetamine abuse. Nevertheless, the enhanced TSH response to TRH will rule out hyperthyroidism.

D. Heparin

Heparin therapy may result in a rise in the serum T_4 concentration and a marked elevation of serum FT_4 and FT_3 concentrations measured by equilibrium dialysis (152,153). Since the RT_3U remains normal, the FT_4I will also be slightly elevated (152). Heparin also induces a concomitant decrease in the serum TSH concentration (153). These changes can be observed within 5 min after intravenous heparin administration with doses as low as 100–500 units (153).

The mechanism underlying these changes is not clear, but increased thyroid hormone secretion can be ruled out, because similar effects have also been observed in heparin-treated athyreotic patients receiving replacement therapy with L-T_4 (153). It is likely that heparin prevents binding of thyroid hormones to serum proteins either by acting as an inhibitor or by altering the stearic configuration of the binding sites. However, TBG- and TBPA-binding capacities are within the normal range during heparin administration (153). Even though addition of free fatty acids (FFA) in vitro inhibits T_4 binding to serum proteins (154), it is unlikely that the heparin effect on tests of thyroid function is secondary to an increase in FFA. This is because protamine, a heparin antagonist, prevents the FFA surge induced by heparin-stimulated lipoprotein lipase activity but does not prevent the heparin-induced increase in serum FT_4 (153). It has also been suggested that treatment with heparin in vivo reduces the cellular binding of T_4 (155).

Since the serum FT_4 is now frequently measured by RIA, it is noteworthy that values are decreased in the plasma of patients receiving heparin when FT_4 is measured by methods that employ a ^{125}I-labeled T_4 analogue (156,157). The effect of heparin on thyroid function tests should be recognized, since many patients are now treated with this drug.

E. Propranolol

The beta-adrenergic blocking agent propranolol is widely used as adjunctive therapy for hyperthyroidism (158). In addition to its beta-blocking property, propranolol and some other beta-adrenergic blocking drugs inhibit the peripheral conversion of T_4 to T_3 (158a,159,160). In hyperthyroid patients, propranolol decreases serum T_3 and increases serum rT_3 concentrations while serum T_4 concentration remains unchanged (161). However, hypothyroid patients receiving T_4 replacement therapy and also taking propranolol often have an elevated serum T_4 concentration (162). Most recently, it has been reported that the serum T_4 concentration is elevated, the serum T_3 concentration is decreased, and the TSH response to TRH is normal or blunted in some euthyroid patients treated with large doses of propranolol (163).

V. Exogenous T_4 Administration

A. L-T_4 Administration

The physiologic replacement dose of L-T_4 for adult hypothyroid patients usually ranges between 0.1 to 0.2 mg daily (164,165). Treatment with doses in this range had been reported to result in normal serum T_4 concentrations (165,166). However, a few reports have indicated that serum T_4 concentrations may be elevated in patients receiving these amounts of L-T_4 as replacement therapy (167–169). Recent studies in our laboratory and elsewhere have demonstrated that the serum T_4 concentrations and the FT_4I are elevated in one-third (170) to two-thirds (167) of clinically euthyroid patients receiving the recommended dose of L-T_4 (either total dose or μg L-T_4/kg body weight) for the treatment of hypothyroidism or suppression of nontoxic goiter. It is important to note that these observations were made during chronic treatment, since the serum T_4 concentration may be transiently elevated during the first month of L-T_4 therapy (171). The serum T_3 concentrations, however, were within the normal range and thus reflected the clinical (euthyroid) status of the patient better than the serum T_4 (167,170). The T_4/T_3 ratio was even higher than one would expect in the absence of thyroid secretion of T_3.

The mechanism underlying these changes is unclear, but a disproportionate delivery of T_4 to the liver after oral administration of L-T_4 may result in less efficient generation of T_3 from T_4. Very recently, Nicoloff and his co-workers reported that the conversion of T_4 to T_3 is more efficient in hypothyroidism and less efficient in hyperthyroidism (172,173). This peripheral autoregulatory control of 5' deiodination may also explain the apparent decrease in the peripheral conversion of T_4 to T_3 when the serum T_4 concentration is elevated. Therefore, hyperthyroxinemia in patients receiving L-T_4 therapy does not necessarily indicate an excessive dose, although some have argued that this might well be the case (174). Normalization or suppression of the serum TSH in hypothyroid patients provides the best indicator of sufficient thyroid hormone replacement therapy but does not provide evidence of overtreatment. Although impractical, a normal TSH response to TRH during thyroid hormone therapy would indicate that the replacement dose is physiologic. Use of the supersensitive TSH assay may also help to determine overreplacement with thyroid hormone. With this assay, many patients who are clinically euthyroid are found to have subnormal serum TSH concentrations suggesting at least mild thyrotoxicosis (175). Patients on L-thyroxine therapy for suppression of goiter or thyroid cancer in all probability require suppression of their TSH secretion. However, patients on thyroid hormone for treatment of primary hypothyroidism should probably have their dose decreased if serum TSH is suppressed using the supersensitive TSH assay. Ideally, the serum TSH concentration should be within the normal range in the latter situation.

B. D-T$_4$ Administration

The D isomer of T$_4$ is used primarily for its cholesterol-lowering effect in the treatment of hyperlipidemia. The metabolic potency of D-T$_4$ is only 10–15% that of L-T$_4$, and it was originally suggested that treatment with D-T$_4$ in amounts that lowered cholesterol would not induce metabolic changes (176,177). This is apparently not the case, since the dose of D-T$_4$ recommended for lowering serum cholesterol also blunts the TSH response to TRH (178,179). Even lower daily doses (1–3 mg) of D-T$_4$ blunt the TSH response to TRH (178,180). This is due to the direct effect of D-T$_4$ on the pituitary, since the serum concentration of L-T$_4$, determined by stereospecific analysis, did not change after 3 days of D-T$_4$ treatment (181). However, the recent finding that D-T$_4$ tablets may be contaminated with L-T$_4$ could partially explain the purported peripheral effects of D-T$_4$ (182).

Since the antibody employed in the RIA for L-T$_4$ cross-reacts 100% with D-T$_4$ and the antibody for L-T$_3$ cross-reacts 100% with D-T$_3$, the serum L-T$_4$ and L-T$_3$ concentrations measured by RIA in the serum of patients receiving D-T$_4$ therapy will be elevated (179, 180). Doses of D-T$_4$ as low as 3 mg result in a fourfold elevation of the serum T$_4$ concentration within 4 h of D-T$_4$ ingestion. The serum T$_4$ remains only slightly elevated 24 h later owing to the rapid clearance of D-T$_4$ from the circulation (178). The serum FT$_4$ measured by equilibrium dialysis and RIA (181) and the FT$_4$I are also elevated during D-T$_4$ therapy.

VI. High Altitude

Exposure to high altitude (12,210–14,000 ft) has been reported to result in an elevation of the serum PBI, T$_4$, FT$_4$I, and FT$_4$ concentrations; a transient elevation of the serum T$_3$ concentration; and no change in the basal and TRH-stimulated TSH concentrations (183–185). Studies from our laboratory (186) have demonstrated that 17 experienced mountain climbers who reached altitudes of 17,000–20,700 ft on Mt. Everest had an increase in the serum T$_4$, at times above the normal range; a small but significant rise in the serum T$_3$ and TSH concentrations; and no change in serum TBG concentrations. Four of 5 subjects tested showed an enhanced TSH response to TRH. These observations suggest that extremely high altitude caused decreased metabolic clearance of T_4 or efflux of T_4 from tissue into plasma and an alteration in the negative feedback set point for pituitary TSH secretion.

VII. Summary

A number of clinical situations have been described in which elevated serum thyroid hormone concentrations do not represent hyperthyroidism. When faced with an elevated serum T_4 or FT_4 concentration or FT_4I value in an otherwise well and clinically euthyroid patient, it is important to consider the possibility of a thyroid hormone-binding abnormality or, far less likely, the presence of a hormone resistance syndrome. Any thyroid hormone values obtained in patients with nonthyroidal illness must be evaluated with cognizance of the frequent presence of abnormal thyroid hormone concentrations in patients with the sick euthyroid syndrome, acute psychiatric illnesses, hyperemesis gravidarum, and, more recently, hyponatremia. Finally, it is important to question the patient specifically about the ingestion of any of the drugs that can cause hyperthyroxinemia. In most situations, determination of whether the patient is hyperthyroid or euthyroid can be made at the time of presentation. However, occasional patients must be followed prospectively (most often those with the sick euthyroid syndrome or acute psychiatric illness) to determine their ultimate thyroid status. The new supersensitive TSH assays hold promise in increasing our ability to distinguish between patients who are euthyroid or thyrotoxic (187). Further work will be needed to evaluate the usefulness of these assays in the clinical assessment of patients with euthyroid hyperthyroxinemia.

Acknowledgments. Supported in part by grant AM-18919 from the NIDDK, NIH, Bethesda, MD.

References

1. Oppenheimer JH (1968) Role of plasma proteins in the binding, distribution and metabolism of thyroid hormones. N Engl J Med 278:1153–1162.
2. Robbins J, Cheng S, Gershengorn MC, Glinoer D, Cahnmann HJ, Edelnoch H (1978) Thyroxine transport proteins of plasma, molecular properties and biosynthesis. Recent Prog Horm Res 34:477–519.
3. Robbins J, Rall JE (1957) Interaction of thyroid hormones and proteins in biological fluids. Recent Prog Horm Res 13:161–208.
4. Dos Remedios LV, Weber PM, Feldman R, Schurr DA, Tsoi TG (1980) Detecting unsuspected thyroid dysfunction by the free thyroxine index. Arch Intern Med 140:1045–1049.

5. Refetoff S (1979) Thyroid hormone transport. In DeGroot LJ, Cahill GF, Martini L, Nelson DH, Odell WD, Potts JT Jr, Steinberger E, Winegard AI (eds): Endocrinology. Grune and Stratton, New York, pp 347–356.
6. Woeber KA, Ingbar SH (1968) The contribution of thyroxine-binding prealbumin to the binding of thyroxine in human serum, as assessed by immunoadsorption. J Clin Invest 47:1710–1721.
7. Azizi F, Vagenakis AG, Portnay GI, Braverman LE, Ingbar SH (1974) Thyroxine transport and metabolism in methadone and heroin addicts. Ann Intern Med 80:194–199.
8. Beierwalts WH, Robbins J (1959) Familial increase in thyroxine binding sites in serum alpha globulin. J Clin Invest 38:1683–1688.
9. Dowling JT, Freinkel N, Ingbar SH (1956) Effect of diethylstibestrol on binding of thyroxine in serum. J Clin Endocrinol Metab 16:1491–1506.
10. Fex G, Adielsson G, Mattson W (1981) Oestrogen-like effects of Tamoxifen on the concentration of proteins in plasma. Acta Endocrinol (Copenh) 97:109–113.
11. Florsheim WH, Dowling JT, Meister L, Bodfish RE (1962) Familial elevation of serum thyroxine binding capacity. J Clin Endocrinol Metab 22:735–740.
12. Galton VA, Ingbar SH, Jimenez-Fonseca J, Hershman JM (1971) Alterations in thyroid hormone economy in patients with hydatidiform mole. J Clin Invest 50:1345–1354.
13. Gardner DF, Carithers RL Jr, Utiger RD (1982) Thyroid function tests in patients with acute and resolved hepatitis B virus infection. Ann Intern Med 96:450–452.
14. Hollander CS, Scott RL, Tschudy DP, Perlroth M, Maxman A, Sterling K (1967) Increased protein bound iodine and thyroxine binding globulin in acute intermittent porphyria. N Engl J Med 227:995–1000.
15. Jones JE, Seal US (1967) X-chromosome linked inheritance of elevated thyroxine-binding globulin. J Clin Endocrinol Metab 27:1521–1528.
16. Kydd DM, Man EB (1951) Precipitable iodine of serum (SPI) in disorders of the liver. J Clin Invest 30:874–878.
17. Lever A, Bird D, Byfield PGH, Lalloz MRA, Webster ADB, Himsworth RL (1983) Increased serum concentration of T4-binding globulin in patients with hypogammaglobulinemia. Clin Endocrinol (Oxf) 18:195–199.
18. Oltman JE, Friedman S (1963) Protein bound iodine in patients receiving perphenazine. JAMA 185:726–727.
19. Refetoff S, Fang VS, Marshall JS, Robin NI (1976) Metabolism of thyroxine binding globulin in man: Abnormal rate of synthesis in inherited thyroxine-binding globulin deficiency and excess. J Clin Invest 57:485–495.
20. Robbins J, Nelson JH (1958) Thyroxine binding by serum protein in pregnancy and in the newborn. J Clin Invest 37:153–159.
21. Ross DS, Daniels GH, Dienstag JL, Ridgway EC (1983) Elevated thyroxine levels due to increased thyroxine-binding globulin in acute hepatitis. Am J Med 74:564–569.
22. Schussler GC, Schaffner F, Korn F (1978) Increased serum thyroid hormone binding and decreased free hormone in chronic active liver disease. N Engl J Med 299:510–515.
23. Vannotti A, Beraud T (1959) Functional relationships between the liver, the thyroxine binding protein of serum and the thyroid. J Clin Endocrinol Metab 19:466–477.
24. Viscardi RM, Shea M, Sriwantanakul K, McCormick K (1983) Hyperthyroxinemia in newborns due to excess thyroxine-binding globulin. N Engl J Med 309:897–899.
25. Webster JB, Coupal JJ, Cushman P Jr (1973) Increased serum thyroxine levels in euthyroid narcotic addicts. J Clin Endocrinol Metab 37:928–934.

26. Glinoer D, McGuire RA, Gershengorn MC, Robbins J, Berman M (1977) Effects of estrogen on thyroxine-binding globulin metabolism in rhesus monkeys. Endocrinology 100:9–17.
27. L'Age M, Meinhold H, Wenzel KW, Schleusener H (1980) Relationships between serum levels of TSH, TBG, T_4, T_3, rT_3 and various histologically classified chronic liver diseases. J Endocrinol Invest 39:501–511.
28. Braverman LE, Abreau CM, Brock P, Kleinmann R, Fournier L, Odstrchel G, Schoemaker HJP (1980) Measurement of serum free thyroxine by RIA in various clinical states. J Nucl Med 21:233–239.
29. Chopra IJ, Van Herle AJ, Chua Teco GN, Nguyen AH (1980) Serum free thyroxine in thyroidal and nonthyroidal illnesses: A comparison of measurements by radioimmunoassay, equilibrium dialysis and free thyroxine index. J Clin Endocrinol Metab 51:135–143.
30. Ingbar SH, Braverman LE, Dawber NA, Lee GY (1965) A new method for measuring the free thyroid hormone in human serum and an analysis of the factors that influence its concentration. J Clin Invest 44:1679–1689.
31. Borst GC, Premachandra BN, Burman KD, Osburne RC, Georges LP, Johnsonbaugh RE (1982) Euthyroid familial hyperthyroxinemia due to abnormal thyroid hormone-binding protein. Am J Med 73:283–289.
32. Docter R, Bos G, Krenning EP, Fekkes D, Visser TJ, Hennemann G (1981) Inherited thyroxine excess: A new serum abnormality due to an increased affinity for modified albumin. Clin Endocrinol (Oxf) 15:363–371.
33. Henneman G, Doctor R, Krenning EP, Bos G, Otten M, Visser TJ (1979) Raised total thyroxine and free thyroxine index but normal free thyroxine: A serum abnormality due to inherited increased affinity of iodothyronine for serum binding protein. Lancet 1:639–642.
34. Lee WNP, Golden MP, Van Herle AJ, Lippe BM, Kaplan SA (1979) Inherited abnormal thyroid hormone-binding protein causing selective increase of total serum thyroxine. J Clin Endocrinol Metab 49:292–299.
35. Ruiz M, Rajatanavin R, Young RA, Taylor C, Brown R, Braverman LE, Ingbar SH (1982) Familial dysalbuminemic hyperthyroxinemia, a syndrome that can be confused with thyrotoxicosis. N Engl J Med 306:635–639.
36. Silverberg JDH, Premachandra BN (1982) Familial hyperthyroxinemia due to abnormal thyroid hormone binding. Ann Intern Med 96:183–186.
37. Stockigt JR, Topliss DJ, Barlow JW, White EL, Hurley DM, Taft P (1981) Familial euthyroid thyroxine excess: An appropriate response to abnormal thyroxine binding associated with albumin. J Clin Endocrinol Metab 53:353–359.
38. Barlow JW, Csicsmann JM, White EL, Funder JW, Stockigt JR (1982) Familial euthyroid thyroxine excess: Characterization of abnormal intermediate affinity thyroxine binding to albumin. J Clin Endocrinol Metab 55:244–250.
39. Ruiz M, Taylor C, Young R, Rajatanavin R, Braverman L, Ingbar SH (1981) Studies of the syndrome of familial euthyroid isolated hyperthyroxinemia (EIH). Proceeding of the 57th Annual Meeting of the American Thyroid Association, Minneapolis, p T-3 (abstract).
40. Rajatanavin R, Fournier L, DeCosimo D, Abreau C, Braverman LE (1982) Elevated serum free thyroxine by thyroxine analog radioimmunoassays in euthyroid patients with familial dysalbuminemic hyperthyroxinemia. Ann Intern Med 97:865–866.
41. Stockigt JR, DeGaris M, Barlow JW (1982) "Unbound analogue" methods for free T_4: A note of caution. N Engl J Med 307:126.
42. Stockigt JR, Degaris M, Csicsmann J, Barlow JW, White EL, Hurley DM (1981) Limitation of a new free thyroxine assay (Amerlex Free T_4). Clin Endocrinol (Oxf) 15:313–318

43. Yabu Y, Amir SM, Ruiz M, Braverman LE, Ingbar SH (1985) Heterogeneity of thyroxine-binding by serum albumins in normal subjects and patients with familial dysalbuminemic hyperthyroxinemia (FDH). J Clin Endocrinol Metab 60:451–459.
44. Lalloz MRA, Byfield PGH, Himsworth RL (1983) Hyperthyroxinaemia: Abnormal binding of T_4 by an inherited albumin variant. Clin Endocrinol 18:11–24.
45. Lalloz MRA, Byfield PGH, Himsworth RL (1985) A new and distinctive albumin variant with increased affinities for both triiodothyronines and causing hyperthyroxinaemia. Clin Endocrinol 22:521–529.
46. Young RA, Stoffer SS, Braverman LE (1987) Familial dysalbuminemic hyperthyroxinemia associated with primary thyroid disease. Am J Med 82:221–223.
47. Byfield PGH, Lalloz MRA, Himsworth RL (1984) Variant prealbumin: Another cause of euthyroid hyperthyroxinemia. Abstracts, 7th International Congress of Endocrinology, Quebec, p 530.
48. Jacobsson B, Petterson T, Sandstedt B, Carlstrom A (1979) Prealbumin in the islets of Langerhans. IRCS Med Sci 7:590.
49. Moses AC, Lawlor J, Haddow J, Jackson I (1982) Familial euthyroid hyperthyroxinemia resulting from increased thyroxine binding to thyroxine binding prealbumin. N Engl J Med 306:966–969.
50. Rajatanavin R, Liberman C, Lawrence GD, D'Arcangues CM, Young RA, Emerson CH (1985) Euthyroid hyperthyroxinemia and thyroxine-binding prealbumin excess in islet cell carcinoma. J Clin Endocrinol Metab 61:17–21.
51. Lalloz MRA, Byfield PGH, Himsworth RL (1984) A prealbumin with increased affinity for T_4 and reverse-T_3. Clin Endocrinol 21:331–338.
52. Skiest D, Braverman LE, Emerson CH (1986) Concentration of free thyroxin in serum of a patient with euthyroid hyperthyroxinemia secondary to increased thyroxin-binding prealbumin: Results by various methods compared. Clin Chem 32:687–689.
53. Moses AC, Lawlor J, Haddow J, Jackson I (1982) Differences between familial hyperthyroxinemic syndrome. N Engl J Med 307:825.
54. Stockigt JR, White EL, Barlow JW (1982) Differences between familial hyperthyroxinemic syndrome. N Engl J Med 307:824–825.
55. Robbins J, Rall JE, Rawson RW (1956) An unusual instance of thyroxine-binding by human serum gamma globulin. J Clin Endocrinol Metab 16:573–579.
56. Premachandra BN, Blumenthal HT (1967) Abnormal binding of thyroid hormone in sera from patients with Hashimoto's disease. J Clin Endocrinol Metab 27:931–936.
57. Ikekubo K, Konishi J, Endo K, Nakajima K, Okuno T, Kasagi K, Mori T, Nagasaka I, Torizuka K (1978) Antithyroxine and antitriiodothyronine antibodies in three cases of Hashimoto's thyroiditis. Acta Endocrinol (Copenh) 89:557–566.
58. Staeheli V, Vallotton MB, Burger A (1975) Detection of human anti-thyroxine and anti-triiodothyronine antibodies in different thyroid conditions. J Clin Endocrinol Metab 41:669–675.
59. Trimarchi F, Benvengar S, Fenzi G, Moriotti S, Consolo F (1982) Immunoglobulin binding of thyroid hormones in a case of Waldenstrom's macroglobulinemia. J Clin Endocrinol Metab 54:1045–1050.
60. Wu SY, Green W (1976) Triiodothyronine (T_3) binding immunoglobulins in a euthyroid woman: Effects on measurement of T_3 (RIA) and T_3 turnover. J Clin Endocrinol Metab 42:642–652.
61. Sakata S, Nakamura S, Miura K (1985) Autoantibodies against thyroid hormone or iodothyronine. Ann Intern Med 103:579–589.

62. Ginsberg J, Segal D, Erlich RM, Walfish PG (1978) Inappropriate triiodothyronine (T_3) and thyroxine (T_4) radioimmunoassay levels secondary to circulating thyroid hormone autoantibodies. Clin Endocrinol (Oxf) 8:133–139.
63. Konishi J, Iida Y, Kousaka T, Ikekubo K, Nakagawa T, Torizuka K (1982) Effect of anti-thyroxine autoantibodies on radioimmunoassay of free thyroxin in serum. Clin Chem 28:1389–1391.
64. Beck-Peccoz P, Romelli PB, Cattaneo MG, Faglia G, White EL, Barlow JW, Stockigt JR (1984) Evaluation of free thyroxine methods in the presence of iodothyronine-binding autoantibodies. J Clin Endocrinol Metab 58:736–739.
65. Mullinger RN, Walker G (1982) Free thyroxine in thyroid disease. Clin Chem 28:1394–1395.
66. Karlsson FA, Wibell L, Wide L (1977) Hypothyroidism due to thyroid hormone binding antibodies. N Engl J Med 296:1146–1148.
67. Refetoff S, Dewind LT, DeGroot LJ (1967) Familial syndrome combining deaf-mutism, stippled epiphyses, goiter and abnormally high PBI: Possible target organ refractoriness to thyroid hormone. J Clin Endocrinol Metab 27:279–294.
68. Weintraub BD, Gershengorn MC, Kourides IA, Fein H (1981) Inappropriate secretion of thyroid stimulating hormone. Ann Intern Med 95:339–351.
69. Lamberg BA, Rosengard S, Leiwendahl K, Saarien P, Evered DC (1978) Familial partial peripheral resistance to thyroid hormones. Acta Endocrinol (Copenh) 87:303–312.
70. Gershengorn MC, Weintraub BD (1975) Thyrotropin-induced hyperthyroidism caused by selective pituitary resistance to thyroid hormone: A new syndrome of "inappropriate secretion of TSH." J Clin Invest 56:633–642.
71. Kaplan MM, Swartz SL, Larson PR (1981) Partial peripheral resistance to thyroid hormone. Am J Med 70:1115–1121.
71a. Refetoff S, DeGroot LJ, Benard B, Dewind LT (1972) Studies of a sibship with apparent hereditary resistance to intracellular action of thyroid hormone. Metabolism 21:723–756.
72. Brooks MH, Barbato AL, Collins S, Garbincus J, Neidballa RG, Hoffman D (1981) Familial thyroid hormone resistance. Am J Med 71:414–421.
73. Gharib H, Klee GG (1985) Familial euthyroid hyperthyroxinemia secondary to pituitary and peripheral resistance to thyroid hormones. Mayo Clin Proc 60:9–15.
74. Refetoff S, Salazar A, Smith TJ, Scherberg NH (1983) The consequences of inappropriate treatment because of failure to recognize the syndrome of pituitary and peripheral tissue resistance to thyroid hormone. Metabolism 32:822–834.
75. Cooper DS, Ladenson PW, Nisula BC, Dunn JF, Chapman EM, Ridgway CE (1982) Familial thyroid hormone resistance. Metabolism 31:504–509.
76. Bode HH, Danon M, Weintraub BD, Maloof F, Crawford JD (1973) Partial target organ resistance to thyroid hormone. J Clin Invest 52:776–782.
77. Liewendahl K (1976) Triiodothyronine binding to lymphocytes from euthyroid subjects and a patient with peripheral resistance to thyroid hormone. Acta Endocrinol (Copenh) 83:64–70.
78. Bernal J, Refetoff S, DeGroot LJ (1978) Abnormality of triiodothyronine binding to lymphocyte and fibroblast nuclei from a patient with peripheral tissue resistance to thyroid hormone action. J Clin Endocrinol Metab 47:1266–1272.
79. Liewendahl K, Rosengard S, Lamberg BA (1978) Nuclear binding of triiodothyronine and thyroxine in lymphocytes from subjects with hyperthyroidism, hypothyroidism and resistance to thyroid hormones. Clin Chim Acta 83:41–48.

80. Kaplowitz PB, D'Ercole JA, Utiger RD (1981) Peripheral resistance to thyroid hormone in an infant. J Clin Endocrinol Metab 53:958–963.
81. Chait A, Kanter R, Green W, Kenny M (1982) Defective thyroid hormone action in fibroblasts cultured from subjects with the syndrome of resistance to thyroid hormones. J Clin Endocrinol Metab 54:767–772.
82. Wortsman J, Premachandra BN, Williams K, Burman KD, Hay ID, Davis PJ (1983) Familial resistance to thyroid hormone associated with decreased transport across the plasma membrane. Ann Intern Med 98:904–909.
83. Menezes-Ferreira MM, Eil C, Wortsman J, Weintraub BD (1984) Decreased nuclear uptake of [^{125}I]triiodo-l-thyronine in fibroblasts from patients with peripheral thyroid hormone resistance. J Clin Endocrinol Metab 59:1081–1087.
84. Wartofsky L, Berman KD (1982) Alterations in thyroid function in patients with systemic illness: The "euthyroid sick syndrome." Endocr Rev 3:164–217.
85. Kaptein LM, MacIntyre SS, Weiner JM, Spencer CA, Nicoloff JT (1981) Free thyroxine estimates in nonthyroidal illness: Comparison of eight methods. J Clin Endocrinol Metab 52:1073–1077.
86. Chopra IJ, Chau Teco GN, Nguyen A, Solomon D (1979) In search of an inhibitor of thyroid hormone binding to serum proteins in nonthyroid illnesses. J Clin Endocrinol Metab 48:63–69.
87. Oppenheimer JH, Schwartz HL, Mariash CN, Kaiser FE (1982) Evidence for a factor in the sera of patients with nonthyroidal disease which inhibits iodothyronine binding by solid matrices, serum proteins and rat hepatocyte. J Clin Endocrinol Metab 54:757–766.
88. Chopra IJ, Solomon DH, Chua Teco GN, Eisenberg JB (1982) An inhibitor of the binding of thyroid hormones to serum proteins is present in extrathyroidal tissues. Science 215:407–409.
89. Chopra IJ, Huang TS, Hurd RE, Beredo A, Solomon DH (1984) A competitive ligand binding assay for measurement of thyroid hormone-binding inhibitor in serum and tissues. J Clin Endocrinol Metab 58:619–628.
90. Wehmann RE, Gregerman RI, Burns WH, Saral R, Santos GW (1985) Suppression of thyrotropin in the low-thyroxine state of severe nonthyroidal illness. N Engl J Med 312:546–552.
91. Chopra IJ, Solomon DH, Hepner GW, Morgenstein AA (1979) Misleading low free thyroxine index and usefulness of reverse triiodothyronine measurement in nonthyroidal illnesses. Ann Intern Med 90:905–912.
92. Bacci V, Schussler GC, Kaplan TB (1972) The relationship between serum triodothyronine and thyrotropin during systemic illness. J Clin Endocrinol Metab 54:1229–1235.
93. Hamblin PS, Dyer SA, Mohr VS, Le Grand BA, Lim C, Tuxen DV, Topliss DJ, Stockigt JR (1986) Relationship between thyrotropin and thyroxine changes during recovery from severe hypothyroxinemia of critical illness. J Clin Endocrinol Metab 62:717–722.
94. Britton KE, Quinn V, Ellis, SM, Miralles JM, Quinn V, Cayley ACD, Brown B, Ekins RP (1975) Is "T$_4$ toxicosis" a normal biochemical finding in elderly women. Lancet 2:141–142.
95. Burrows AW, Shakespear RA, Hesch RD, Cooper E, Aicken CM, Burke CW (1975) Thyroid hormones in elderly sick: "T$_4$ euthyroidism." Br Med J 4:437–439.
96. Gavin LA, Rosenthal M, Cavalieri RR (1979) The diagnostic dilemma of isolated hyperthyroxinemia in acute illness. JAMA 242:251–253.
97. Gooch BR, Isley WL, Utiger RD (1982) Abnormalities in thyroid function tests in patients admitted to a medical service. Arch Intern Med 142:1801–1805.
98. Caplan RH, Pagliara AS, Wickus G (1980) Thyroxine thyrotoxicosis: A common variant of hyperthyroidism. JAMA 244:1934–1938.

98a. Turner TG, Browlie BEW, Sadder WA (1975) Does T_4 toxicosis exist? Lancet 1:407–408.
99. Birkhauser M, Busset R, Burer T, Burger A (1977) Diagnosis of hyperthyroidism when serum thyroxine alone is raised. Lancet 2:53–56.
100. Engler D, Donaldson EB, Stockigt JR, Taft P (1978) Hyperthyroidism without triiodothyronine excess: An effect of severe non-thyroidal illness. J Clin Endocrinol Metab 46:77–82.
101. Maturlo SJ, Rosenbaum RL, Pan C, Surks MI (1980) Variable thyrotropin response to thyrotropin releasing hormone after small decrease in plasma free thyroid hormone concentrations in patients with nonthyroidal disease. J Clin Invest 66:451–456.
102. McLarty DG, Ratcliffe WA, Ratcliffe JG, Shimmins JG, Goldberg A (1978) A study of thyroid function in psychiatric in patients. Br J Psychiatry 133:211–218.
103. Whybrow PC, Prange AJ, Treadway CK (1969) Mental changes accompanying thyroid gland dysfunction. Arch Gen Psychiatry 20:48–63.
104. Goldberg IJL, Lawton K, Ridges AP (1977) The effect of depression and its treatment on serum thyroxine. Postgrad Med J 53:211–215.
105. Whybrow PC, Coppen A, Prange AJ, Noguera R, Bailey JE (1972) Thyroid function and the response to liothyronine in depression. Arch Gen Psychiatry 26:242–245.
106. Kirkegaard C, Faber J (1981) Altered serum levels of thyroxine, triiodothyronines and diiodothyronines in endogenous depression. Acta Endocrinol (Copenh) 96:199–207.
107. Pecknold JC, Ban TA, Lehmann HE, Nair NPV, Orbach L (1976) Thyrotropin releasing hormone in depression: Clinical and endocrinological findings. Psychopharmacol Bull 12:44–48.
108. Cohen LK, Swigar ME (1979) Thyroid function screening in psychiatric patients. JAMA 242:254–257.
109. Morley JE, Shafer RB (1982) Thyroid function screening in new psychiatric admissions. Arch Intern Med 142:591--593.
110. Spratt DI, Pont A, Miller MB, McDougall IR, Bayer MF, McLaughlin WT (1982) Hyperthyroxinemia in patients with acute psychiatric disorders. Am J Med 73:41–48.
111. Kramlinger KG, Gharib H, Swanson DW, Maruta T (1984) Normal serum thyroxine values in patients with acute psychiatric illness. Am J Med 76:799–801.
112. Brunn T, Kristoffersen K (1978) Thyroid function during pregnancy, with special reference to hydatidiform mole and hyperemesis. Acta Endocrinol (Copenh) 88:383–389.
113. Bouillon R, Naesens M, Van Assche (1982) Thyroid function in patients with hyperemesis gravidarum. Am J Obstet Gynecol 143:922–926.
114. Kenimer JG, Hershman JM, Higgens PM (1975) The thyrotropin in hydatiform moles is human chorionic gonadotropin. J Clin Endocrinol Metab 40:482–491.
115. Fairweather DVI (1968) Nausea and vomiting in pregnancy. Am J Obstet Gynecol 102:135–175.
116. Cogan E, Abramow M (1986) Transient hyperthyroxinemia in symptomatic hyponatremic patients. Arch Intern Med 146:545–547.
117. Cavalieri RR, Pitt-Rivers R (1981) The effects of drugs on the distribution and metabolism of thyroid hormones. Pharmacol Rev 32:55–80.
118. Burgi H, Wimpfheimer C, Burger A, Zaunbauer W, Rosler H,

88   M. Safran and L. E. Braverman

Lemarchand-Beraund T (1976) Changes of circulatory thyroxine, triiodothyronine and reverse triiodothyronine after radiographic contrast agents. J Clin
Endocrinol Metab 43:1203–1210.

119. Kleinmann RE, Vagenakis AG, Braverman LE (1980) The effect of iopanoic
acid on the regulation of thyrotropin secretion in euthyroid subjects. J Clin
Endocrinol Metab 51:399–403.

120. Larsen PR, Dick TE, Markovitz BP, Kaplan MM, Gard TG (1979) Inhibition
of intrapituitary thyroxine to 3,5,3'-triiodothyronine conversion prevents the
acute suppression of thyrotropin released by thyroxine in hypothyroid rats. J
Clin Invest 64:117–128.

121. Obregon MJ, Pascual A, Mallol J, Morreale de Escobar G, Escobar del Rey F
(1980) Evidence against a major role of L-thyroxine at the pituitary level:
Studies in rats treated with iopanoic acid (Telepaque). Endocrinology
106:1827–1836.

122. Felicetta JV, Green WL, Nelp WB (1980) Inhibition of hepatic binding of
thyroxine by cholecystographic agents. J Clin Invest 65:1032–1040.

123. Larsen PR (1982) Thyroid-pituitary interaction: Feedback regulation of
thyrotropin secretion by thyroid hormone. N Engl J Med 306:23–32.

124. Kleinmann RE, Sternthal E, Starobin O, Braverman LE (1982) Cardiac
catheterization dye does not affect serum thyroid hormone concentrations or
TSH secretion. Cathet Cardiovasc Diagn 8:261–265.

125. Fairhurst BJ, Nagui N (1975) Hyperthyroidism after cholecystography. Br
Med J 3:630–631.

126. Heger JJ, Prystowsky EN, Jackman WM, Naccarelli GV, Warfel KA,
Rinkenberger RL, Zipes DP (1981) Amiodarone. Clinical efficacy and
electrophysiology during long-term therapy for recurrent ventricular tachy-
cardia or ventricular fibrillation. N Engl J Med 305:539–545.

127. Nademanee K, Singh B, Hendrickson J, Intarachot V, Lopez B, Feld G,
Cannon DS, Weis JL (1983) Amiodarone in refractory life threatening
ventricular arrhythmias. Ann Intern Med 98:577–586.

128. Rosenbaum MB, Chiale PA, Halpern MS, Nau GJ, Brzybyiski J, Levi R,
Lazzari JO, Elizar IM (1976) Clinical efficacy of amiodarone as an antiarrhythmic agent. Am J Cardiol 38:936–966.

129. Singh BN, Vaughan Williams EM (1970) The effect of amiodarone, a new
antianginal drug, on cardiac muscle. Br J Pharmacol 39:657–667.

130. Toubol P, Huerta R, Porte J, Delahaye JP (1976) Bases electrophysiologiques
de l'action antiarytmique de l'amiodarone chez l'homme. Arch Mal Coeur
8:845–853.

131. Vastesaeger M, Gillot P, Rasson G (1967) Etude clinique d'une nouvelle
medication antiangioreuse. Acta Cardiol 22:483–500.

132. Balsam A, Ingbar SH (1978) The influence of fasting, diabetes and several
pharmacological agents on the pathways of thyroxine metabolism in rat liver.
J Clin Invest 62:415–424.

133. Burger A, Dinichert D, Nicod P, Jenny M, Lemarchand-Beraud T, Vallotton
MB (1976) Effect of amiodarone on serum triiodothyronine, reverse triiodothyronine, thyroxin and thyrotropin. A drug influencing peripheral metabolism of thyroid hormone. J Clin Invest 58:255–259.

134. Jonckheer MH, Blockx P, Broeckaert I, Cornette C, Beckers C (1978) "Low
T₃ syndrome" in patients chronically treated with an iodine-containing drug,
amiodarone. Clin Endocrinol (Oxf) 9:27–35.

135. Kannan R, Ookhtens M, Chopra IJ, Singh BN (1984) Effects of chronic
administration of amiodarone on kinetics of metabolism of iodothyronines.
Endocrinology 115:1710–1716.

136. Melmed S, Nademanee K, Reed AW, Hendrickson JA, Singh BN, Hershman

JM (1981) Hyperthyroxinemia with bradycardia and normal thyrotropin secretion after chronic amiodarone administration. J Clin Endocrinol Metab 53:997–1001.

137. Pritchard DA, Singh BN, Hurley PJ (1975) Effects of amiodarone on thyroid function in patients with ischaemic heart disease. Br Heart J 37:856–860.

138. Lambert MJ, Burger AG, Galeazzi RL, Engler D (1982) Are selective increases in serum thyroxine (T_4) due to iodinated inhibitors of T_4 monodeiodination indicative of hyperthyroidism? J Clin Endocrinol Metab 55:1058–1065.

139. Martino E, Safran M, Aghini-Lombardi F, Rajatanavin R, Lenziardi M, Fay M, Pacchiarotti A, Aronin N, Macchia E, Haffajee C, Odoguardi L, Love J, Bigalli A, Baschieri L, Pinchera A, Braverman L (1984) Environmental iodine intake and thyroid dysfunction during chronic amiodarone therapy. Ann Intern Med 101:28–34.

140. Hershman JM, Sugawara M, Pekary AE, Reed AW, Ross R, Nademanee K, Singh BN, DiStefano JJ (1983) T_4 and T_3 kinetics in cardiac patients taking amiodarone. In Program of the 59th Meeting of the American Thyroid Association, New Orleans, p T-29.

141. Jaggarao NSV, Sheldon J, Grundy EN, Vincent R, Chamberlain DA (1982) The effects of amiodarone on thyroid function. Postgrad Med J 58:693–696.

142. Amico JA, Richardson V, Alpert B, Klein I (1984) Clinical and chemical assessment of thyroid function during therapy with amiodarone. Arch Intern Med 144:487–490.

143. Baillet J (1977) Amiodarone et Dysthyroide. Colloque ser l'Amiodarone, Paris 1977. Documentation Labaz, Paris, p 238.

144. Bugugnani MJ, Bailly M, DeSoutter P, Fouye H, Haiat R (1980) Surveillance de la fonction thyroidienne au cours des troitements prolonges par l'amiodarone. Ann Cardiol Angeiol 29:375–378.

145. Chevigne-Brancart M, Vandalem JL (1982) Thyroid function during and after amiodarone therapy. Proceedings of the 64th Annual Meeting of the Endocrine Society, San Francisco, p 111 (abstract).

146. Fradkin JE, Wolff J (1983) Iodine-induced thyrotoxicosis. Medicine 62:1–20.

147. Posner J, Sobel R, Glick S (1984) Effect of amiodarone on thyroid hormone economy. Isr J Med Sci 20:113–117.

148. Singh BN, Nademanee K (1983) Amiodarone and thyroid function: Clinical implications during antiarrhythmic therapy. Am Heart J 106:857–869.

149. Borowski GD, Garofano CD, Rose LI, Spielman SR, Rotmensch HR, Greenspan AM, Horowitz LN (1985) Effect of long-term amiodarone therapy on thyroid hormone levels and thyroid function. Am J Med 78:443–450.

150. Safran M, Fang SL, Bambini G, Pinchera A, Martino E, Braverman LE (1986) Effects of amiodarone and desethylamiodarone on pituitary deiodinase activity and thyrotropin secretion in the rat. Am J Med Sci 292:136–141.

151. Morley JE, Shafer RB, Elson MK, Flag ML, Raleigh MJ, Brammer G, Yuwiler A, Hershman J (1980) Amphetamine-induced hyperthyroxinemia. Ann Intern Med 93:707–709.

152. Hershman JM, Jones CM, Bailey AL (1972) Reciprocal changes in serum thyrotropin and free thyroxine produced by heparin. J Clin Endocrinol Metab 34:574–579.

153. Schatz DL, Sheppard RH, Steiner G, Chandarlapaty CS, De Vebur GA (1969) Influence of heparin on serum free thyroxine. J Clin Endocrinol Metab 29:1015–1022.

154. Hollander CS, Scott RL, Burgess JA, Rabinowitz D, Merimee TJ, Oppenheimer JH (1967) Free fatty acids: A possible regulation of free thyroid hormone levels in man. J Clin Endocrinol Metab 27:1219–1223.

155. Schwartz HL, Schadlow AR, Faierman D, Surks MI, Oppenheimer JH (1973) Heparin administration appears to decrease cellular binding of thyroxine. J Clin Endocrinol Metab 36:598–600.
156. Boss M, Kingstone D, Chan MK, Varghese Z (1982) Contradictory findings in the measurements of free thyroxine after administration of heparin. Clin Chem 28:1238–1239.
157. Lundberg PA, Jagenburg K, Lindstedt G, Nystrom E (1982) Heparin in vivo effect on free thyroxine. Clin Chem 28:1241–1242.
158. Shanks RB, Hadden DR, Lowe DC, McDermitt DG, Montgomery DAD (1969) Control trial of propranolol in thyrotoxicosis. Lancet 1:993–994.
158a. Heyma P, Larkins RG, Campbell DG (1980) Inhibition by propranolol of 3,5,3'-triiodothyroxine formation from thyroxine in isolated rat renal tubules: An effect independent of β-adrenergic blockade. Endocrinology 106:1437–1441.
159. Harrower ADB, Fyffe JA, Horn DB, Strong JA (1977) Thyroxine and triiodothyronine levels in hyperthyroid patients during treatment with propranolol. Clin Endocrinol (Oxf) 7:41–44.
160. Braverman LE, Chiovato L (1985) Thyroid storm. In Rippe JM, Irwin RS, Alpert JS, Dalen JE (eds): Intensive Care Medicine. Little, Brown, Boston, pp 798–801.
161. Verhoeven RP, Visser TJ, Docter R, Hennemann G, Schalekamp MADH (1977) Plasma thyroxine, 3,3',5-triiodothyronine and 3,3',5'-triiodothyronine during β-adrenergic blockade in hyperthyroidism. J Clin Endocrinol Metab 44:1002–1005.
162. Wiersinga WM, Touber JL (1977) The influence of β-adrenoreceptor blocking agents on plasma thyroxine and triiodothyronine. J Clin Endocrinol Metab 45:293–298.
163. Cooper DS, Daniels GH, Ladenson PW, Ridgway EC (1982) Hyperthyroxinemia in patients treated with high-dose propranolol. Am J Med 73:867–871.
164. Braverman LE, Vagenakis AG, Downs P, Foster AE, Sterling K, Ingbar SH (1973) Effects of replacement doses of sodium L-thyroxine on the peripheral metabolism of thyroxine and triiodothyronine in man. J Clin Invest 52:1010–1017.
165. Stock JM, Surks MI, Oppenheimer JH (1974) Replacement dosage of L-thyroxine in hypothyroidism: A re-evaluation. N Engl J Med 290:529–533.
166. Saberi M, Utiger RD (1974) Serum thyroid hormone and thyrotropin concentrations during thyroxine and triiodothyronine therapy. J Clin Endocrinol Metab 39:923–927.
167. Ingbar JC, Borges M, Iflah S, Kleinmann RE, Braverman LE, Ingbar SH (1982) Elevated serum thyroxine concentration in patients receiving "replacement" doses of levothyroxine. J Endocrinol Invest 5:77–85.
168. Kahn A (1973) Serum thyroxine levels in patients receiving L-thyroxine suppression or replacement therapy. Can Med Assoc J 109:279–281.
169. Wong MM, Volpe R (1981) What is the best test for monitoring level of thyroxine therapy? Can Med Assoc J 124:1181–1183.
170. Pearce J, Himsworth RL (1985) Total and free thyroid hormone concentrations in patients receiving maintenance replacement treatment with thyroxine. Br Med J 288:693–695.
171. Brown ME, Refetoff S (1980) Transient elevation of serum thyroid hormone concentration after initiation of replacement therapy in myxedema. Ann Intern Med 92:491–495.
172. Lo Presti JS, Warren DW, Kaptein EM, Croyson MJ, Nicoloff JT (1982)

Urinary immunoprecipitation method for estimation of thyroxine to triiodothyronine conversion in altered thyroid states. J Clin Endocrinol Metab 55:666–670.

173. Lum SMC, Nicoloff JT, Spencer CA, Kaptein EM (1984) Peripheral tissue mechanism for maintenance of serum triiodothyronine values in a thyroxine-deficient state in man. J Clin Invest 73:570–575.

174. Jennings PE, O'Malley BP, Griffin KE, Northover B, Rosenthal FD (1984) Relevance of increased serum thyroxine concentrations associated with normal serum triiodothyronine values in hypothyroid patients receiving thyroxine: A case for "tissue thyrotoxicosis." Br Med J 289:1645–1647.

175. Ross DS (1986) New sensitive immunoradiometric assays for thyrotropin. Ann Intern Med 104:718–720.

176. Greene R, Ferran HE (1958) The physiological activity of D-thyroxine. Br Med J 2:1057–1060.

177. Oliver MF, Boyd GS (1961) Reduction of serum cholesterol by dextrothyroxine in men with coronary heart disease. Lancet 1:783–785.

178. Bantle JP, Oppenheimer JH, Schwartz HL, Hunninghake DB, Propbsfield JL, Hanson RF (1981) TSH response to TRH in euthyroid, hypercholesterolemic patients treated with graded dose of dextrothyroxine. Metabolism 30:63–66.

179. Gorman CA, Jiang N-S, Ellefson RD, Elveback LR (1979) Comparative effectiveness of dextrothyroxine and levothyroxine in correcting hypothyroidism and lowering blood lipid levels in hypothyroid patients. J Clin Endocrinol Metab 49:1–7.

180. Yosha S, Fay M, Longcope C, Braverman LE (1984) Effect of D-thyroxine on serum sex hormone binding globulin (SBHG), testosterone, and pituitary thyroid function in euthyroid subjects. J Endocrinol Invest 7:489–494.

181. Leb G, Lankmayer FP, Goebel R, Pristautz H, Nachtmann F, Knapp G (1981) Stereospecific determination of D-thyroxin and L-thyroxin in serum. Klin Wochenschr 59:861–863.

182. Young WF Jr, Gorman CA, Jiang NS, Machacek D, Hay ID (1984) L-thyroxine contamination of pharmaceutical D-thyroxine: Probable cause of therapeutic effect. Clin Pharmacol Ther 36:781–787.

183. Kotchen TA, Mongey EH, Hogan KP, Boyd AE III, Pennington LL, Mason JW (1973) Thyroid response to simulated altitude. J Appl Physiol 34:165–168.

184. Rastogi GK, Malhotra MS, Srivastava MC, Sawhney RC, Dua GL, Sridharan K, Hoon RS, Singh I (1977) Study of the pituitary-thyroid functions at high altitude in men. J Clin Endocrinol Metab 44:447–452.

185. Surks MI (1966) Elevated PBI, free thyroxine, and plasma protein concentration in man at high altitude. J Appl Physiol 21:1185–1190.

186. Mordes JP, Blume FD, Boyer S, Zheng MR, Braverman LE (1983) High-altitude pituitary-thyroid dysfunction on Mount Everest. N Engl J Med 30:1135–1138.

187. Martino E, Bambini G, Bartalena L, Mammoli C, Aghini-Lombardi F, Baschieri L, Pinchera A (1986) Human serum thyrotrophin measurement by ultrasensitive immunoradiometric assay as a first-line test in the evaluation of thyroid function. Clin Endocrinol (Oxf) 24:599–606.

4

Pseudohypoparathyroidism: Target Organ Resistance to Parathyroid Hormone and Other Metabolic Defects

ARNOLD S. BRICKMAN and HAROLD E. CARLSON

Pseudohypoparathyroidism (PsHP) is the generic term for a clinical disorder characterized by target tissue resistance to the action(s) of parathyroid hormone (PTH). The biochemical expressions of PTH resistance include hypocalcemia, hyperphosphatemia, and increased circulating levels of PTH appropriate to the degree and duration of hypocalcemia. Altered vitamin D metabolism may be reflected by low circulating levels of $1,25(OH)_2D_3$. These biochemical changes reflect PTH resistance at both the kidney and skeleton sites of action. Administration of exogenous PTH fails to elicit normal responses at either or both of these target organs.

Although the first described cases of PsHP showed distinctive somatic anomalies, other forms of PsHP lacking these somatic features and with different metabolic etiologies have been identified subsequently. This review summarizes our knowledge and understanding of various disorders associated with PTH resistance and commonly described as "variants" or types of PsHP.

I. A Selective Historical Review

In 1942, Albright and his associates (1) first described 3 cases of a disorder they named PsHP and proposed that it represented a form of resistance to PTH. The patients reported by Albright and other early investigators had chemical features of hypoparathyroidism (hypocalcemia and hyperphosphatemia) associated with a characteristic phenotypic appearance consisting of short stature with a thickset body habitus; shortening of metacarpals, metatarsals, and phalanges (brachydactyly); and subcutaneous calcifications (1–5). This triad became known as Albright's hereditary osteodystrophy (AHO), and this term is still used today. Ten years after their first report, Albright et al. described a second syndrome, in which the features of AHO were present but the patients were normocalcemic and lacked evidence of PTH resistance (6); they referred to this disorder as pseudopseudohypoparathyroidism (PsPsHP).

In the initially described cases of PsHP, demonstration of blunted calcemic and phosphaturic responses to infusion or injection of crude preparations of bovine parathyroid extract suggested that PsHP was a disorder of hormone resistance. Examination of parathyroid tissue obtained at surgery from patients with PsHP revealed enlarged hyperplastic glands, suggesting that hormone secretion was intact (1,3,7). Subsequently, Tashjian et al. (8) and others demonstrated that circulating levels of immunoreactive PTH are indeed increased in patients with PsHP (9). At this time, patients with PsHP and resistance to exogenous PTH, but without the AHO phenotype, were also described (5,7).

In 1969, Chase et al. (10) provided the first insight into the molecular basis for PsHP, with the report that in patients with PsHP, urinary cyclic 3′,5′-adenosine monophosphate (cAMP) response to exogenous PTH was blunted. This work provided evidence that the metabolic defect in PsHP might involve the PTH receptor-adenylate cyclase complex. It also provided a very useful procedure for the diagnosis of renal resistance to PTH. Indeed, using this diagnostic aid, Drezner et al. (11) were able to identify a third type of PsHP (PsHP type II), lacking the AHO features and characterized by normal PTH-stimulated cAMP generation but without phosphaturic response.

In 1971, Marcus et al. (12) demonstrated the presence of PTH-sensitive adenylate cyclase in renal cortical tissue from a patient with PsHP. Because this enzyme was activated in vitro by saturating concentrations of PTH, it could be concluded that the basic abnormality was not attributable to a depletion of either the hormone-binding or the catalytic activity of adenylate cyclase. Drezner and Burch (13), also working with renal cortical tissue from a patient with PsHP, provided evidence that the defect in PTH-induced cyclic AMP production could be explained by a defect of the GTP-binding protein of the receptor, resulting in a partial uncoupling of the receptor-cyclase complex. This laid the foundation for our current understanding of the basic defect in one form of PsHP.

In 1980, Farfel et al. (14) and other investigators (15–17) demonstrated that the activity of the guanine nucleotide receptor cyclase coupling protein (termed N_s or G_s protein subunit) is reduced in erythrocyte and other tissue membranes from some patients with PsHP. This membrane protein couples the hormone binding to the catalytic subunit of adenylate cyclase, bringing about stimulation of the enzyme (18). Thus, N_s protein functions as a transducer, conveying a hormonal signal from the external surface of the cell (Fig. 4.1). Reduced N_s protein activity in various target tissues can explain impaired cellular generation of cAMP. Measurement of N_s protein activity has been useful in confirming the heterogeneity of PsHP and in providing help in understanding the complex inheritance of AHO with and without PTH resistance.

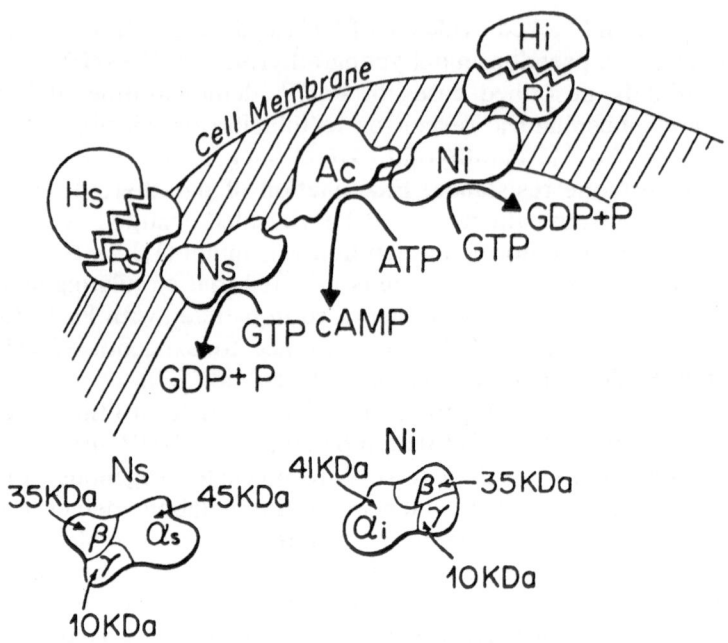

FIGURE 4.1. Components of hormonal stimulatory (H_s) and inhibitory (H_i) pathways for activation or inactivation of catalytic adenylate cyclase (AC). R_s, stimulatory pathway receptor; N_s, stimulatory guanine nucleotide-binding protein; R_i, inhibitory pathway receptor; N_i, inhibitory guanine nucleotide-binding protein; α_s, α subunit of N_s; α_i, α subunit of N_i; β and γ, β and γ subunits of N_s and N_i.

II. Classification

A. Pseudohypoparathyroidism Type I

Individuals with this type of PTH resistance manifest impaired renal response to the phosphaturic and cAMP excretory effects of exogenous PTH. Skeletal resistance to PTH may be present but is hard to measure, since factors other than primary PTH resistance may influence the calcemic response to exogenous PTH (the standard marker for skeletal responsiveness to PTH). Furthermore, not all patients with renal resistance appear to manifest skeletal resistance to the hormone, and hence the latter has not been considered a prerequisite for the diagnosis of PsHP type I. Patients with PsHP type I usually have hypocalcemia (and hyperphosphatemia), at least intermittently (Table 4.1).

By current convention, PsHP type Ia is characterized by the presence

TABLE 4.1. Classification of types of pseudohypoparathyroidism.

Type	Features	N_s protein activity	PTH resistance	Mode of genetic transmission
Ia	AHO; multiple hormone resistance	Reduced	+	Autosomal-dominant Autosomal-recessive
Pseudo-PsHP	AHO; ?hormone resistance other than PTH	Reduced	−	Affected individuals often in kindreds with PsHP Ia
Ib	Normal somatic features; kindreds with additional syndromes or metabolic defects reported; probably comprises a heterogeneous group of disorders:	(Normal)	+	Autosomal-dominant
	PTH receptor defect	Normal	+	?
	Abnormality at catalytic adenylate cyclase	Normal	+	?
II	Normal somatic features; probably comprises a heterogeneous group of disorders; PTH resistance characterized by normal cAMP response but absent phosphaturic response; defect may be reversible with normalization of calcium	Normal	+	?

of AHO features—PTH resistance, and deficient cell membrane N_s protein (generally reduced by approximately 50%) (14–17). For purposes of patient screening, N_s protein activity has been measured in erythrocyte membrane preparations. Two types of assay procedures have been employed. In one (complementation assay), activity is determined by the ability of extracts of patient erythrocyte membranes to complement the N_s protein deficiency of membrane preparations from the S49 mouse lymphoma cyc⁻ strain in the stimulation of adenylate cyclase activity (cAMP generation) (14). The assay system requires the presence of an agonist, such as isoproterenol and guanosine-5'-0-3-thiotriphosphate. The second assay (ribosylation assay) involves the use of cholera toxin to catalyze the transfer of ^{32}P-ADP-ribose from ^{32}P-NAD to the 41-kd subunit of N_s protein in erythrocyte membrane preparations (14). Results from both assays are expressed as a percent of the activity measured in erythrocyte membrane preparations from normal subjects (Fig. 4.2).

In contrast to individuals with PsHP type Ia, patients with PsHP type Ib have normal somatic features and normal levels of N_s protein activity (14). PsHP type Ib appears to include a heterogeneous group of conditions composed of several subtypes; although the pathogenesis is unclear, separate disorders with defects at the level of the PTH receptor or the catalytic adenylate cyclase have been described (19,20). Among them are

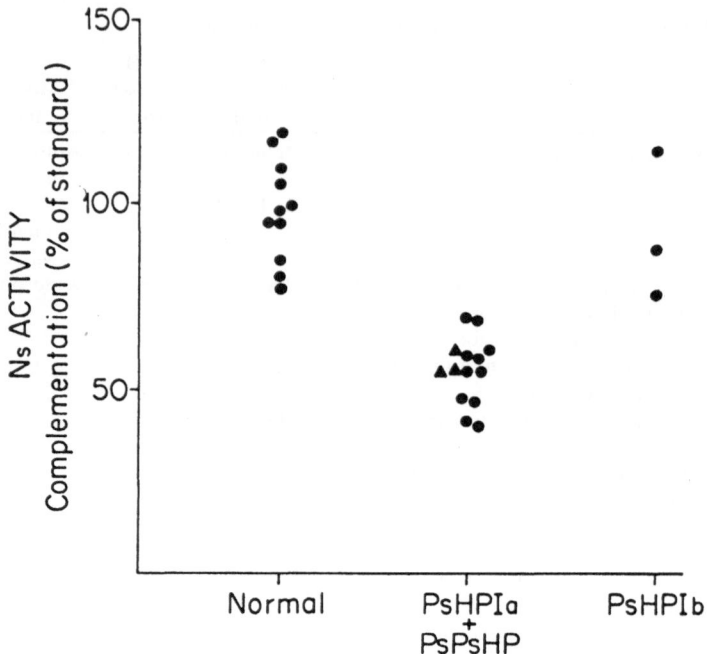

FIGURE 4.2. Erythrocyte N$_s$ protein activity measured by complementation assay in normal controls, patients with pseudohypoparathyroidism (PsHP) types Ia and type Ib (●), and patients with pseudopseudohypoparathyroidism (▲).

the rare cases of patients with AHO and PTH resistance but with normal N$_s$ protein activity (at least by current technology) (14).

B. Pseudohypoparathyroidism Type II

Individuals with this type of PTH resistance manifest a normal urinary cAMP response but absent phosphaturic response to infusion of PTH under hypocalcemic conditions (11,21). It has been reported that normalization of serum calcium corrects the phosphaturic defect (21–24). Although in some individuals PsHP type II may be a primary form of PTH resistance, the uncoupling of the pathway between intracellular generation of cAMP and activation of luminal transport of phosphate may be a feature of this heterogeneous group of disorders (22–24). Insufficient information is available concerning possible skeletal resistance to PTH in patients with PsHP type II to permit conclusions at the present time. Finally, this disorder has been described both in individuals with AHO and in those with normal somatic features.

C. Pseudopseudohypoparathyroidism

The designation PsPsHP was initially given to patients with somatic features of AHO, persistent normocalcemia, and normal responses to PTH infusion (6,7). Unfortunately, the use of this designation to indicate the presence of normocalcemia in patients with demonstrated renal resistance to PTH has created some confusion (25). Periods of normocalcemia have been described in both PsHP type Ia and PsHP type Ib, adding to the difficulty in the absence of more careful diagnostic characterization. Because it is now recognized that PsHP type Ia is usually associated with resistance to other hormones, it remains possible that although PTH resistance is not present (at least as characterized by the current methods of evaluation), other, perhaps subtle expressions of hormone resistance (or alterations in other cAMP-mediated processes) may occur in patients with PsPsHP. Many patients with PsPsHP are relatives of patients with overt PsHP type Ia, and recent studies have reported that patients with PsPsHP usually have reduced N_s protein activity (26). Thus, patients currently designated as having PsPsHP may represent one end of a spectrum of diseases more fully expressed in patients with PsHP type Ia.

In this regard, Fisher et al. (27) have recently reported the occurrence of increased TSH responsivity to TRH in some individuals with PsPsHP (e.g., with AHO, reduced N_s protein activity, and lacking PTH resistance). However, from available reports that some patients with PsPsHP have no evidence of hormone resistance, it is suggested that deficiency of N_s protein activity alone is not sufficient to explain the hormonal and metabolic abnormalities of PsHP. Rather, the N_s protein abnormality seems to be more closely (but not perfectly) correlated with the occurrence of AHO. More extensive testing for evidence of defects in cAMP-mediated physiologic processes in such patients will be required before it can be concluded that individuals with PsPsHP lack such metabolic abnormalities.

Finally, evidence has also been provided that some individuals with PsHP type Ib may have target tissue resistance to several hormones, including PTH, TSH, and glucagon (28). At present it is not known whether multiple hormone resistance in this syndrome shares a common underlying defect analogous to that seen in PsHP type Ia. It is possible that within kindreds of patients with PsHP type Ib, individuals may be identified who lack PTH resistance but who manifest other expressions of hormone resistance. Indeed, Brickman et al. (28) have reported 2 individuals in a family affected with PsHP type Ib who have normal responses to PTH infusion but primary hypothyroidism and glucagon resistance. Current nomenclature suggests that these individuals could be designated PsPsHP type Ib.

III. Pathogenesis

A. Pseudohypoparathyroidism Type Ia

It now seems clear that patients with this form of PTH resistance do indeed have approximately 50% reduction of N_s protein activity in cell membranes (Fig. 4.2). This defect has been demonstrated in erythrocytes (14–17), transformed lymphoblasts (29), fibroblasts (30), platelets (31), and renal tubular cells (32). N_s protein is composed of three subunits; the apparent site of defect in PsHP type Ia is in the 45 kD α subunit (α_s), which has GTP-binding capacity (Fig. 4.1). Levine et al. (33) recently reported results of examination of genomic DNA from cultured fibroblasts in eight kindreds of patients with PsHP type Ia and in some individuals with PsPsHP. Their preliminary studies indicate that the defect in N_s protein is not due to a major deletion or rearrangement of the gene for the α subunit. However, results of measurements of levels of pretranslational mRNA identified two subgroups of patients with low N_s protein activity: those with reduced and those with normal levels of mRNA for the α subunit. These workers also found that patients with PsPsHP may have reduced levels of α subunit mRNA. Although preliminary, these intriguing findings underscore the genetic heterogeneity of these disorders. It is also possible that PsHP patients will be found with disorders of the β and γ subunits of N_s protein.

A second transducing protein, the inhibitory N protein (N_i), that inhibits adenylate cyclase activity (Fig. 4.1), is normal in PsHP type Ia (16,32,33). N_i protein also contains an α subunit (α_i) with a molecular weight of 41 kD which apparently differs from that of the N_s protein.

Binding of GTP to N_s brings about its dissociation from the hormone-binding moiety and the dissociation of the α, β, and γ subunits and, consequently, the stimulation of the catalytic moiety of the receptor by α-GTP and an increase in the production of cAMP. A similar mechanism activates N_i, reducing the production of cAMP. These phenomena are self-limiting, because the binding of GTP confers GTPase activity to the N proteins reconverting active α-GTP to inactive α-GDP and allowing the reassociation of the α, β, and γ subunits. The GTPase activity is inhibited when N_s is ribosylated by the transfer of the ADP-ribose moiety of NAD, a process catalyzed by the toxin of γ·cholerae, while the GTPase activity of N_i is similarly inhibited by the toxin of β·pertussis. Thus, these bacterial toxins bring about a persistent activation of N_s and N_i and a persistent stimulation or inhibition of cAMP production, respectively.

B. Pseudohypoparathyroidism Type Ib

Information concerning the pathogenesis of PsHP type Ib is limited. Reports by several investigators have documented normal complements of membrane N_s and N_i protein (14,17). This is consistent with the observation that the majority of patients with PsHP type Ib lack evidence of multiple hormone resistance, although Brickman et al. (28) have reported concomitant hypothyroidism in two PsHP type Ib kindreds (and a third sporadic case) and prolactin deficiency in several members of one of these families. Fisher et al. (27) have also described concomitant PTH resistance and primary hypothyroidism in a kindred of individuals lacking features of AHO and with normal erythrocyte N_s protein activity.

Silve et al. (19) found decreased cAMP responses to PTH in cultured fibroblasts in 7 of 10 patients with PsHP type Ib. The remaining 3 patients had normal or increased cAMP generation. Fibroblast cAMP responses to prostaglandin E_1 were normal, suggesting that these patients may have a defect limited to PTH action—perhaps an abnormality of the PTH-binding moiety of the receptor. Another form of PsHP type Ib has been described in a patient whose cultured fibroblasts showed decreased cAMP responses to several agonists including prostaglandin E_1, forskolin, isoproterenol, and PTH. The findings suggest a generalized defect in adenylate cyclase signal transduction in this patient (20). Together, these various observations point to the pathogenetic heterogeneity of several disorders currently classified as PsHP type Ib.

C. Expressions of PTH Resistance and Hypocalcemia in PsHP Type I

At the present time, it is not possible to distinguish between PsHP type Ia and type Ib on the basis of renal or putative skeletal resistance to PTH. In both types of the disorder, renal resistance is expressed by impaired cAMP and blunted phosphaturic responses to PTH (Fig. 4.3); moreover, in both types, infusion of dibutyryl cAMP produces a phosphaturic response similar to that observed in normal subjects (36). In both conditions, effects of endogenous and exogenous PTH on renal handling of other electrolytes including calcium, magnesium, sodium, potassium, and bicarbonate appear to be similar (37). Finally, in both conditions, treatment with vitamin D may (in some cases) result in improvement or normalization of the phosphaturic response to exogenous PTH (38–40).

Another marker of renal resistance to PTH in PsHP type Ia and type Ib is the reduced production of $1,25(OH)_2D_3$ (41–46). Several reports have documented low circulating levels of $1,25(OH)_2D_3$ in patients with both conditions (45); however, this does not appear to be a constant phenomenon, as some individuals have normal levels of vitamin D metabolites (Fig. 4.4). Similarly, the expected stimulatory effect of exogenous PTH infusion on $1,25(OH)_2D_3$ may be absent or blunted in some but not all individuals with these disorders (Fig. 4.5) (43,44). The explanation for this variable expression of altered vitamin D metabolism is unclear.

Although hypocalcemia per se is present in the majority of patients with PsHP type Ia or Ib, this is not invariably the case. Spontaneous

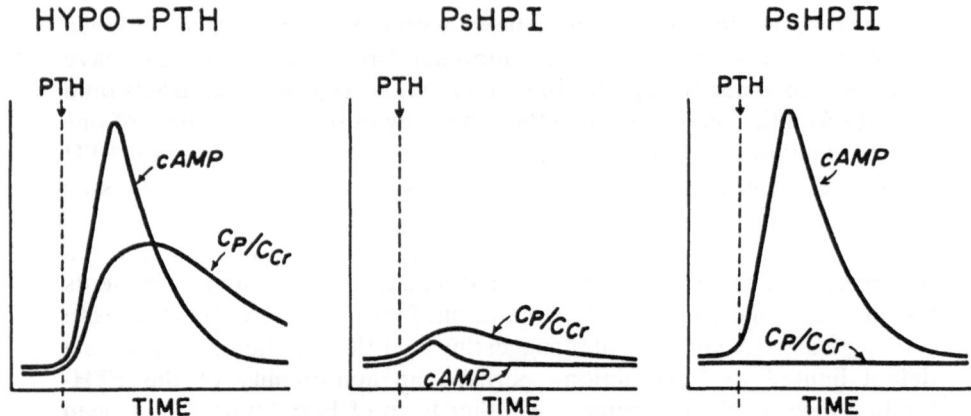

FIGURE 4.3. Schematic depiction of renal responses to infusion of exogenous PTH in typical patients with hypoparathyroidism (HYPO-PTH), pseudohypoparathyroidism type I (PsHP I), and type II (PsHP II). Depending on the molecular species, dose, and rate of infusion of PTH, the typical rise in urinary cAMP is 50- to 100-fold over baseline in normal subjects or patients with HYPO-PTH. The response in PsHP is usually less than a fivefold rise in urinary cAMP. The typical maximum fractional excretion of phosphorus (Cp/CCr) is 0.4–0.5 in normals or HYPO-PTH. The phosphaturic response in PsHP I is variable but usually markedly blunted. Patients with PsHP II under conditions of hypocalcemia demonstrate a normal cAMP response but a blunted or absent phosphaturic response.

normocalcemia may occur in both disorders even in the presence of renal PTH resistance, with or without low circulating levels of 1,25(OH)$_2$D$_3$ (45,46). This observation strongly suggests that blunted effects of PTH on skeletal mineral homeostasis play a central role in development of hypocalcemia. The site(s) of such a defect(s) is unclear at present but may involve alterations in effects of PTH on bone remodeling and homeostatic maintenance of extracellular calcium pool size. At present, the evidence for primary resistance to PTH in bone cells (in terms of cAMP generation) is inferential. Thus, skeletal PTH resistance generally has been defined in terms of an absent or blunted calcemic response to administration of exogenous PTH. However, several studies have shown that long-term treatment with various forms of vitamin D analogues may normalize the calcemic response to exogenous PTH (38,39) (Fig. 4.6). Although 1,25(OH)$_2$D$_3$ appears to play a role in modulating the calcemic response to exogenous PTH, short-term (up to 21 days) replacement therapy in patients with PsHP type Ia or Ib has no effect on this response to exogenous PTH (Fig. 4.7). These observations suggest that in some individuals altered effects of PTH on the skeleton may be attributable (at

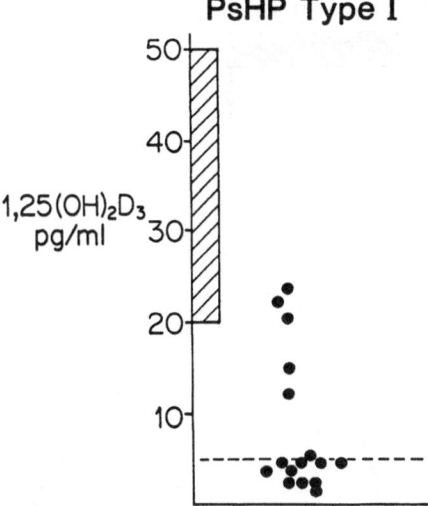

FIGURE 4.4. Serum levels of calcitriol in patients with untreated pseudohypoparathyroidism type I (PsHP). The hatched bar indicates the normal range.

least in part) to acquired factors. Further evidence in support of this concept is the prolonged maintenance of a normocalcemic state following long-term treatment with vitamin D. We have observed several patients treated with either vitamin D_2 or calcitriol who have remained normocalcemic for periods of 3–7 years following discontinuation of therapy. None of these patients had experienced vitamin D intoxication; moreover, the duration of normocalcemia far exceeded that attributable to subclinical vitamin D excess.

Other factors that may affect the development of hypocalcemia in PsHP type I include increased calcitonin secretion (47–49), hyperphosphatemia (5,7,45), reduced intestinal absorption of calcium (50), and acquired conditions such as magnesium or vitamin D depletion (41,46). Morimoto et al. (49) have reported increased urinary secretion of calcitonin in PsHP type I; other studies have shown increased calcitonin secretory reserve as assessed by calcium infusion in such patients, although the magnitude of the response was not different from that seen in individuals with other chronic hypocalcemic disorders such as idiopathic hypoparathyroidism (47). At present, it is unknown whether either PsHP type Ia or type Ib is associated with an innate defect in response to or secretion of calcitonin. However, it seems unlikely that increased calcitonin secretion in these conditions plays a significant role in the generation or maintenance of hypocalcemia.

A number of reports have described an apparent paradoxic association of radiologic and histologic features of hyperparathyroidism and the

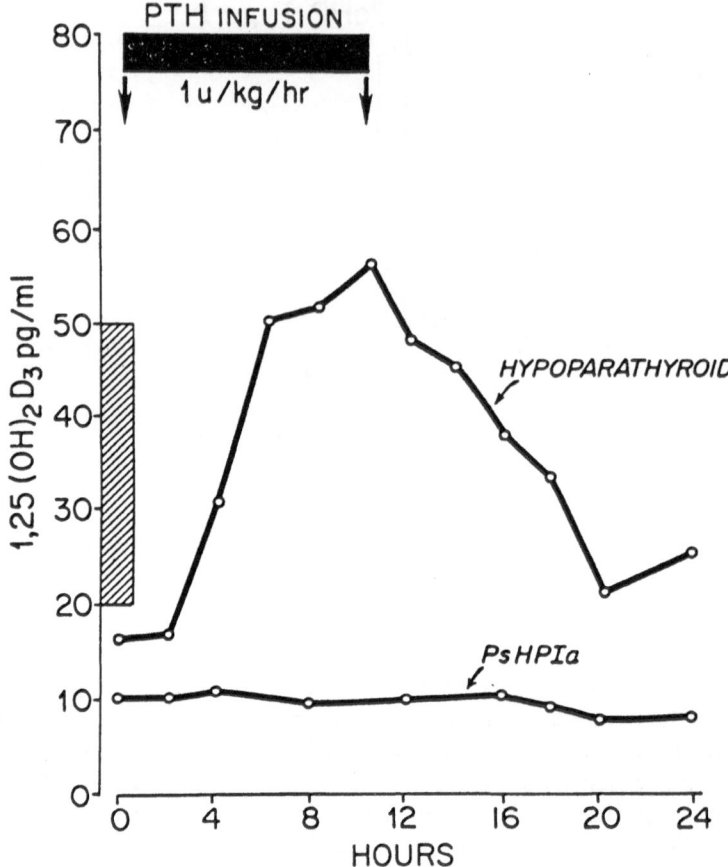

FIGURE 4.5. Changes in serum levels of calcitriol during a 10-h infusion of parathyroid extract (1 unit/kg/h) in patients with hypoparathyroidism and pseudo-hypoparathyroidism type Ia, respectively. The hatched bar indicates the range of baseline serum calcitriol values in normal subjects.

presence of hypocalcemia and presumed skeletal resistance to PTH, with established renal PTH resistance in patients with PsHP type I (50–56). However, such an association appears to be relatively uncommon, possibly occurring more frequently in patients with various forms of PsHP type Ib (55). In both types of PsHP, reported bone biopsy features range from marked osteitis fibrosa with increased osteoclastic resorption to markedly reduced resorptive activity and/or features of osteomalacia (45). Again, such findings underscore the possibility of selective defects in PTH action or extracellular calcium concentration, in contrast to actions involving skeletal-remodeling processes.

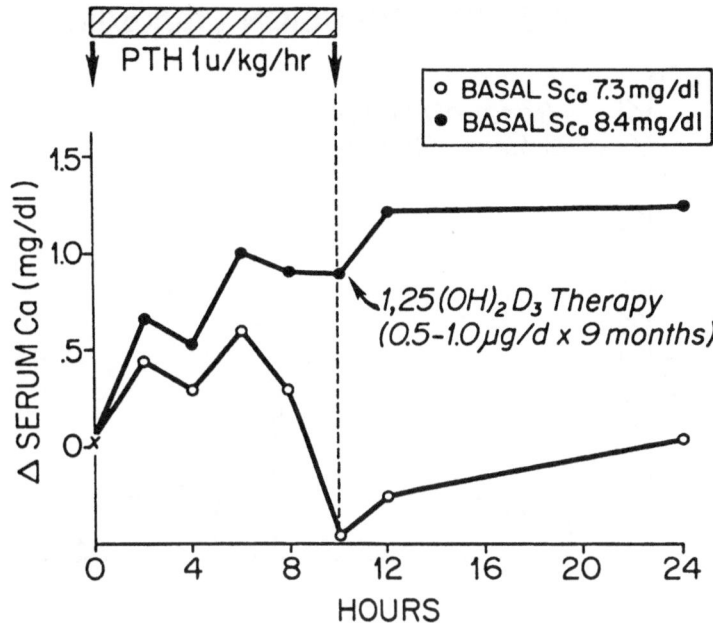

FIGURE 4.6. Changes in serum calcium during 10-h infusions of parathyroid extract (1 unit/kg/h) in a patient with PsHP type Ib, prior to (○) and after (●) 9 months of treatment with 1,25(OH)₂D₃ (0.5–1.0 μg/day). Therapy had been discontinued for 1 week prior to performing the second infusion study. Note that the latter response is within the range of responses observed in 17 normal subjects receiving the same infusion (range of calcemic responses 0.8–2.4 mg/dl).

D. Pseudohypoparathyroidism Type II

As in the case of other forms of PsHP, PsHP type II may encompass a group of disorders with diverse etiologies. The important physiologic marker in this condition is a dissociation between PTH-induced cAMP generation and its subsequent effect on phosphorus transport. Although PsHP may represent a primary metabolic disorder in some individuals, there is evidence in other cases that it may be an acquired condition. At present, there is only one possible example of familial occurrence of PsHP type II (57); this clearly contrasts with both types of PsHP type I. Recently, Rao et al. (24) reported features of PsHP type II in individuals with hypocalcemia due to vitamin D depletion and associated osteomalacia. Other conditions associated with PsHP type II include anticonvulsant therapy (58) and Sjogren's syndrome (59). In the latter report, the patient's serum contained immunoglobulin which blocked PTH-induced phosphaturia. These observations have expanded our concept of the pathogenesis of PsHP type II. Clearly, exclusion of altered vitamin D

10 HOUR PTH INFUSION (1u/kg/hr)

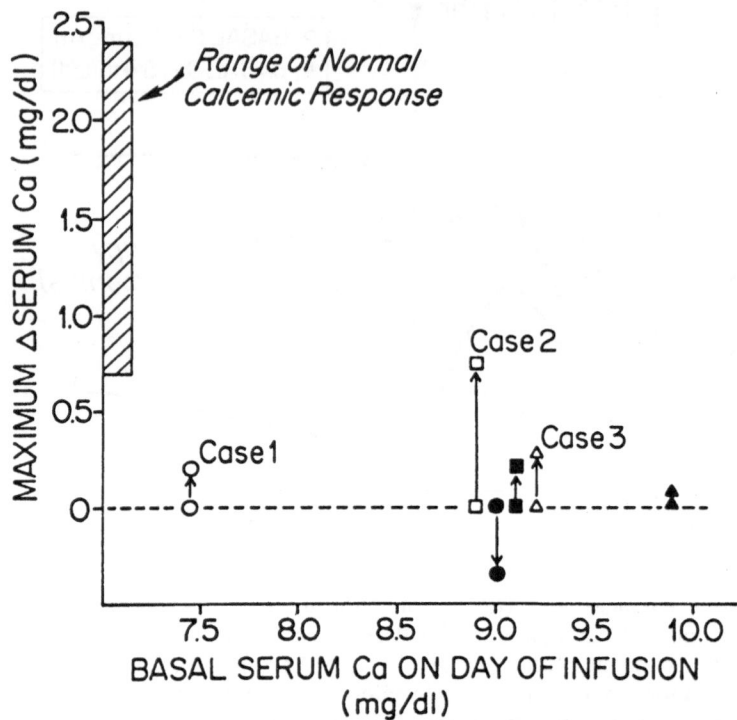

FIGURE 4.7. Effects of 21 days of therapy with 1,25(OH)$_2$D$_3$ (1 μg/day) on maximum calcemic responses during a 10-h infusion of parathyroid extract (1 unit/kg/h) in 3 patients with pseudohypoparathyroidism. Open symbols indicate levels of serum calcium at the time of onset of therapy; closed symbols indicate levels of serum calcium after 21 days of treatment. Arrows indicate the direction of change in serum calcium during the infusions. Case 1 is a patient with PsHP type Ib, and cases 2 and 3 are patients with PsHP type Ia with spontaneous normocalcemia at the time of study.

metabolism is a diagnostic requirement in establishing the pathogenesis of this disorder.

   Studies by several investigators have demonstrated that acute and chronic correction of hypocalcemia can restore or normalize PTH-induced phosphaturia in PsHP type II (21–23). This has prompted the speculation that the defect in this disorder can be attributed to an impairment of PTH-mediated calcium transport processes in the kidney (and possibly the skeleton). The significance of the observation of Yamada et al. (23) that infusion of dibutyryl cAMP induced phosphaturia in a patient with PsHP type II (identical to responses observed in PsHP type I and normal subjects) independent of changes in extracellular calcium is uncertain at this time.

E. Other Theories of Pathogenesis

1. CIRCULATING INHIBITORS

Two groups of investigators have now reported that plasma from PsHP patients with AHO (thus, probably with PsHP type Ia) inhibits the action of PTH in vitro in a cytochemical bioassay system (60–62). This inhibitor is suppressed by prolonged (days), but not brief (hours), hypercalcemia. The nature of the inhibitor is unknown. Because it is suppressed by hypercalcemia and because it disappears after total parathyroidectomy, it has been suggested that is originates in the parathyroid glands (62); one group of investigators has suggested that it could be an abnormal form of PTH (61). The proposal that a circulating inhibitor of PTH action might be important in the pathogenesis of PsHP has merit, because the mere presence of a 50% deficiency of N_s protein appears insufficient to explain the occurrence of PsHP, given that individuals with PsPsHP have a similar N_s deficiency without endocrine abnormalities (26,34). Nevertheless, the cAMP response to PTH generally does not improve when normocalcemia is restored in PsHP type Ia; the phosphaturic and calcemic responses also do not invariably improve. If the inhibitor can also induce resistance to other hormones besides PTH, it could also play a role in other features of this disorder. It is also possible (although there are no data yet available to support or refute the hypothesis) that such inhibitors could occur in PsHP type Ib and perhaps account for the occurrence of prolactin deficiency, TSH resistance, and resistance to beta-adrenergic agonists in some of these patients (see below). In only 1 patient, however, has deficient pituitary, thyroid, and gonadal function been corrected by restoration of normocalcemia (63).

2. CIRCULATING ANTIBODIES

In some of our patients with PsHP type Ia or type Ib, we have noted the presence of antithyroid or other autoantibodies (64). Although the link between PsHP and endocrine autoimmunity is obscure, it is possible that autoimmune thyroiditis may explain thyroid dysfunction in some patients. "Autoantibodies" to the PTH receptor have been described in uremic patients with secondary hyperparathyroidism (65), although the inhibiting material has not been identified.

IV. Function of Other Endocrine Systems in PsHP (Table 4.2)

A. Thyroid

Many patients with PsHP type Ia have either overt hypothyroidism or compensated hypothyroidism (euthyroid with elevated serum TSH), usually without a goiter. Although this is most common in PsHP type Ia, it also occurs in type Ib. Early studies suggesting that the hypothyroidism was central in origin have been refuted by the finding of elevated serum TSH in PsHP patients, as summarized below:

	Number of patients	Percent with elevated TSH	Reference
Type Ia	13	85	(67)
	8	88	Present authors
Type Ib	16	0	(67)
	12	33	Present authors

In 2 patients with PsHP type Ia, in vitro studies of excised thyroid tissue have demonstrated a blunted cAMP response to TSH (67). The explanation for thyroid dysfunction in patients with PsHP type Ib remains unknown. Since PsPsHP patients have N_s protein deficiency but often normal thyroid function, it is not clear what role the N_s protein deficiency plays in the hypothyroidism of PsHP type Ia subjects.

B. Glucagon

Most patients with PsHP type Ia demonstrate a blunted plasma cAMP response to intravenous glucagon, whereas most patients with PsHP type Ib show normal responses (28,67) (Fig. 4.8). Nevertheless, we have studied 2 relatives of a patient with the type Ib disorder, both of whom showed blunted cAMP responses to glucagon, although their PsHP type Ib relative had a normal response (28).

Although the plasma cAMP response is blunted, the glucose and insulin responses to injected glucagon are normal in PsHP type Ia (28); clinically, the patients do not suffer from hypoglycemia. Thus, it is possible that cAMP is more efficient in promoting glycogenolysis in PsHP type Ia; alternatively, movement of cAMP from hepatocytes to plasma might be impaired in this disorder. A third, albeit speculative possibility is that

TABLE 4.2. Metabolic disorders described in pseudohypoparathyroidism type I.

Tissue	Hormone or metabolic abnormality	Nature of disorder	PsHP Ia	PsHP Ib
Kidney; skeleton	PTH	Kidney: impaired cAMP-mediated phosphaturia; reduced 1,25(OH)$_2$D$_3$ generation	+ +/N	+ +/N
		Skeleton: impaired remodeling? calcemic response to PTH infusion	+	+
Pituitary	PTH	Impaired prolactin secretory response to PTH infusion	+	+
Kidney	ADH	Possible impaired urinary-concentrating capacity	±	−
Thyroid	TSH	Hyperresponsive TSH secretion to TRH infusion; possible primary hypothyroidism	+	±
Liver	Glucagon	Impaired hepatic cAMP release in response to glucagon	+	±/−
Ovary	LH/FSH	Amenorrhea	+	−
Pituitary	Prolactin	Hypoprolactinemia; reduced prolactin secretory reserve (see PTH above)	+	+
CNS	Intellectual impairment	Nature of metabolic defect requires clarification	+	−
?	Hypertension	Nature of metabolic defect(s) requires clarification	+	+
Adipose	Obesity	Nature of metabolic defect(s) requires clarification	+	+
Olfactory tissue	Olfaction	Impaired smell	+	−

these patients have larger glycogen stores than normal owing to resistance to their own endogenous glucagon.

C. Beta-Adrenergic Agonists

We have recently demonstrated that the plasma cAMP response to intravenous infusion of isoproterenol is blunted in both PsHP type Ia and Ib (68). Despite this, cardiovascular, glucose, and free fatty-acid responses were normal in both groups (69). Thus, as in the case of glucagon, enhanced activity of cAMP may compensate for its decreased production in these patients. Heisimer et al. (70) have reported that PsHP type Ia patients have alterations in erythrocyte β-adrenergic receptors, apparently related to the N$_s$ protein deficiency. Although the deficient cAMP production in PsHP type Ia is presumably due to the N$_s$ protein deficiency, there is so far no explanation for the observed decrease in cAMP responses to isoproterenol in PsHP type Ib.

GLUCAGON 2 μg/kg

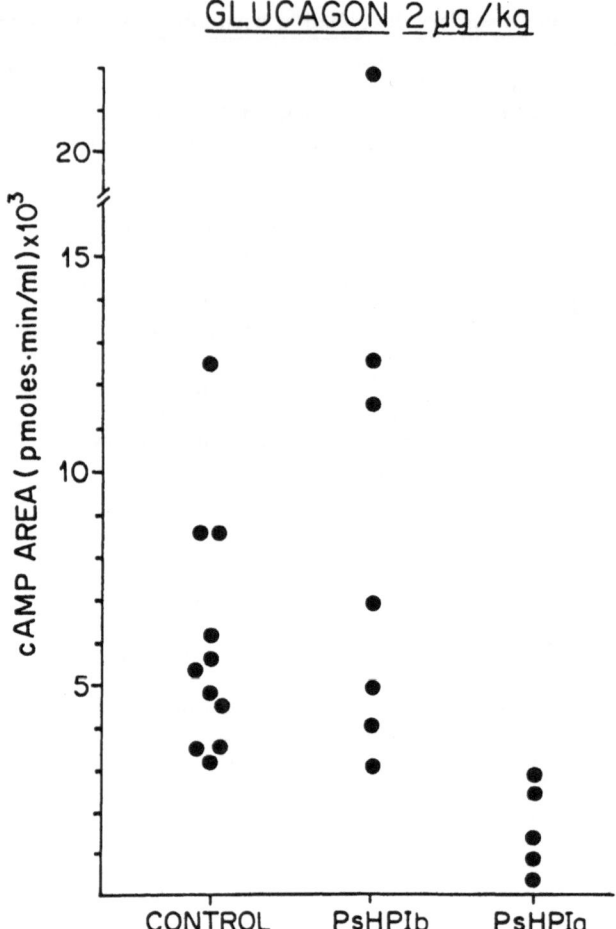

FIGURE 4.8. Plasma cAMP response to infusion of glucagon (2 μg/kg) in normal (control) subjects and patients with pseudohypoparathyroidism type Ia (PsHP Ia) and type Ib (PsHP Ib). (Reprinted with permission from Brickman AS, Carlson HE, Levin SR. Responses to glucagon infusion in pseudohypoparathyroidism. J Clin Endocrinol Metab 63:1354–1360, © The Endocrine Society (1986).)

D. Gonadal Function

Clinical abnormalities of ovarian function (e.g., oligomenorrhea, infertility) are common in women with PsHP type Ia, although most of these patients do not have elevated serum gonadotropins (67). Two patients with type Ia disease and 1 with type Ib did have elevated serum gonadotropins and an exaggerated response to GnRH (63,71,72); 2 of these 3 subjects were not clinically hypogonadal. In 1 patient, gonadotropin secretion normalized when serum calcium was corrected (63).

E. Prolactin Deficiency

Prolactin deficiency clearly occurs in patients with PsHP type Ia, type Ib, and type II disorders (64,72–78). Except for 1 case of failure to lactate (77), all cases have been clinically "silent," an observation consistent with the belief that prolactin has no important functions in nonlactating females and males. The deficiency in secretion of prolactin can be demonstrated using a variety of stimuli (TRH, dopamine antagonists, hypoglycemia, sleep, PTH) and is not corrected by restoration of normocalcemia (64). Prolactin deficiency also did not correlate with any other feature of the disorder (e.g., AHO, hypothyroidism, glucagon resistance). Although PTH is a weak stimulator of prolactin secretion, it is unlikely that deficient PTH action accounts for the prolactin deficiency, since patients with surgical or idiopathic hypoparathyroidism have normal prolactin responses to TRH (75). The exact frequency of prolactin deficiency in PsHP is unknown; we have found it in about half of our patients (5/9 PsHP type Ia patients; 7/12 PsHP type Ib patients). Others have reported a few cases, but prolactin deficiency has not been observed in any other large reported group of PsHP subjects (67).

The cause of prolactin deficiency is unknown. Since it is found in PsHP type Ia and type Ib as well as type II patients, it is probably not directly related to deficient N_s protein activity.

F. Antidiuretic Hormone (AVP) Function

The occurrence of renal resistance to AVP in patients with PsHP type Ia remains controversial. The physiologic expression of AVP resistance would be a form of nephrogenic diabetes insipidus. Moses et al. (79) have examined responses to water deprivation and administration of exogenous AVP in 7 patients with PsHP type Ia and 5 patients with PsHP type Ib. Measuring urinary AVP and/or plasma and urinary osmolality, these authors were unable to demonstrate abnormalities in antidiuretic responses to endogenous or exogenous AVP. They concluded that a putative reduction in intracellular cAMP generation in response to AVP in patients with PsHP type Ia may be insufficient to impair renal-concentrating ability in this disorder. However, Brickman and Weitzman (80) found increased plasma levels of AVP and impaired urine-concentrating capacity following overnight water deprivation testing in 2 of 4 patients with PsHP type Ia and reduced N_s protein activity. Neither of the patients showed evidence of polyuria. In contrast, plasma levels of AVP and concentrating capacity were normal in each of 6 patients with PsHP type Ib. These findings suggest that renal resistance to AVP may occur in rare cases of PsHP type Ia and that it is clinically silent.

G. Other Pituitary Hormones

Although short stature is frequent, especially in PsHP type Ia, available data suggest that the secretion of growth hormone and the production of somatomedin are normal (64,72,74,81–83). Interestingly, 2 patients with gonadotropin resistance have shown deficient GH secretion (63,71); in 1 case, normal GH responses were restored by the attainment of normocalcemia (63). It has been suggested that the brachydactyly (and perhaps the short stature) is due to premature epiphyseal closure (81,84).

Secretion of ACTH and cortisol has been found normal in essentially all subjects studied (64,67,73,74,81), a puzzling phenomenon in view of the fact that the action of ACTH is mediated by cAMP.

H. Miscellaneous Defects

A number of less well characterized nonendocrine abnormalities have been described in patients with PsHP. These include a high incidence of mental retardation (5,7,45,85), disturbances in taste and olfaction (86), and a possible increased incidence of hypertension (76).

Henkin (86) described varying degrees of impairment in tastes of sour and bitter as well as an increased threshold for the smell of pyridine, thiophene, and nitrobenzene in 6 patients with PsHP and AHO. The similarity of these defects to those found in patients with chromatin-negative gonadal dysgenesis was noted. However, in a recent report, Winestock et al. (87) observed that defects in olfaction occurred in patients with N_s protein deficiency but not in those with PsHP type Ib and isolated resistance to PTH. Since olfaction is mediated by metabolic processes involving activation of adenylate cyclase, impaired intracellular generation of cAMP may explain these observations.

Mental retardation has been recognized as a common feature of patients with PsHP since the original description by Albright et al. (1). Its causes are uncertain and may include hypothyroidism and untreated or poorly managed hypocalcemia in infancy and/or early childhood. Farfel and Friedman (85) have recently reported that mental deficiency occurred in 9 of 14 patients with N_s deficiency but that intellectual function was normal in 11 patients with PsHP type Ib, calling attention to the recently recognized involvement of the N_s protein–adenylate cyclase–cAMP pathway in learning processes (88,89). As in the case of other cAMP-mediated metabolic abnormalities in PsHP type Ia, since mental retardation is not invariably present in patients with N_s protein deficiency, the latter appears to be a necessary but not sufficient etiologic factor.

Our observations suggest that the incidence of hypertension is increased in adult patients with PsHP (76). Approximately 50% of our patients with either type of PsHP have elevated arterial blood pressure. The hypertension appears to be associated with low renin levels but normal circulating levels of aldosterone. It is possible that hypothyroidism and obesity, two well-recognized factors in other forms of endocrine hypertension, may also contribute to the pathogenesis of hypertension in PsHP type I.

V. Genetics of PsHP

The mode of genetic transmission of PsHP type I is poorly understood. Subdivision of kindreds of affected individuals into type Ia and type Ib disorders and better characterization of the biochemical features of these disorders may facilitate our understanding. Substantial evidence now appears to refute the earlier notion that PsHP type Ia is inherited as an X-linked dominant disorder (90). Males with PsHP frequently produce unaffected females, and male-to-male transmission has been demonstrated (82,90).

With the use of N_s protein measurements to characterize and classify type Ia patients, examples of both autosomal-dominant and autosomal-recessive transmission have been reported (91,92). However, as the biochemical heterogeneity of PsHP type Ia has become apparent, the likelihood of a complex mode of inheritance has strengthened. Clearly, the features of AHO may be inherited independently of those of hormone resistance, although the two are closely linked, as suggested by the presence of PsHP and PsPsHP within the same kindred (25,26). The answers to the questions of genetic transmission may be provided by a better understanding of the quantitative structure-function relationships of the subunits of N_s protein. Although the N_s protein defect may be ubiquitous among patients with PsHP type Ia or PsPsHP, the quantitative distribution of the defect in different tissues may vary. Johnson (93) has offered the concept of "metabolic interference" in which two nonallelic genes interact, resulting in a mode of inheritance that does not follow a classic Mendelian pattern (e.g., kindreds appear to inherit through either autosomal-recessive or autosomal-dominant transmission). Available information suggests that this form of genetic expression may be applicable to the disorders encompassed under the diagnoses PsHP type Ia and PsPsHP. Information concerning inheritance of PsHP type Ib (also apparently a heterogeneous group of disorders) is even more limited. Studies in three large kindreds suggest either autosomal or X-linked patterns of transmission (14,92,94).

VI. Treatment of PsHP

The medical management of the abnormal mineral metabolism in patients with PsHP involves modalities not usually considered in the treatment of *PTH*-deficient hypoparathyroidism. Patients with PsHP type I may have clinically latent or overt metabolic bone disease. The types of bone disease reported range from mild, generalized osteopenia (detected with noninvasive quantitative procedures) to severe osteitis fibrosa cystica (and other features of severe secondary hyperparathyroidism), or osteomalacia obvious on routine X-rays and confirmed with bone biopsy. In this setting there are two goals of therapy: correction of hypocalcemia, and treatment of the underlying metabolic bone disease.

Patients with PsHP type I may have reduced levels of $1,25(OH)_2D_3$ attributable to impaired renal production of this sterol. Effective oral doses of calcitriol range from 0.5 to 2.0 μg/day, with lower doses sometimes required in children and occasional higher doses of 4–5 μg daily in adults. Patients with these disorders appear to respond to doses of ergocalciferol, dihydrotachysterol, or 25-hydroxyvitamin D_3 similar to those effective in PTH-deficient hypoparathyroidism (95). Dose requirements for calcitriol vary with the amount and form of calcium supplementation and with the nature of the underlying metabolic bone disease. As in the case of other forms of metabolic bone disease, higher doses of vitamin D analogues may be tolerated or required in the early phase of treatment; in the later stages of therapy, when the underlying metabolic bone disease has been healed and bone calcium stores have been repleted, lower maintenance doses of vitamin D are indicated. Failure to adjust doses of vitamin D analogues or the amount of calcium supplements may result in development of hypercalcemia (and other features of vitamin D intoxication), especially in patients treated with analogues with prolonged tissue storage and physiologic half-life.

Serial measurement of several parameters may be useful in following the course of vitamin D treatment in PsHP. When possible, serial evaluation of skeletal histomorphometry from bone biopsies can provide detailed information concerning changes in bone-remodeling dynamics. Measurement of serum alkaline phosphatase and urinary hydroxyproline may provide similar information. The normalization of a previously elevated level of serum alkaline phosphatase has been useful as an indicator of a need for reducing the dosage. Serial measurement of serum PTH may be useful in providing evidence for suppression of secondary hyperparathyroidism. Measurements of morning levels of serum calcium do not provide information concerning the degree of perturbation of serum calcium levels; demonstration of an elevated serum PTH level may indicate the occurrence of intermittent hypocalcemia and inadequate vitamin D and calcium therapy.

At the present time, information concerning the required duration of vitamin D therapy is not available. Cessation of long-term treatment in some individuals may result in prolonged periods of normocalcemia, lasting for several years. As discussed previously, such observations provide support for the view that, in some individuals, acquired factors (e.g., vitamin D deficiency) may be important for the development of hypocalcemia. A rise in serum levels of PTH or serum alkaline phospha-tase, or the development of hypocalciuria, is evidence for the recurrence of secondary hyperparathyroidism, even in the presence of apparent persistent normocalcemia. Similarly, patients with PsHP and spontane-ous normocalcemia may have elevated levels of PTH, perhaps contri-buting to maintenance of a normocalcemic state in the presence of renal PTH resistance. At the present time, there is insufficient information available concerning the need or benefits of reinstituting or instituting vitamin D therapy.

Control of hyperphosphatemia has not proved to be a problem in adults with PsHP, since serum phosphorous levels invariably normalize with the correction of hypocalcemia. However, in young children and infants, control of hyperphosphatemia may prove to be difficult, requiring severe dietary restriction of phosphorous and use of phosphate-binding antacids or calcium carbonate.

Although magnesium depletion may complicate the management of hypoparathyroidism, resulting in resistance to the actions of vitamin D, it does not appear to be a complication in management of PsHP (96). This probably reflects the rarity of this disorder and the fact that such patients are relatively protected against magnesium depletion in the presence of underlying secondary hyperparathyroidism. Theoretically, magnesium depletion might be particularly likely to develop in PsHP in patients being treated with magnesiuric drugs (thiazides) or in whom hypercalciuria is persistent. Thus, the occasional monitoring of serum levels of magnesium will reduce the likelihood of its occurrence.

References

1. Albright F, Burnett CH, Smith PH, Parson W (1942) Pseudohypoparathy-roidism—an example of "Seabright-Bantam syndrome." Report of three cases. Endocrinology 30:922–932.
2. Peterman H, Garvey JL (1949) Pseudohypoparathyroidism: Case report. Pediatrics 4:790–797.
3. Elrick H, Albright F, Bartter FC, Forbes AP, Reeves J (1950) Further studies on pseudo-hypoparathyroidism: Report of four new cases. Acta Endocrinol (Copenh) 5:199–225.
4. Reynolds TB, Jacobson G, Edmondson HA, Martin HE, Nelson CH (1952) Pseudohypoparathyroidism: Report of a case showing bony demineralization. J Clin Endocrinol Metab 12:560–573.
5. Bronsky D, Kushner DS, Dubin A, Snapper I (1958) Idiopathic hypoparathy-roidism and pseudohypoparathyroidism: Case reports and review of the literature. Medicine 37:317–352.
6. Albright F, Forbes AP, Henneman PH (1952) Pseudo-pseudohypoparathy-roidism. Trans Assoc Am Phys 65:337–350.
7. Bartter FC (1966) Pseudohypoparathyroidism and Pseudopseudohypopara-thyroidism. In Stanbury JB, Wyngaarden JB, Frederickson DS (eds): The Metabolic Basis of Inherited Disease. McGraw Hill, New York, pp 1024–1031.

8. Tashjian AH Jr, Frantz AG, Lee JB (1966) Pseudohypoparathyroidism: Assays of parathyroid hormone and thyrocalcitonin. Proc Natl Acad Sci USA 56:1138–1142.
9. Lee JB, Tashjian AH Jr, Streeto JM, Frantz AG (1968) Familial pseudohypoparathyroidism: Role of parathyroidism and thyrocalcitonin. N Engl J Med 279:1179–1184.
10. Chase LR, Melson GL, Aurbach GD (1969) Pseudohypoparathyroidism: Defective excretion of 3'5'-AMP in response to parathyroid hormone. J Clin Invest 48:1832–1842.
11. Drezner M, Neelon FA, Lebovitz HE (1973) Pseudohypoparathyroidism type II: A possible defect in the reception of the cyclic AMP signal. N Engl J Med 289:1056–1060.
12. Marcus R, Wilber JF, Aurbach GD (1971) Parathyroid hormone-sensitive adenyl cyclase from the renal cortex of a patient with pseudohypoparathyroidism. J Clin Endocrinol Metab 33:537–541.
13. Drezner MK, Burch WM Jr (1978) Altered activity of the nucleotide regulatory site in the parathyroid hormone-sensitive adenylate cyclase from a patient with pseudohypoparathyroidism. J Clin Invest 62:1222–1227.
14. Farfel Z, Brickman AS, Kaslow HR, Brothers VM, Bourne HR (1980) Defect of receptor-cyclase coupling protein in pseudohypoparathyroidism. N Engl J Med 303:237–242.
15. Levine MA, Downs RW Jr, Singer MJ, Marx SJ, Spiegel AM (1980) Deficient activity of guanine nucleotide regulatory protein in erythrocytes from patients with pseudohypoparathyroidism. Biochem Biophys Res Commun 94:1319–1324.
16. Akita Y, Saito T, Yajima Y, Sakuma S (1985) The stimulatory and inhibitory guanine nucleotide-binding proteins of adenylate cyclase in erythrocytes from patients with pseudohypoparathyroidism type I. J Clin Endocrinol Metab 61:1012–1017.
17. Radeke HH, Auf'mkolk B, Juppner H, Krohn H-P, Keck E, Hesch R-D (1986) Multiple pre- and postreceptor defects in pseudohypoparathyroidism (a multicenter study with twenty four patients). J Clin Endocrinol Metab 62:393–402.
18. Gilman AG (1984) Guanine nucleotide-binding regulatory proteins and dual control of adenylate cyclase. J Clin Invest 73:1–4.
19. Silve C, Santora A, Breslau N, Moses A, Spiegel A (1986) Selective resistance to parathyroid hormone in cultured skin fibroblasts from patients with pseudohypoparathyroidism type Ib. J Clin Endocrinol Metab 62:640–644.
20. Barrett DA, Spiegel AM, Downs RW Jr (1986) Reduced adenylate cyclase activity of cultured skin fibroblasts from an atypical patient with pseudohypoparathyroidism. Clin Res 34:540A.
21. Rodriquez HJ, Villarreal H Jr, Klahr S, Slatopolsky E (1974) Pseudohypoparathyroidism type II: Restoration of normal renal responsiveness to parathyroid hormone by calcium administration. J Clin Endocrinol Metab 39:693–701.
22. Renier J-C, Boasson M, Bernat M, Basle M (1977) Vitamin resistant osteomalacia due to defect of hepatic hydroxylation of vitamin D. Secondary reversible pseudohypoparathyroidism by administration of calcium. Sem Hop Paris 53:1723–1529.
23. Yamada K, Tamura Y, Yamamoto M, Kumagai A (1979) Effect of calcium administration on renal responsiveness to parathyroid hormone in pseudohypoparathyroidism type I and II in comparison with normal, idiopathic and surgical hypoparathyroidism. Endocrinol Jpn 26:147–157.

24. Rao DS, Parfitt AM, Kleerekoper M, Pumo BS, Frame B (1985) Dissociation between the effects of endogenous parathyroid hormone on adenosine 3'5'-monophosphate generation and phosphate reabsorption in hypocalcemia due to vitamin D depletion: An acquired disorder resembling pseudohypoparathyroidism type II. J Clin Endocrinol Metab 61:285–290.
25. Mann JB, Alterman S, Hills AG (1962) Albright's hereditary osteodystrophy comprising pseudohypoparathyroidism and pseudo-pseudohypoparathyroidism. Ann Intern Med 56:315–342.
26. Levine MA, Jap T-S, Mauseth RS, Downs RW, Spiegel AM (1986) Activity of the stimulatory guanine nucleotide-binding protein is reduced in erythrocytes from patients with pseudohypoparathyroidism and pseudopseudohypoparathyroidism: Biochemical, endocrine and genetic analysis of Albright's hereditary osteodystrophy in six kindreds. J Clin Endocrinol Metab 62:497–502.
27. Fisher JA, Bourne HR, Dambacher MA, Tschopp F, DeMeyer R, Devogelaer J-P, Werder EA, De Deuxchaisnes CN (1983) Pseudohypoparathyroidism: Inheritance and expression of deficient receptor-cyclase coupling protein activity. Clin Endocrinol (Oxf) 19:747–754.
28. Brickman AS, Carlson HE, Levin SR (1986) Responses to glucagon infusion in pseudohypoparathyroidism. J Clin Endocrinol Metab 63:1354–1360.
29. Farfel Z, Abood ME, Brickman AS, Bourne HR (1982) Deficient activity of receptor-cyclase coupling protein in transformed lymphoblasts of patients with pseudohypoparathyroidism, type I. J Clin Endocrinol Metab 55:113–117.
30. Bourne HR, Kaslow HR, Brickman AS, Farfel Z (1981) Fibroblast defect in pseudohypoparathyroidism, type I: Reduced activity of receptor-cyclase coupling protein. J Clin Endocrinol Metab 53:636–640.
31. Farfel Z, Bourne HR (1980) Deficient activity of receptor-cyclase coupling protein in platelets of patients with pseudohypoparathyroidism. J Clin Endocrinol Metab 51:1202–1204.
32. Downs RW Jr, Levin MA, Drezner MK, Burch WM Jr, Spiegel AM (1983) Deficient adenylate cyclase regulatory protein in renal membranes from a patient with pseudohypoparathyroidism. J Clin Invest 71:231–235.
33. Levine MA, Ahn TG, Van Dop C, Kaufman K, Smallwood P, Bourne HR, Sullivan K (1986) Molecular basis for genetic deficiency of G_s in pseudohypoparathyroidism. J Bone Mineral Res 1 (Suppl 1):160.
34. Motulsky HJ, Hughes RJ, Brickman AS, Farfel Z, Bourne HR, Insel PA (1982) Platelets of pseudohypoparathyroid patients: Evidence that distinct receptor-cyclase coupling proteins mediate stimulation and inhibition of adenylate cyclase. Proc Natl Acad Sci USA 79:4193–4197.
35. Downs RW, Sekura RD, Levine MA, Spiegel AM (1985) The inhibitor adenylate cyclase coupling protein in pseudohypoparathyroidism. 61:351–354.
36. Bell NH, Avery S, Sinha T, Clark CM Jr, Allen DO, Johnston C Jr (1972) Effects of dibutyryl cyclic adenosine 3',5'-monophosphate and parathyroid extract on calcium and phosphorus metabolism in hypoparathyroidism and pseudohypoparathyroidism. J Clin Invest 51:816–823.
37. Moses AM, Breslau N, Coulson R (1976) Renal responses to PTH in patients with hormone-resistant (pseudo) hypoparathyroidism. Am J Med 61:184–189.
38. Suh SM, Fraser D, Kooh SW (1970) Pseudohypoparathyroidism: Responsiveness to parathyroid extract induced by vitamin D_2 therapy. J Clin Endocrinol Metab 30:609–614.

39. Strogmann W, Fisher JA (1975) Pseudohypoparathyroidism. Disappearance of the resistance to parathyroid extract during treatment with vitamin D. Am J Med 59:140–144.
40. Brickman AS, Norman AW, Coburn JW (1976) Restoration of PTH-dependent phosphaturia by 1,25(OH)$_2$-vitamin D$_3$ in pseudohypoparathyroidism type I. Kidney Int 10:488 (abstract).
41. Drezner MK, Neelon FA, Haussler M, McPherson HT, Lebovitz HE (1976) 1,25-Dihydroxycholecalciferol deficiency: The probable cause of hypocalcemia and metabolic bone disease in pseudohypoparathyroidism. J Clin Endocrinol Metab 42:621–628.
42. Sinha TK, DeLuca HF, Bell NH (1977) Evidence for a defect in the formation of 1,25-dihydroxyvitamin D in pseudohypoparathyroidism. Metabolism 26:731–738.
43. Lambert PW, Hollis BW, Bell NH, Epstein S (1980) Demonstration of a lack of change in serum 1,25-dihydroxyvitamin D in response to parathyroid extract in pseudohypoparathyroidism. J Clin Invest 66:782–791.
44. Aksnes L, Aarskog D (1980) Effect of parathyroid hormone on 1,25-dihydroxyvitamin D formation in type I pseudohypoparathyroidism. J Clin Endocrinol Metab 51:1223–1226.
45. Drezner MK, Nellon FA (1985) Pseudohypoparathyroidism. In Stanbury JB, Wyngaarden JB, Frederickson DS, Goldstein JL, Brown MS (eds): Metabolic Basis of Inherited Disease, 5th Ed. McGraw-Hill, New York, pp 1508–1528.
46. Drezner MK, Haussler MR (1979) Normocalcemic pseudohypoparathyroidism: Association with normal vitamin D$_3$ metabolism. Am J Med 66:503–508.
47. Deftos LJ, Powell D, Parthemore JG, Potts JT Jr (1973) Secretion of calcitonin in hypocalcemic states in man. J Clin Invest 52:3109–3114.
48. Wagar G, Lehtivuori J, Salven I, Backman R, Sivula A (1980) Pseudohypoparathyroidism associated with hypercalcitoninaemia. Acta Endocrinol (Copenh) 93:43–48.
49. Morimoto S, Onishi T, Kumahara Y, Avioli L (1983) Differentiation of pseudo- and idiopathic hypoparathyroidism by measuring urinary calcitonin. J Clin Endocrinol Metab 57:1616–1220.
50. Bell NH, Gerard ES, Bartter FC (1963) Pseudohypoparathyroidism with osteitis fibrosa cystica and impaired absorption of calcium. J Clin Endocrinol Metab 23:759–772.
51. Singleton EB, Teng CT (1962) Pseudohypoparathyroidism with bone changes simulating hyperparathyroidism: Report of a case. Radiology 78:388–393.
52. Kolb FO, Steinbach HL (1962) Pseudohypoparathyroidism with osteitis fibrosa. J Clin Endocrinol Metab 22:59–70.
53. Frame B, Hanson CA, Frost HM, Block M, Arnstein AR (1972) Renal resistance to parathyroid hormone with osteitis fibrosa: "Pseudohypohyperparathyroidism." Am J Med 52:311–321.
54. Connors MH, Irias JJ, Golabi M (1977) Hypo-hyperparathyroidism: Evidence for a defective parathyroid hormone. Pediatrics 60:343–348.
55. Kidd GS, Schaaf M, Adler RA, Lassman MN, Wray HL (1980) Skeletal responsiveness in pseudohypoparathyroidism: A spectrum of clinical disease. Am J Med 68:772–781.
56. Wilson JD, Hadden DR (1980) Pseudohypoparathyroidism presenting with rickets. J Clin Endocrinol Metab 51:1184–1189.
57. Pallardo-Sanchez LFP, Montero A, Vidal O, Sicilia LS, Cerdan A (1977) Seudohipoparatiroidismo tipo II de presentacion familiar. Rev Clin Esp 145:369–374.
58. Matsuda I, Takekoshi Y, Tanaka M, Matsuura N, Nagai B, Seino Y (1979)

Pseudohypoparathyroidism type II and anticonvulsant rickets. Eur J Pediatr 132:303–308.

59. Yamada K, Tamura Y, Hisao T, Kumagai A, Yoshida S (1984) Possible existence of anti-renal tubular plasma membrane auto-antibody which blocked parathyroid hormone-induced phosphaturia in a patient with pseudohypoparathyroidism type II and Sjogren's syndrome. J Clin Endocrinol Metab 58:339–343.

60. Loveridge N, Fisher JA, Devogelaer J-P, Nagant de Deuxchaisnes C (1986) Suppression of parathyroid hormone inhibitory activity of plasma in pseudohypoparathyroidism type I by calcium. Clin Endocrinol (Oxf) 24:549–554.

61. Mitchell J, Goltzman D (1985) Examination of circulating parathyroid hormone in pseudohypoparathyroidism. J Clin Endocrinol Metab 61:328–334.

62. Loveridge N, Fisher JA, Nagant de Deuxchaisnes C, et al. (1982) Inhibition of cytochemical bioactivity of parathyroid hormone by plasma in pseudohypoparathyroidism type I. J Clin Endocrinol Metab 54:1274–1275.

63. Suzuki H, Kasai K, Shimoda S, Mori K, Miyaska M (1982) Improvement in abnormal secretion of thyrotropin and gonadotropin after restoration of serum calcium in pseudohypoparathyroidism. Endocrinol Jpn 29:69–75.

64. Carlson HE, Brickman AS, Botazzo GF (1977) Prolactin deficiency in pseudohypoparathyroidism. N Engl J Med 296:140–144.

65. Juppner H, Bialasiewicz AA, Hesch RD (1978) Autoantibodies to parathyroid hormone receptor. Lancet 2:1222–1224.

66. Mallet E, Carayon P, Amr S, Brunelle P, Ducastelle T, Basuyau JP, De Menibus CH (1982) Coupling defect of thyrotropin receptor and adenylate cyclase in a pseudohypoparathyroid patient. J Clin Endocrinol Metab 54:1028–1032.

67. Levine MA, Downs RW Jr, Moses AM, Breslau NA, Marx SJ, Lasker RD, Rizzoli RE, Aurbach GD, Spiegel AM (1983) Resistance to multiple hormones in patients with pseudohypoparathyroidism. Association with deficient activity of guanine nucleotide regulatory protein. Am J Med 74:545–556.

68. Carlson HE, Brickman AS, Williams A (1983) Blunted plasma cyclic adenosine monophosphate response to isoproterenol in pseudohypoparathyroidism. 57:1323–1326.

69. Carlson HE, Brickman AS, Burns TW, Langley PE (1985) Normal free fatty acid response to isoproterenol in pseudohypoparathyroidism. J Clin Endocrinol Metab 61:382–384.

70. Heinsimer JA, Davies AO, Downs RW, Levine MA, Spiegel AM, Drezner MK, DeLean A, Wreggett KA, Caron MG, Lefkowitz RJ (1984) Impaired formation of beta-adrenergic receptor-nucleotide regulatory complexes in pseudohypoparathyroidism. J Clin Invest 73:1335–1343.

71. Wolfsdorf JI, Rosenfield RL, Fang VS, Kobayashi R, Razdan AK, Kim MH (1978) Partial gonadotropin-resistance in pseudohypoparathyroidism. Acta Endocrinol (Copenh) 88:321–328.

72. Shapiro MS, Bernheim J, Gutman A, Arber I, Spitz IM (1980) Multiple abnormalities of anterior pituitary hormone secretion in association with pseudohypoparathyroidism. J Clin Endocrinol Metab 51:483–487.

73. Windeck R, Menken U, Benker G, Reinwein D (1981) Basal ganglia calcification in pseudohypoparathyroidism type II. Clin Endocrinol (Copenh) 15:57–63.

118   A. S. Brickman and H. E. Carlson

74. Kruse K, Gutekunst B, Kracht U, Schwerda K (1981) Deficient response to parathyroid hormone in hypocalcemic and normocalcemic pseudohypoparathyroidism. J Clin Endocrinol Metab 52:1099–1105.
75. Brickman AS, Carlson HE, Deftos LJ (1981) Prolactin and calcitonin responses to parathyroid hormone infusion in hypoparathyroid, pseudohypoparathyroid and normal subjects. J Clin Endocrinol Metab 53:661–664.
76. Sowers JR, Brickman AS (1982) Circadian blood pressure and renin, aldosterone, cortisol and prolactin levels in hypertensive pseudohypoparathyroid patients. J Clin Endocrinol Metab 55:1202–1208.
77. Devogelaer JP, Huaux JP, Docquier C, Crabbe J, Nagant de Deuxchaisnes C (1984) Pseudohypoparathyroidism: A case report with studies on the pathogenesis of the condition. Acta Clin Belg 39:228–247.
78. Weisman Y, Golander A, Spirer Z, Farfel Z (1985) Pseudohypoparathyroidism type Ia presenting as congenital hypothyroidism. J Pediatr 107:413–415.
79. Moses AM, Weinstock RS, Levine MS, Breslau NA (1986) Evidence for normal antidiuretic responses to endogenous and exogenous arginine vasopressin in patients with guanine nucleotide-binding stimulatory protein-deficient pseudohypoparathyroidism. J Clin Endocrinol Metab 62:221–224.
80. Brickman AS, Weitzman RE (1978) Renal resistance to arginine vasopressin in pseudohypoparathyroidism. Clin Res 26:164A (abstract).
81. Werder EA (1979) Pseudohypoparathyroidism. Adv Intern Med Pediatr 42:191–221.
82. Urdanivia E, Mataverde A, Cohen MP (1975) Growth hormone secretion and sulfation factor activity in pseudohypoparathyroidism. J Lab Clin Med 86:772–776.
83. Van Dop C, Bourne HR, Neer RM (1984) Father to son transmission of decreased N$_s$ activity in pseudohypoparathyroidism type Ia. J Clin Endocrinol Metab 59:825--828.
84. De Wijn EM, Steendijk R (1982) Growth and maturation in pseudohypoparathyroidism; a longitudinal study in 5 patients. Acta Endocrinol (Copenh) 101:223–226.
85. Farfel Z, Friedman E (1986) Mental deficiency in pseudohypoparathyroidism type I is associated with N$_s$-protein deficiency. Ann Intern Med 105:197–199.
86. Henkin RI (1986) Impairment of olfaction and of the tastes of sour and bitter in pseudohypoparathyroidism. J Clin Endocrinol Metab 38:624–628.
87. Weinstock RS, Wright HN, Spiegel AM, Levine MA, Moses AM (1986) Olfactory dysfunction in humans with deficient guanine nucleotide-binding protein. Nature 322:635–636.
88. Aceves-Pina EO, Booker R, Duer JS (1983) Learning and memory in Drosophila, studies with mutants. Cold Spring Harbor Symp Quant Biol 48:831–840.
89. Tully T (1984) Drosophila learning: Behavior and biochemistry. Behav Genet 14:527–557.
90. Fitch N (1982) Albright's hereditary osteodystrophy: A review. Am J Med Genet 11:11–29.
91. Cederbaum SD, Lippe BM (1973) Probable autosomal recessive inheritance in a family with Albright's hereditary osteodystrophy and an evaluation of genetics in the disorder. Am J Hum Genet 25:638–645.
92. Farfel Z, Brothers VM, Brickman AS, Conte F, Neer R, Bourne HR (1981) Pseudohypoparathyroidism: Inheritance of deficient receptor-cyclase coupling activity. Proc Natl Acad Sci USA 78:3098–3102.
93. Johnson WG (1980) Metabolic interference and the +− heterozygote. A hypothetical form of simple inheritance which is neither dominant or recessive. Am J Hum Genet 32:374–386.

94. Winter JSD, Hughes IA (1980) Familial pseudohypoparathyroidism without somatic anomalies. Can Med Assoc J 123:26–31.
95. Brickman AS (1985) Pseudohypoparathyroidism. In Krieger DT, Bardin CW (eds): Current Therapy in Endocrinology and Metabolism. Mosby, Philadelphia, pp 334–337.
96. Rosler A, Rabinowitz D (1973) Magnesium-induced reversal of vitamin-D resistance in hypoparathyroidism. Lancet 1:803–805.

5
Syndromes of Vitamin D Resistance

JAMES F. MCLEOD and JOHN G. HADDAD JR.

Mellanby, over 50 years ago, recognized the significance of the fat-soluble substance, vitamin D, as a nutritional antirachitic agent (1). Rickets and osteomalacia, persistent problems in the smoke-ladened skies and dank sweat shops of the industrialized world of Western Europe and America, rapidly declined over the decade following this discovery. Importation of cod liver oil rapidly rose, and its consumption became a wintertime ritual of youth. With the decline in the frequency of rickets, physicians began to note patients who failed to respond to prescribed regimens of cod liver oil. Albright and his colleagues presented the first detailed report of such a patient who had the typical radiographic features of rickets and no improvement during vitamin D supplementation (2). This patient differed from the usual case of nutritional rickets by having hypophosphatemia without hypocalcemia. Prader et al., in 1961, described a second variant of vitamin D resistance in several patients with hypocalcemia, hypophosphatemia, and the radiographic features of rickets (3). During the ensuing two decades, advances in the understanding of vitamin D metabolism and of skeletal and mineral homeostasis helped elucidate the molecular basis for rickets and the disruption of the pathways involved in vitamin D action that result in syndromes of resistance to vitamin D in man.

I. Vitamin D Metabolism

The family of vitamin D sterols encompasses a group of related seco-steroids possessing antirachitic activity. The naturally occurring vitamin is cholecalciferol, vitamin D_3 (D_3), which is synthesized photochemically in the epidermal layer of the skin by UV irradiation of 7-dehydrocholesterol to form a previtamin (4) (Fig. 5.1). Over several days, the previtamin undergoes isomerization to D_3 which is apparently removed from the skin bound to the serum vitamin D–binding protein (DBP). The major dietary source of vitamin D is ergocalciferol, vitamin D_2 (D_2), which is synthesized by UV irradiation of the plant cell membrane sterol, ergosterol. Oral

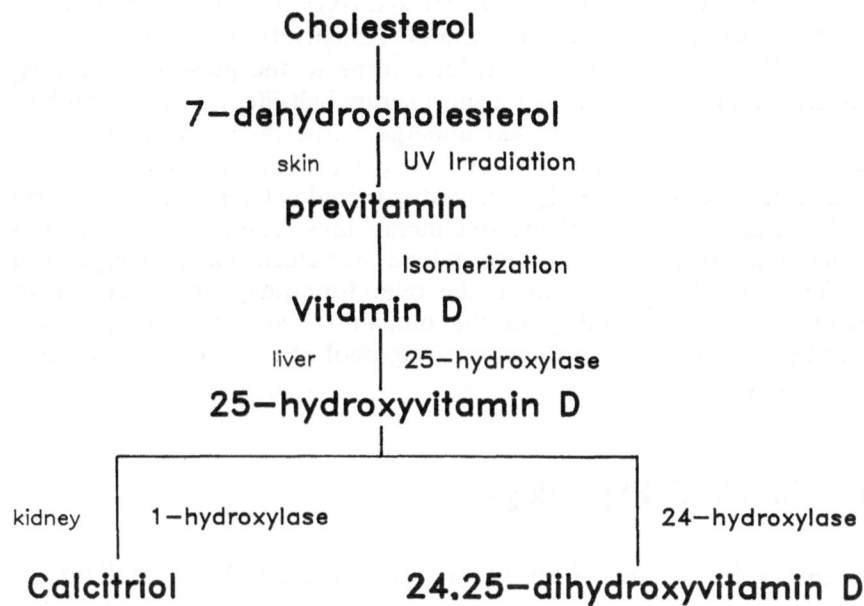

FIGURE 5.1. Synthesis and metabolism of major members of the family of vitamin D secosteroids.

vitamin D is absorbed from the intestinal lumen along with bile salts and fats in the small intestine (5) and is subsequently incorporated into chylomicrons. Vitamin D, associated with the metabolites of chylomicrons or by diffusion from the serum pool, is delivered to the liver, where it undergoes further hydroxylation by mitochondrial and microsomal enzymes to form 25-hydroxyvitamin D (25-OH D) (6). The 25-OH D returns to plasma and circulates bound to DBP in a long-lived serum pool ($t_{\frac{1}{2}} \approx 14$ days). As a result, the plasma level of 25-OH D is an excellent index of the total body vitamin D stores (Table 5.1).

TABLE 5.1. Clinical features of common calciferols.

Calciferol	Plasma concentration	Free plasma[a] concentration	Plasma half-life	Relative[b] potency	Daily requirement
Vitamin D	2.6–13 nM (1–5 ng/ml)	?	1 day	—	10 μg (400 IU)
25-OH D	25–75 nM (10–30 ng/ml)	8–24 pM	14 days	1	>1 μg
Calcitriol	50–150 pM (20–60 pg/ml)	0.28–0.30 pM	½ day	500–1000	0.5–1.0 μg

[a] Determined by plasma ultrafiltration.
[b] Bioactivity measured by ^{45}Ca release from fetal rat long-bone rudiments in organ culture.

In the proximal renal tubule, further hydroxylation of 25-OH D occurs to form either $1\alpha,25$-dihydroxyvitamin D (calcitriol), or 24,25-dihydroxy-vitamin D (24,25-$(OH)_2$D). Calcitriol returns to the plasma associating with DBP and albumin and has a short serum half-life ($t_{\frac{1}{2}} \approx 12$ h). Both of the dihydroxylated D sterols may undergo further hydroxylation to form $1\alpha,24,25$-trihydroxyvitamin D_3. The di- and trihydroxylated vitamin D metabolites are subsequently inactivated by side-chain cleavage to form calcitroic acid and other bioinactive metabolites. The final disposition of the metabolites is not clearly determined, but glucuronate conjugates of the polar metabolites appear in the bile. Enterohepatic circulation of calcitriol, noted previously in the human (7) and rat (8), does not contribute significantly to the circulating pool of calcitriol (9), as previously proposed.

II. Vitamin D Physiology

A. General Features of the Control of Vitamin D Metabolism

Regulation of the production of the active vitamin D sterols is controlled to varying degrees at the hydroxylation steps. Vitamin D_3 production in the epidermis is limited by the period of sunlight exposure, availability of the 7-dehydrocholesterol substrate, extent of dermal pigmentation, increasing age, and an increased conversion of the previtamin to bioinactive metabolites with intensive exposure to sunlight (10).

In animals, the 25-hydroxylation of vitamin D by the liver is inhibited by vitamin D (11). A decrease in the 25-hydroxylase activity by vitamin D has been noted in both animals and man; however, the decrease in enzymatic activity is more than offset by the increased concentration of sterol substrate, so that the total 25-OH D production is increased. The substrate inhibition of the 25-hydroxylase enzyme has been confirmed in the perfused rat liver (12).

Calcitriol and 25-OH D also may inhibit the hepatic 25-hydroxylase system. In rachitic rat liver homogenates, 25-OH D is reported to competitively inhibit vitamin D conversion in the liver, and supraphysiologic quantities of calcitriol are reported to produce noncompetitive inhibition (13); however, only a minimal inhibitory effect by 25-OH D and calcitriol can be demonstrated during perfusion of the rat liver (14). The very limited quantities of free plasma 25-OH D and calcitriol that are available to the liver parenchyma under physiologic conditions militates against a direct effect of these sterols on 25-OH D synthesis. The chronic administration of calcitriol to the rat lowers the plasma level of 25-OH D by increasing metabolic conversion of 25-OH D to 24,25-$(OH)_2$D and other metabolites (15). An accelerated metabolism of 25-OH D and possibly other vitamin D sterols by calcitriol may account for many of the reported changes in plasma levels of vitamin D sterols with calcitriol administration.

A similar inhibition of the rise in plasma levels of 25-OH D by vitamin D (16), 25-OH D (17), and calcitriol (18,19) has been observed in man. A single dose of 2 μg of calcitriol (4–6 times the physiologic amount) prevents the increased plasma level of 25-OH D after vitamin D loading in normal adults (18). This effect of calcitriol may be important in disease states in which calcitriol production is poorly regulated (e.g., sarcoidosis), by limiting the substrate for l-hydroxylation (20).

In normal humans, calcitriol levels are closely regulated. However, when exogenous 25-OH D was administered to a group of patients with primary hyperparathyroidism in whom calcitriol levels were normal, a progressive elevation in serum calcitriol level was observed (21). In this group of patients, 25-OH D access to the lα-hydroxylase was important in limiting the levels of calcitriol production.

B. Effects of Parathyroid Hormone and Calcitriol

Parathyroid hormone (PTH) and calcitriol are the primary regulators of the pair of renal mitochondrial mixed-function oxidases responsible for the formation of calcitriol and 24,25-$(OH)_2$D. Elevation of serum calcitriol levels has been noted after PTH administration to hyperparathyroid, normal, and hypoparathyroid adult humans. In kidney homogenates (22,23) and in dispersed renal tubule epithelial cells (24), PTH stimulation in vitamin D–replete animals increases lα-hydroxylase activity with a concomitant decrease in 24-hydroxylase activity. This stimulation is associated with activation of the adenylate-cyclase system in the proximal renal tubule. Indeed, cyclic adenosine-monophosphate (cAMP) and cAMP analogues also stimulate lα-hydroxylase activity.

In vitamin D–replete normal adult humans (25) and animals, administration of vitamin D or 25-OH D does not usually increase calcitriol levels in plasma, which is consistent with tight control of calcitrol production (26). In the kidney, calcitriol suppresses its synthesis by lowering lα-hydroxylase activity while increasing 24-hydroxylase activity in the kidney and other tissues. In primary cell cultures of the chicken kidney, the addition of calcitriol to the media results in a diminution of calcitriol production and a reciprocal increase in the production of 24,25-$(OH)_2$D after several hours (22). The readdition of media without calcitriol reverses the alterations in enzymatic activity, again after several hours. The changes in enzymatic activity are abolished by the coaddition of the RNA transcription inhibitor, actinomycin D, or the RNA translational inhibitor, cyclohexamide, to culture media with calcitriol. This indicates that messenger RNA (mRNA) and new protein synthesis are required for the calcitriol-induced effects on enzyme activities.

C. Effects of Calcium and Phosphorus

Diets deficient in calcium or phosphorus affect calcitriol levels and calcium transport across the intestinal lumen. The major effect of manipulation of dietary calcium content appears to be the alteration of 1α-hydroxylase activity, which occurs indirectly by stimulation or suppression of PTH secretion. Evidence for a direct effect of extracellular calcium on 1α-hydroxylase activity has been sought using isolated mitochondria, but the results have been conflicting (6). There is little evidence that, under physiologic conditions, calcium itself plays a significant role in the regulation of 1α-hydroxylase.

The effect of low-phosphorus diets and hypophosphatemia on plasma calcitriol levels, 1α-hydroxylase activity, and the intestinal transport of calcium is multifaceted. In the rat and chick, low-phosphorus diets (0.3–0.4% by weight) result in plasma and intestinal accumulation of calcitriol and increased intestinal calcium transport, without detectable changes in 1α-hydroxylase activity. More severe restriction of phosphorus intake (0.15–0.2%), with the development of hypophosphatemia, increases 1α-hydroxylase activity in the chick (27,28), and parathyroidectomy does not prevent this effect (29). Nevertheless, the effect of phosphorus deprivation is small in comparison to that of calcium deficiency. Over the physiologic range, increasing the phosphorus concentration from 0 to 1.2 mM suppresses calcitriol synthesis in isolated chick renal tubules, and higher concentrations increase the production rate (30). In humans, dietary restriction of phosphorus increases the production rate and the plasma levels of calcitriol; high-phosphate diets suppress both the production rate and the plasma levels (31). Although in hypophosphatemia, 1α-hydroxylase activity is clearly elevated in the chicken, and serum calcitriol levels and production rate are elevated in humans (32), the contribution of serum phosphate to regulation of 1α-hydroxylase activity under conditions of normal mineral homeostasis is unclear.

D. Extrarenal Production of Calcitriol

It has been reported that several nonrenal tissues metabolize 25-OH D to polar compounds that, upon analysis with high-pressure liquid chromatography (HPLC) and calcitriol receptor binding, appear to be similar or identical to calcitriol. These tissues include the rat placenta (33), bone (34), peritoneal macrophages (35), human alveolar macrophages (36), and transformed leukocytes (37). Using mass spectroscopy for unequivocal identification, production of calcitriol has been firmly established in the rat placenta (38), human sarcoid alveolar macrophages (39), chick bone cells (40), and human transformed T lymphocytes (41). During pregnancy, the placenta contributes to the plasma pool of calcitriol (42). However, in the nonpregnant normal adult rat, there is no evidence of a significant contribution by extrarenal sources (43). A local role for small quantities of calcitriol and other biologically active metabolites has yet to be determined.

Hypercalcemia and, more frequently, hypercalciuria have been recognized in sarcoidosis associated with increased plasma levels and production of calcitriol (44,45). Pulmonary alveolar macrophages and lymph node homogenates (46) from patients with sarcoidosis converted 25-OH D_3 to a compound indistinguishable from calcitriol by HPLC, receptor-binding studies, and mass spectroscopy. Hypercalcemia, in conjunction with elevation of serum calcitriol levels, has been reported in other granulomatous diseases including tuberculosis (47), silicon-induced granulomas (48), and lymphoma (49).

III. Hormone Action

A. General Features of the Calcitriol Receptor

The major target tissues for vitamin D are the intestine and bone. Although these organs are prominent in mineral homeostasis, a large number of other tissues and cell culture lines possess the calcitriol receptor and have tissue- or cell-specific responses to the hormone.

The initial reports indicating that vitamin D is not a simple cofactor described the distribution of vitamin D metabolites in the intestinal mucosa (50) and the requirement of new RNA synthesis for action (51) and suggested that the mode of action of D sterols was similar to that of other steroid hormones. The subsequent identification of a cytosolic protein with high affinity for the most biologically active metabolites added further support to this hypothesis. Since then, the receptor has been partially characterized, and some of its cellular effects have been identified. However, its interaction with the genome, and the features controlling protein expression, remain largely unexplored.

Using sucrose gradient density ultracentrifugation analysis, ligand saturation analysis, and DNA-cellulose chromatography, the calcitriol receptor has been identified in a large number of tissues, including bone marrow, kidney, parathyroid, placenta, yolk sac, breast, pituitary, testis, uterus, and peripheral mononuclear cells (52). In avian species, the receptor has a similar distribution and is also found in the ovarian shell gland and chorioallantoic membrane. The receptor has been largely localized in the nucleus, using careful cell disruption and nuclear extraction techniques (53). Stumpf and co-workers, using in situ autoradiography after infusion of tritiated calcitriol, confirmed this localization (54). This method has allowed the receptor to be identified in a limited number of cell types in complex organs. The abundance of calcitriol receptor is less than that of other steroid receptors (55) and varies considerably from tissue to tissue and in different cultured cell lines. The intestine has the highest receptor content, with other tissues possessing a lower concentration of receptor or fewer receptor-positive cell types. The concentration of receptor in responsive tissues also varies during development, being undetectable in the neonatal rat duodenum and abundant by 21 days of age (56). As conventional assay methods improve and receptor complementary DNA (cDNA) probes become available, all tissue may eventually prove to possess the calcitriol receptor in at least a few cell types.

The receptor binds calcitriol with high affinity (Kd $\approx 10^{-10}$ to 10^{-11} M) and sediments in sucrose gradients with a density of 3.0 to 3.7 S. By gel filtration, the avian receptor has been found to have a molecular weight of 60 to 70 kD, increasing to 95 kD in the presence of sterol and protease inhibitors (57). Purification of avian receptor has been achieved by two groups of researchers (58,59). The production of polyclonal rat antisera and monoclonal antibodies (60) has aided in the characterization of the receptor. Using sodium dodecyl-sulfate polyacrylamide gel electrophoresis (SDS-PAGE) under reducing conditions and immunoblotting with the monoclonal antibodies, the molecular weight of the avian receptor has been found to be approximately 60 kD (61). The mammalian receptor is slightly smaller (54 kD) and is recognized by antibodies directed against the avian as well as the porcine receptor (62).

Receptor occupancy by the ligand is low (10–15%) but normally increases rapidly following the intravenous administration of exogenous calcitriol to the rachitic chick, peaking at 2 h and declining over the next 10 h (63). The total number of receptors declines over the same time course by 30%, with return to pretreatment levels by 36 h. The serum level of calcitriol, the level of receptor occupancy, and the subsequent synthesis of vitamin D–dependent proteins are tightly correlated.

With ligand occupancy, the receptor undergoes a physicochemical change resulting in an increased affinity for chromatin. Thus, calcitriol binding stabilizes the aposterol receptor, which is susceptible to reduction in the presence of sulfhydryl blocking agents and organomercurials, and causes the holosterol receptor to acquire an increased affinity for DNA, as evidenced by its increased resistance to salt extraction from the nucleus. Recently, the binding of calcitriol has been associated with phosphorylation of the receptor (64). Many of these properties of the calcitriol receptor are shared by other nuclear steroid receptors.

B. Cellular Effects of Calcitriol

The major biologic effects of vitamin D are the normalization of serum calcium and phosphate and the healing of bony lesions of vitamin D–deficient rickets. Early investigations revealed an enhancement of calcium and phosphorus transport from the intestinal lumen of rachitic animals treated with vitamin D. A family of vitamin D–dependent calcium-binding proteins (CaBPs), calbindins, has been identified in chick and mammalian tissues (65). Four classes of CaBPs have been isolated: one class with molecular weights of 29 kD was found in avian and mammalian tissues; one with molecular weights of 8–11 kD in mammalian intestines; one with a molecular weight of 9 kD from mammalian skin (immunologically distinct from the mammalian intestinal CaBP); and one with a molecular weight of 6 kD from avian and mammalian bone. All CaBPs, except the bone CaBP, have been shown to have high affinity for calcium (Kd $\approx 10^{-7}$ M), similar to the affinities displayed by regulatory elements of the cytoskeleton and by calmodulin. Cloning of the cDNAs of the CaBPs has demonstrated little homology between the 9-kD and 28-kD CaBPs. The 28-kD CaBP family, with a 6-domain structure replicating the calcium-binding domain, has a high degree of nucleotide and amino acid homology with several vitamin D–independent calcium-binding proteins, including calmodulin (66).

1. INTESTINE

In the mammalian and chick intestine, CaBP is located exclusively in the mature columnar epithelial cell of the villus and constitutes 1–3% of the total cellular protein. Subcellular localization demonstrates CaBP associated with the terminal web and the basilar cytosolic region, with smaller amounts in the perinuclear region and minimal amounts in the microvilli (52). CaBP is absent in the intestine of the rachitic animal, but it appears rapidly after exposure to calcitriol and new RNA synthesis (67). In the chick, a second 76-kD peptide also increases in quantity after vitamin D treatment.

Calcium transport in the rachitic animal remains unchanged until CaBP appears in the cytosol, but in the partially vitamin D–replete animal, calcium transport rapidly increases prior to new protein synthesis (68,69). Based on cellular localization of CaBP and the rapid activation of calcium transport in D-replete animals, some investigators have hypothesized that calcitriol may have direct cellular effects not mediated through the CaBP gene activation. Supporting this hypothesis is the observation that calcitriol directly alters phospholipid metabolism in intestinal mucosal cells (70,71). It has been proposed that the CaBP protein acts as an intracellular calcium buffer, allowing for high rates of transcellular calcium flux without significant alteration in the intracellular calcium concentration (≈ 1–10μM). In spite of extensive characterization of the cellular responses to calcitriol and the purification of CaBP, the mechanism by which calcitriol influences flux across the intestine remains to be elucidated.

2. KIDNEY

Calcitriol is synthesized by cells in the proximal renal tubule, which respond to calcitriol by synthesizing both 28-kD CaBP and 24-hydroxylase. In the chick, the clear cells of the distal convoluted tubule, which are the site of calcium reclamation, contain the CaBP. The 28-kD CaBP mRNA accumulates in the rat and chick kidney in response to calcitriol treatment. In contrast to the renal 28-kD and the intestinal 9-kD CaBPs, the 9-kD CaBP mRNA level in the kidney is strongly suppressed by exposure to calcitriol (72). Nuclear receptors have been localized in the CaBP-containing cells; however, no receptors have been found in the cells of the proximal tubule, even using sensitive autoradiography methods. Since the 1α-hydroxylases and 24-hydroxylase activities in the proximal tubule are responsive to calcitriol, further investigations of this finding and of the differential regulation of renal CaBP mRNA species are required.

Calcium and phosphate reabsorption occurs throughout the nephron with no net transport in the loop of Henle. The bulk of the transport of calcium and phosphate occurs in the proximal tubule. Calcium transport is regulated predominantly by the filtered load and by PTH. Phosphate transport is likewise regulated by PTH and serum phosphate concentration, and it is influenced by urinary sodium load, phosphate intake, insulin, glucose, acid-base balance, and serum calcium. Dietary phosphate restriction markedly reduces urinary phosphate loss within 24 h; nevertheless, fecal losses are unaffected, and phosphate depletion soon develops (73). The accompanying slight reductions of PTH secretion and serum phosphate cannot account for these rapid transport changes, and the mechanism by which this adaptation occurs is unknown. When acutely thyroparathyroidectomized rats are infused with a phosphate load, the reabsorption of phosphate markedly decreases within 40 min, and isolated renal tubule membrane vesicles also demonstrate a marked decrease in Na^+-dependent phosphate uptake (74). This clearly demonstrates the rapidity and the PTH independence of this aspect of phosphate transport.

Phosphate is removed from the glomerular filtrate by an electrochemical gradient that is sodium- and energy-dependent. The reabsorbed phosphate ion readily crosses the cell and freely diffuses across the basolateral surface membrane. In vitro studies of isolated membrane vesicles have suggested that the capacity and not the V_{max} of the phosphate transport system is the limiting factor in ion transport (75). Calcitriol stimulates phosphate uptake by renal brush border vesicles, whereas chronic administration slightly suppresses phosphate uptake, possibly a result of alteration in serum calcium, filtered phosphate load, and other regulators of serum calcium and phosphate (76). The direct contribution of calcitriol to renal phosphate transport under physiologic conditions is unclear, although indirect action by increasing intestinal transport of calcium and phosphorus and suppression of PTH is significant. The common site of phosphate transport and calcitriol synthesis, and the potent stimulation of calcitriol production by phosphate deprivation, suggests an intriguing interrelationship.

3. BONE

Bone, as a biologic unit and a structural element of the skeleton, is compromised by both vitamin D excess and vitamin D deficiency. Early studies in the chick demonstrated that diets deficient in calcium or phosphate produce skeletal abnormalities similar to those seen in the D-deficient animal, and that calcium and phosphate supplementation corrected the mineralization defects. It was suggested that the impairment of calcium transport in the intestine, and the accompanying hypocalcemia and hypophosphatemia, solely accounted for the bony changes of vitamin D deficiency.

Subsequent work has demonstrated that the calcitriol receptor is present in the nucleus of osteoblasts, osteocytes, chondrocytes, and monocytes (the putative osteoclast progenitor); that calcitriol stimulates the synthesis of specific proteins by osteoblasts and chondrocytes; and that osteoclasts in mixed marrow cell populations are activated by calcitriol (52). However, osteoclasts lack both PTH and calcitriol receptors. When stimulated by calcitriol, bone marrow–derived mononuclear cells from several species, including primates, undergo fusion to form multinucleated cells with several histologic and immunocytochemical characteristics of osteoclasts (77). Calcitriol may contribute to the recruitment of osteoclast precursors from the marrow and to the regulation of other aspects of bone formation and resorption (78).

Treatment of rachitic animals with vitamin D results in increased synthesis and incorporation of bone matrix proteins along with mineralization of excessive osteoid. Collagen, hyaluronic acid, phosphoproteins, and the 6-kD bone-specific CaBP, osteocalcin, are increased in bone after D treatment in vivo. In cultured osteoblastlike cells, calcitriol stimulates the production of collagen and alkaline phosphatase activity at physiologic concentrations, whereas higher concentrations suppress synthesis (79). In bone organ culture systems, similar changes have been noted; however, the dominant effects of calcitriol at supraphysiologic concentrations are the inhibition of collagen synthesis and the stimulation of bone resorption.

Osteocalcin, with three vitamin K–dependent calcium-binding gamma-carboxyglutamic acid residues, has a much lower affinity ($Kd \approx 10^{-4}$ M) for calcium than the other CaBPs in bone and other tissues (80). It is secreted in large quantities into the bone matrix and into the plasma in response to calcitriol in vivo and by osteosarcoma cells in vitro. The treatment of rats with warfarin or a D-deficient, high-calcium, high-phosphate diet diminishes the serum levels of osteocalcin and depletes bone of osteocalcin; however, no abnormality in mineralization is observed. Although osteocalcin is an excellent marker of bone turnover, its role in bone homeostasis remains to be defined. The physiologic role that calcitriol plays, in comparison to PTH and local factors, is obscured by the lack of basic knowledge of bone cell biology and interactions.

4. PARATHYROID GLANDS

The indirect suppression of PTH secretion by vitamin D, mediated by serum calcium, has long been recognized. Initial experiments in animals subjected to dietary and surgical manipulations failed to provide consistent evidence for a proposed direct-feedback loop. The identification of calcitriol receptors in the parathyroid (81,82) led to a reevaluation of this problem using parathyroid tissue cultures. In this system, calcitriol suppresses PTH accumulation in culture media in a dose-dependent manner, and the cells continue to exhibit calcium suppressibility (83). Calcitriol treatment of dispersed D-replete bovine parathyroid cells reduces prepro-PTH mRNA levels by 30–40% (84). The acute administration of calcitriol transiently suppresses the levels of prepro-PTH mRNA in the parathyroid glands of intact rats (85), and long-term intravenous administration of calcitriol to patients with renal osteodystrophy and severe secondary hyperparathyroidism who are undergoing dialysis suppresses plasma PTH levels (86). Unfortunately, hypercalcemia develops in some of these patients, preventing continued treatment. The proposed long-term regulatory role of calcitriol in parathyroid activity and cell mass will likely be defined by further research.

5. IMMUNE SYSTEM

Immunocytes are capable of synthesizing small quantities of calcitriol and of other biologically active vitamin D metabolites. The effects of these small amounts, which may locally exceed the circulating calcitriol of nephrogenous origin, is difficult to establish in vivo. Vitamin D deficiency has not been associated with profound immunologic dysfunction, suggesting that the role of these active metabolites is modulatory and not necessary for overall immunologic competency. Cells of the lymphopoietic system respond to calcitriol by alteration of secretion of potent cell mediators by granulocytic and mononuclear cells, inhibition of cellular proliferation, and cellular differentiation (87). Peripheral blood monocytes possess calcitriol receptors, and B and T lymphocytes acquire detectable receptors after lectin activation. Receptor can be detected in lymphocytes from the thymus and tonsils of normal children. Approximately 80% of adults with rheumatoid arthritis have calcitriol receptors in their peripheral lymphocytes (88).

Activated lymphocytes respond in vitro to physiologic concentrations of calcitriol (10^{-12} to 10^{-10} M) with a 100-fold reduction in interleukin-2 (IL-2) secretion, the expression of different cell surface markers, and a suppression of cellular proliferation. Cellular proliferation can only be partially restored by exogenous IL-2. Some observers believe that the production and binding of calcitriol by immunocytes indicate an autocrine or paracrine system for this sterol.

IV. Vitamin D and Elemental Deficiency

A. General Considerations

Nutritional rickets remains a major problem in countries in which dietary sources of vitamin D and exposure to sunlight are limited. In Western societies, nutritional rickets is infrequently encountered in the general childhood population, although it may be associated with chronic illness, malnutrition, parental neglect, and dietary faddism. Thus, a significant percentage of children presenting the rachitic findings are afflicted with inherited forms of rickets.

The premature infant is at particular risk of developing rickets. Deprived of active placental transport of calcium, these infants may be unable to achieve the requisite positive calcium balance (up to 300 mg daily) necessary for mineralizing the developing skeleton. The absorption of calcium in the infants is hindered by immaturity of the intestine, which limits mineral transport, and by immaturity of the kidney and parathyroid glands, which limits mineral homeostasis and calcitriol synthesis. Supplementation with vitamin D and calcium improves the calcium homeostasis, with resulting mineralization of the skeleton. Although human colostrum is a comparatively poor source of calcium ($\approx$ 33 mg/L) (89), rickets is uncommon in exclusively breast-fed infants and, when present, is usually associated with insufficient (maternal) exposure to sunlight.

Malabsorption, and drugs that induce the hepatic microsomal oxidase system, cause rickets in children and osteomalacia in adults. These patients display varying degrees of insensitivity to the usual therapeutic dose of vitamin D. The increased requirement for vitamin D does does not indicate D resistance; plasma calcitriol levels are low in the face of decreased serum 25-OH D concentrations resulting from vitamin D malabsorption or the increased hepatic catabolism of calciferols. Other causes of vitamin D–deficient rickets or osteomalacia that are insensitive to treatment with calciferols include phosphate depletion from antacid use, urinary loss, aluminum-associated osteomalacia of renal insufficiency, and calcium-deficient diets. The nutritional causes of rickets are usually responsive to treatment by increasing the dietary intake of the deficient mineral and vitamin D or by treatment of the underlying disease.

B. Clinical Features of Vitamin D Deficiency

The clinical picture of patients with rickets or osteomalacia is influenced by the severity of the deficiency, the underlying etiology, and the patient's age and general nutritional status. In most patients, the signs and symptoms of rickets develop insidiously over several months or years before clinical diagnosis is made. In children, the cessation of linear growth is usually asymptomatic and is not noted until the defective ossification of endochondral and membranous bone and poor callus formation are radiographically manifest. In infants, there is a generalized softening of all bones with widening of the cranial sutures and flaring and widening of the long bones in the area of the endochondral growth plates. Muscular weakness is commonly noted, along with irritability upon movement of the child. Most affected infants fail to walk at the appropriate age. Chest wall abnormalities occur in cases of severe disease, and a rachitic rosary forms owing to widening growth plates in the developing ribs and pectus carinature.

These bony changes, in conjunction with muscular weakness, lead to compromised pulmonary function. Rachitic children are also reported to have an increased rate of infection, and pneumonia may be a life-threatening event in children with restrictive pulmonary disease. Older children who are ambulatory prior to the onset of rickets develop bony deformities of the limbs and pelvis. Bowing of the tibia and femur are the most striking changes noted, but upper extremities are also affected. A coxa varum deformity and a waddling gait are apparent on examination, and surgical correction may be required in severely affected children. Children with other forms of rickets, including end-organ resistance to calcitriol, have similar signs and symptoms.

The abnormalities in biochemical parameters reflecting mineral metabolism correlate well with the severity of clinical disease. Hypocalcemia, hypophosphatemia, and hyperphosphatasia are common, and secondary hyperparathyroidism is uniformly observed. Radiographic changes of diffuse osteopenia are noted in conjunction with a poorly mineralized, widened, and broadened epiphyseal plate and flaring of metaphyses. In older children and adults with more protracted or severe deficiency states, pseudofractures of the extremities and pelvis may occur. Serum levels of 25-OH D are uniformly low in vitamin D deficiency rickets, and plasma calcitriol levels are inappropriately low (low or normal) for the amount of calcium needed and in relation to the concomitant hyperparathyroidism. Urinary calcium excretion is very low.

The rachitic animal responds to treatment with vitamin D, 25-OH D, or calcitriol, given in appropriate doses (90). Treatment of rachitic children with physiologic quantities of vitamin D results in a rapid rise of the calcitriol level to 4–5 times the normal plasma level. The elevated calcitriol level slowly declines over several months, although it may persist for up to 9 months (91,92). The serum calcium level rises after the increase in calcitriol and becomes normal after several months. Serum alkaline phosphatase concentration, a marker of osteoblastic activity, parallels the decline in calcitriol levels after the initiation of treatment. Radiographic improvement is usually noted after 1–2 months with calcification of widened epiphyseal margins, and occasionally a "bone-within-a-bone" appearance is noted along the metaphyseal surfaces. There is minimal skeletal deformity, and a period of rapid growth occurs when treatment is initiated in the early phase of deficiency.

V. Phosphate-Wasting Syndromes

A. .Familial X-Linked Hypophosphatemic Rickets

Familial X-linked hypophosphatemic rickets (XLH), also known as X-linked vitamin D–resistant rickets, is the most common form of hereditary rickets. The disease has an X-linked dominant inheritance pattern with variable penetrance, usually more severely affecting the male members of a kindred. A sporadic autosomal-dominant form is also recognized. Short stature, rachitic skeletal deformities, and hypophosphatemia with renal phosphate wasting are the hallmarks of this disease. In contrast to nutritional rickets, the serum calcium is normal, and there is no evidence of severe secondary hyperparathyroidism or myopathy (Table 5.2). The plasma calcitriol level is usually normal in untreated adults and children (93) but low in patients receiving vitamin D and oral phosphate (94). The renal proximal tubule, with a reduced threshold for phosphate transport, is the site of phosphate wasting in these patients (95). Intestinal transport of calcium and phosphate is slightly impaired (96); however, normal phosphate transport by jejunal mucosal specimens from patients with XLH (97) and the improvement of intestinal transport with calciferol treatment militate against a primary intestinal defect in these patients.

Patients with XLH display mild derangement of the regulation of PTH secretion and calcitriol synthesis. PTH administration to XLH patients results in a normal generation of nephrogenous cAMP, but plasma calcitriol levels increase only to 35% of peak levels attained by normal individuals (98). In normal adults, dietary phosphate restriction results in a threefold increase in plasma calcitriol levels; in contrast, dietary restriction in patients with XLH does not elevate plasma calcitriol levels (99). A calcitriol receptor defect had been proposed as the basis for these observations. However, cultured dermal fibroblasts from XLH patients displayed normal receptor number, ligand affinity, and production of 24-hydroxylase in response to calcitriol stimulation (100). The common

TABLE 5.2. Biochemical profiles of patients with rachitic syndromes.

Symptom	Serum calcium	Serum inorganic phosphate	25-OH D	Calcitriol	PTH	U_{Ca}	TRP
Vitamin D deficiency	↓	↓/nl	↓↓	↓/nl	↑	↓↓	↓/nl
Calcium deficiency	↓	↓/nl	↓	↑	↑	↓↓	↓/nl
XLH	nl	↓	nl	nl	nl	nl/↓	↓
XLH with hypercalciuria	nl	↓	—	↑	nl	↑	↓
Oncogenic osteomalacia	nl	↓	nl	↓	nl	—	↓
1α-Hydroxylase deficiency (VDDR I)	↓	↓	↑	↓↓	↑↑	↓↓	↓/nl
Receptor defects (VDDR II)	↓	↓	nl/↑	↑↑↑	↑↑	↓↓	↓/nl

Abbreviations: TRP, tubular reabsorption of phosphate; nl, normal.

anatomic location of the putative defects of XLH, involving calcitriol synthesis, PTH renal activity, and phosphate handling, suggests that there is an interrelated abnormality that determines the phenotypic expression of the disease (101).

A murine model of XLH, the Hyp mouse, with similar genetic, biochemical, and physiologic findings, has been extensively studied (102). This small-bodied, short-nosed, short-tailed mouse has hypophosphatemia with decreased phosphate transport in the proximal tubule. Other tissues do not share this impairment of phosphate transport. Hyp mice also have decreased intestinal transport of calcium and phosphate and low levels of intestinal vitamin D–dependent CaBP. Plasma calcitriol levels are normal in spite of persistent hypophosphatemia, and these levels do not increase with phosphate deprivation. Hyp mice present an excellent animal model for the investigation of the pathophysiologic bases of XLH; nonetheless, we remain ignorant of the cellular basis of the phosphate transport defect. It has not been established whether an intrinsic defect of phosphate transport capacity and rate in the proximal tubule or a humoral phosphaturic factor is the cause of the hypophosphatemia.

B. Hereditary Hypophosphatemic Rickets with Hypercalciuria

A recently recognized syndrome of hypophosphatemic rickets with hypercalciuria provides a clear contrast to XLH with respect to plasma calcitriol levels and PTH regulation (103). The disease has been identified only in a single Bedouin kindred, in which an autosomal-recessive pattern of inheritance was proposed. The kindred had the usual clinical findings of XLH, with the addition of nephrolithiasis and renal impairment in affected individuals. These patients have increased calcitriol levels (150–550 pg/ml), decreased serum PTH levels, and suppression of nephrogenous cAMP (Table 5.2). Calcium loading in affected individuals resulted in marked hyperabsorption of calcium, and urinary calcium excretion was normalized by fasting. The associated findings of phosphate wasting and an inappropriately low calcitriol production rate in response to hypophosphatemia that are observed in XLH patients are separated in this syndrome. Kindred members displayed the increased serum calcitriol levels in response to phosphate restriction observed in normal individuals. The affected members had severe rachitic bony deformity despite their exuberant calcitriol responses; thus, the relative contribution of the abnormalities in vitamin D metabolism to pathogenesis of XLH remains unresolved. Oral phosphate alone improved the rickets in the hypercalciuric, hypophosphatemic patients. In contrast, patients with XLH require the addition of calciferols, in conjunction with phosphate supplementation, for improvement in bony disease.

Most authors have suggested that patients with XLH receive either 1α-hydroxyvitamin D or calcitriol in preference to vitamin D, in the hope of avoiding D intoxication which occasionally results from prolonged administration (104,105). A regimen of phosphate supplementation alone results in the development of secondary hyperparathyroidism; however, it has been reported that the addition of an active calciferol results in nearly normal growth over several years when treatment is initiated in early childhood.

C. Oncogenic Hypophosphatemic Osteomalacia

Almost 50 patients with a hypophosphatemic syndrome, in association with a mesenchymal tumor, that resolves upon successful resection of the tumor have been described (106). Their age varied from 5 to 62 years, and their symptoms included bone pain, fractures, and myopathy. In general, the tumors are small (<1 cm), occasionally multiple, and difficult to detect. Many years usually elapse between the onset of symptoms, the localization of the tumor, and its successful resection. The patients have osteomalacia or rickets associated with hypophosphatemia, hyperphosphatasia, normal plasma concentrations of PTH and 25-OH D, low plasma concentration of calcitriol, and increased urinary excretion of hydroxyproline (107) (Table 5.2). Additional patients with these findings, who showed no improvement with partial resection of the associated malignancy, have also been described.

Histologic examination of the tumors from patients cured by complete surgical excision has revealed a wide variety of rare, usually benign tumors including ossifying mesenchymal tumors, osteoblastomas, hemangiopericytoma-like tumors, nonossifying fibromas, and 11 other types (108). In general, the tumors have had a mixed, complex architecture with multinucleated giant cells and are highly vascularized.

A few patients with nonresectable tumors have been treated with vitamin D metabolites and phosphate supplementation, similar to the regimens used for the treatment of XLH, and the osteomalacia has responded, given sufficient time for the therapeutic trial. Resection of the tumor has resulted in restoration of plasma calcitriol levels in the majority of patients and resolution of the phosphate wasting. Tumor homogenates from several patients have produced phosphaturia in laboratory animals, but no further characterization of the proposed phosphaturic factor has been reported. This syndrome may be analogous to XLH, but it is clearly humorally mediated. The factor(s) produced by this lesion may be important in the normal regulation of renal phosphate handling.

D. Fanconi Syndrome

Fanconi syndrome, a heterogeneous group of disorders with generalized proximal tubule dysfunction, is frequently associated with osteomalacia and rickets. This syndrome, characterized by amino aciduria, glycosuria, phosphaturia, kaliuresis, and, usually, metabolic acidosis, may present in childhood associated with a variety of enzymatic deficiencies or as an acquired disease associated with heavy-metal intoxication, dys-proteinemic states, drug use, or tubulointerstitial disease (109). PTH and plasma calcitriol levels are reported to be normal in this syndrome (110), although vitamin D administration appears to suppress renal phosphate losses and increase phosphate retention. In general, patients have re-sponded to calciferols, alkaline salts, and phosphates similar to XLH patients, with improvement of symptoms and osteomalacia.

VI. Vitamin D–Dependent Rickets

Patients with end-organ resistance to vitamin D develop the aforementioned signs and symptoms of deficiency rickets. The failure of these patients to respond to treatment with supplemental vitamin D prompted a search for a more complex etiology than the usual nutritional deficiency, including end-organ resistance to calcitriol in the intestine or other responsive organs, or disorders of sterol and mineral absorption or metabolism.

Steroid hormones interact with intranuclear receptors, ultimately re-sulting in the production of new proteins and the alteration of cellular function. The diagnosis of end-organ resistance requires the presence of sufficient or excessive quantities of the active hormone in the target tissue, the presence of the necessary tissue elements for a specific response, and the failure of competent cells to respond in a tissue-specific manner. The phenotypic findings may be highly variable and may include resistance to androgens, as noted above, and glucocorticoids (111), or an unidentified and probably lethal mutation, with resistance to estrogens. The outcome of end-organ resistance is complete or partial failure of activation of the steroid-responsive genes; since normal activation involves many cellular elements (e.g., receptors, specific nucleoplasmic proteins and DNA regulator elements, mRNA transcription and processing, and receptor-binding site on responsive genes) and multiple steps, defects in any of these processes could theoretically occur. Indeed, several defects have been identified in different patients with true resistance to vitamin D sterols.

A. Vitamin D–Dependent Rickets Type I

Among the first patients to be identified with true resistance to vitamin D were those described by Fraser and co-workers (112). These patients had severe rickets and hypocalcemia, and they required large doses of vitamin D (1.25–2.5 mg) or 25-OH D_3 (0.4–0.9 mg) for a skeletal response, but showed rapid normalization of serum calcium and amelioration of rickets after two brief courses of calcitriol in physiologic doses (113). Individuals with this type of rickets have hypocalcemia, hypophosphatemia, and hyperphosphatasia with elevated levels of PTH, nephrogenous cAMP, and 25-OH D (Table 5.2). There is a subnormal urinary cAMP response to PTH infusion. The basal renal tubular reabsorption of phosphates is greater than 90%, and it is unresponsive to the PTH infusion. In spite of the evidence of secondary hyperparathyroidism and adequate concentration of 25-OH D, plasma calcitriol levels are subnormal. Calcitriol treatment results in rapid reversal of the clinical and biochemical findings of rickets.

Severe impairment of 1α-hydroxylase activity and successful treatment with modest quantities of 1α-hydroxyvitamin D or calcitriol are characteristic of the disease. It has yet to be determined whether these patients have a decreased concentration of the normal hydroxylase enzyme, an abnormal hydroxylase enzyme, or an absent or abnormal component of this multiple enzyme system. In this syndrome, known as vitamin D–dependent rickets type I (VDDR I), normal growth can be anticipated with adjustment of calciferol dose to the developing child's weight and mineral requirements.

B. Vitamin D–Dependent Rickets Type II

1. CLINICAL FEATURES

Twenty-one kindreds with 36 affected individuals with true resistance to calcitriol have been described (Table 5.3). Patients have been characterized to a varying extent by several groups of investigators using a variety

TABLE 5.3. Clinical profiles of patients with vitamin D–dependent rickets type II.

Author designation	Sibs	Consan-guinity	Age at onset (years)	Alo-pecia	25-OH D[a] (ng/ml)	Calcitriol pre[a] (pg/ml)	Calcitriol post[b] (pg/ml)	Reference
—	1	?	15	0	14	137	297	114
1A	2	–	2	0	25	213	270	115
1B	2	–	0.5	0	44	280	189	115
—(*)	?	?	2	0	19	212	?	116
A	2	–	9	0	?	93	?	117
B	2	–	2	0	?	54	?	117
—	1	?	45	0	?	?	?	118
—	1	–	12	0	?	143	?	119
2A	2	+	2	+	142	169	?	120
2B	2	+	1	+	?	142	800	120
3	2	+	1	+	20	710	?	121
—	1	?	1.5	+	32–900	>200	?	122
KN	?	+	1.5	+	48	?	4800	123
IK	2	+	1	+	68	108	?	124,128
RK	2	+	1	+	29	83	?	128
4	2	+	0.8	+	?	66	?	125
5	1	–	0.6	+	124	?	1400	125
6	2	–	3	+	32	916	?	126
—	2	+	2	+	12	500	?	127
SH	2	–	1	+	36	112	1500	128
RH	2	–	0.7	+	16	118	1500	128
D₁	3	+	?	+	?	≈150	?	129
D₂	3	+	?	+	?	≈150	?	129
A*	1	–	1.5	+	26	319	4000	132,133
B*	?	?	?	–	?	?	468	133

Table layout based on S. J. Marx et al. (1984) (Recent Prog Horm Res 40:589–620).
[a] Pretreatment plasma levels of 25-OH D and calcitriol reflect previous treatment with vitamin D.
[b] Calcitriol level at time of smallest dose required to attain and maintain normocalcemia.

of techniques (114–134). Patients with this syndrome, referred to as VDDR II, have ranged in age from 6 months to 45 years. The younger patients usually have more profound hypocalcemia and more severe rachitic deformities. Older patients have less severe disease with osteomalacia and mild hypocalcemia. Within a kindred, the characteristics of the syndrome have been consistent with respect to age of onset, presence of alopecia, severity of bone disease, success of treatment, and pharmacologic requirements of calciferols and calcium.

All patients were asymptomatic at birth and had no congenital anomalies or evidence of rickets. In spite of a presumed defect in placental calcitriol receptors, calcium transport to the fetus is apparently sufficient for the modest amount of bone mineralization that occurs in utero. In contrast, full-term infants with neonatal rickets have vitamin D or mineral deficiency derived from a maternal deficiency state. The youngest VDDR

II patients to develop rickets presented during or near the end of the first year of life with hypocalcemia, hypophosphatemia, and the physical and radiographic findings of rickets. These patients also had evidence of secondary hyperparathyroidism with elevated serum PTH, urinary nephrogenous cAMP, and alkaline phosphatase (2.5–6 times normal) (Table 5.2).

2. VDDR II with Alopecia Totalis

VDDR II is characterized by a wide spectrum of receptor dysfunction and clinical response to treatment; however, a few coherent features are evident. All of the patients with alopecia presented for evaluation by the age of 3 years and were severely affected at that time. Marx and colleagues suggested that the children who develop alopecia totalis, after loss of lanugo hair, have a particularly severe form of vitamin D resistance (135). Supporting this hypothesis is the earlier age of onset in this group of VDDR II patients (1.3 vs. 11 years), the lower success rate in achieving normocalcemia with therapy, and the higher plasma calcitriol levels in patients attaining normocalcemia (2330 vs. 349 pg/ml) compared to patients without alopecia (Table 5.3).

Marx et al. postulated that the alopecia is a direct consequence of the calcitriol receptor defect, not an associated but unrelated finding. The identification of calcitriol receptor in the cells of hair follicular sheaths, the presence of CaBP in the hair follicle and the dermis, and the accumulation of 7-dehydrocholesterol in dermis with calcitriol treatment are cited as evidence that calcitriol has a physiologic role in the dermis and the hair follicle. Furthermore, the heterogeneity of receptor defects identified in these patients, each possibly representing a different defective allele, militates against the idea that alopecia results from linkage of the calcitriol receptor gene with a contiguous but unrelated gene. In addition, nutritional rickets and VDDR I are not associated with alopecia. However, the function of calcitriol in the skin and hair follicle is unknown, as is the role that the receptor defects play in alopecia.

Patients with alopecia vary in their ability to achieve and maintain normocalcemia. Fewer than half of them responded to 1 year of appropriate treatment with calciferols and oral calcium. Some patients in whom this treatment regimen failed had persistent plasma calcitriol concentrations in excess of several thousand pg/ml, and as high as 19,000 pg/ml. Among those who responded to treatment, a variable clinical course was noted. The following summaries illustrate several of the difficulties encountered in the evaluation and care of these individuals:

An Algerian infant (patient 5 in Table 5.3), who initially failed to respond to calcium and calciferols, had a belated clinical improvement at the end of 1 year. During the year of treatment with calcitriol (5–10) μg/day), all biochemical and radiographic features of rickets had resolved; however, after 2½ years of continuous therapy, there was a progressive decline in serum calcium and phosphorus. Radiographic evidence of rickets and findings of secondary hyperparathyroidism recurred, and the child again became refractory to doses of up to 20 μg of calcitriol daily.

A young female (patient RK in Table 5.3) with alopecia and extensive radiographic changes of rickets, who had no calcemic responses to appropriate treatment, had progressive improvement of rickets beginning at age 7. Over the ensuing 1½ years, with a serum calcium of 7.5–8.4 mg/dl and a phosphorus level of 4.0–5.0 mg/dl, the alkaline phosphatase declined, the rachitic changes completely resolved, and only mild changes of rickets were observed on bone biopsy.

The clinical response to therapy in this group of patients is far more variable than in patients without alopecia. A minority of these patients achieve normocalcemia, whereas a few will relapse despite continuing treatment, and spontaneous healing of rickets may occur in the face of persistent hypocalcemia. The lack of a calcemic response has been attributed to inadequate induction of intestinal calcium transport by calcitriol. This supposition has been directly confirmed in a 4-year-old patient who failed to have a calcemic response to 50 μg of calcitriol daily (134). Intestinal calcium absorption, measured by dual tracer studies with stable isotopes, remained at the low, pretreatment level (≈ 12%).

Among the affected siblings of the reported patients, several deaths have occurred prior to the age of 3 years. The exact cause of death was undetermined, but pulmonary insufficiency associated with severe rickets is likely to have contributed directly or indirectly. Treatment of affected children may ameliorate this aspect of the disease process, indirectly by improvement in pulmonary performance or, possibly, directly by improving leukocyte and mononuclear cell function. Since clinical study of VDDR II patients with alopecia has been possible for only the past 8 years, the clinical course and the requirement for treatment during adolescence and adulthood are unknown.

3. VDDR II Patients without Alopecia Totalis

The initial case report of VDDR II described a patient with the mildest recognized form of this disease. The minimal hypocalcemia, mild secondary hyperparathyroidism, and osteomalacia responded to treatment with 4000 IU of vitamin D, although the plasma calcitriol level remained elevated after remission of the bone disease. All other patients without alopecia have responded to treatment with calcium and various calciferols and have similarly attained normocalcemia (135).

The response may be very slow, as illustrated by a girl who was evaluated initially at 2 years of age and treated with a daily dose of 2 mg of vitamin D, with gradual improvement over several years (patient 4 in Table 5.3). At age 28, after an uneventful pregnancy, she developed hip pain and pseudofractures of the public ramus and femurs. The plasma calcitriol concentration was elevated (4.5 times normal), and treatment with 25-OH D initially did not alter the serum calcium or phosphorus. However, the patient was in net positive calcium balance (48 mg/day) and became normocalcemic after 6 months of continuous therapy. The placental transfer of calcium to the fetus during the third trimester may have far exceeded the maximum intestinal transport capacity in this patient and caused the postpartum osteomalacia (136). This same discrepancy may have impeded a rapid restoration of skeletal mineral content, as noted at both age 2 and age 28. This group of patients requires continuous treatment to maintain a normal serum calcium and phosphorus concentration.

4. Proposed Treatment

Most patients have been treated with vitamin D, ranging from the usual replacement dose to several milligrams daily, without clinical improvement prior to referral and evaluation. Many of the initial patients identified with VDDR II had received a multitude of doses and types of calciferol for variable lengths of time, with varying clinical improvement. The failure of some patients to attain normocalcemia may have been a result of both the severity of the resistance to calcitriol and the choice and dose of calciferol and oral calcium. In considering this problem, and based on their experience with many of the reported patients, Marx et al. (137) recommend the following regimen: (a) oral calcium supplements, adjusted for age and weight; (b) very high doses of either 1α-OH D_3 or calcitriol (20–60 μg/day) to achieve consistent serum calcitriol levels greater than 2000 pg/ml; and (c) treatment of sufficient duration (>3 months) to overcome any hypocalcemic effects arising from rapid initial mineralization of preformed osteoid.

Since approximately one-half of the patients with alopecia fail to respond to the recommended regimen, an alternative approach has been developed and used in a single patient. A 4-year-old child, whose intestinal absorption of calcium could not be augmented by calcitriol, received nightly intravenous calcium infusion via a permanent central venous catheter. This regimen allowed the patient to walk for the first time, corrected the radiographic rachitic findings, and improved the biochemical indices of rickets (134). Such a regimen, although successful in this patient, would be difficult to maintain throughout childhood. Calcium requirements decline after puberty, so that in some unresponsive patients the low rate of intestinal calcium transport may prove sufficient for near mineralization of the fully developed skeleton. Therapeutic alternatives are clearly needed for all patients with VDDR II.

C. In Vitro Studies

Although the major target issues of calcitriol (intestine, bone, and kidney) are generally difficult to obtain for in vitro studies of the receptor, the recognition that calcitriol receptors are present in a variety of other tissues and the development of techniques for primary cultures of human tissues have allowed evaluation of calcitriol receptor in cells from VDDR II patients. Thus, the calcitriol receptor is demonstrable after several weeks of continuous cell culture in fibroblasts obtained from skin biopsy of normal individuals (55,138). Circulating monocytes can also be used for such studies (146). With skin biopsies from VDDR II patients, several steps in the pathway of calcitriol-receptor binding and gene activation have been investigated (139–145). In spite of methodological differences, the results obtained by different groups of investigators can be compared, and some general conclusions can be reached. The general method employed by all investigators requires the proliferation of skin fibroblasts for several weeks in cell culture, harvesting the receptor in a soluble extract (or associated with nuclear chromatin), and determining the affinity and capacity of the receptor for calcitriol, using steroid saturation analysis.

Binding of calcitriol to its receptor induces a physicochemical change in the receptor, increasing its affinity for nuclear chromatin, with which it interacts to accomplish the next step in activation of responsive genes. To evaluate this process, fibroblasts are incubated with ^{3}H-calcitriol, the cell membrane is then gently disrupted, and receptors unbound to chromatin are extracted. Nuclear chromatin binding of the ^{3}H-calcitriol-receptor complex can be quantified after separation of the nucleus and the cellular extract. This procedure allows distinction between cells with receptors that display apparently normal nuclear chromatin binding from those with receptors that have normal calcitriol binding but an altered DNA binding site.

Interaction of the calcitriol-receptor complex with chromatin ultimately alters the expression of multiple genes. Among them is the gene for 24-hydroxylase, whose expression is markedly increased by calcitriol in a variety of tissues, including fibroblasts. The activity of this enzyme represents the final step in calcitriol action and can be evaluated by measuring the conversion of ^{3}H-25-OH D_3 to ^{3}H-24,25-$(OH)_2D_3$ in cultured fibroblasts from normal individuals or from patients with VDDR II.

The radioligand immunoassay (RLIA) employs monoclonal antibodies generated against the receptor to evaluate for the presence of a receptor unable to bind calcitriol. The antibody is mixed with the fibroblast extract and a small excess of preformed ^{3}H-calcitriol-avian-receptor complexes. The amount of receptor protein can be estimated by its competition with the labeled avian receptor for the binding to the antibody.

D. Results of the In Vitro Testing in VDDR II Patients

The in vitro studies described above have been conducted with cultured fibroblasts from only a limited number of vitamin D–resistant patients,

TABLE 5.4. Results of in vitro testing from patients with VDDR type II.

Author desig- nation	Alo- pecia	Receptor extraction assay		Nuclear binding assay		24-Hy- droxylase activity (% control)	RLIA (fmol/mg prot)	Cal- cemic response
		Kd (pM)	Capacity (fmol/mg prot)	Kd (pM)	Capacity (% control)			
1A	−	170	19	UM		<10	58	+
1B	−	380	18	UM		—	—	+
—(*)	−	200	26	?	100	5–10	—	+
2B	+	40	28	UM		<10	69	+
3	+	110	46	300	50	—	—	−
KN	+	UM		UM		†	—	+
IK	+	UM		UM		NR	—	−
RK	+	UM		UM		NR	—	−
4	+	150	4.2	500	8	—	48	−
5	+	UM		UM		—	53	±
6	+	UM		UM		—	47	−
D_1	+	55	32	—		NR	—	+
D_2	+	63	52	—	75	NR	—	+
A*	+	1200	10	—		NR	—	+
B*	−	2200	30	—		6%	—	+

† No suppression of lectin-stimulated proliferation of peripheral mononuclear cells by calcitriol.
Abbreviations: UM, unmeasurable; NR, no response.

and in only 2 patients have all four analyses been performed (Table 5.4). Nevertheless, the receptor defects appear to be hetergeneous, although in those kindreds in which 2 members were studied, the defects in receptor sterol binding, nuclear association, and 24-hydroxylase activation have been found to be consistent within the kindred. Such consistency correlates with the clinical observation that members of the same kindred have similar calcemic responses to treatment, requirements for calciferols and calcium, and affliction with alopecia totalis.

1. ABNORMALITIES IN CALCITRIOL BINDING

In vitro studies with cultured cells from patients with VDDR have demonstrated altered sterol-binding affinity and binding capacity, abnormal receptor-chromatin binding, and ineffective activation of responsive genes. Patients KN, IK, RK, 5, and 6 (Table 5.4) had no demonstrable binding of ³H-calcitriol, suggesting either an absence of receptor protein or a dysfunctional sterol binding site of the receptor. The RLIA showed that the amount of receptor protein in 2 of these patients (Nos. 5 and 6) was normal. Receptor protein from patient 6 had a normal sedimentation rate during sucrose gradient ultracentrifugation, suggesting a defect in the sterol-binding region without a major change in the physical features of the protein.

Extracts of cells from patients A* and B* showed a 10- to 40-fold reduction in affinity for calcitriol but a normal binding capacity. The clear difference between these 2 patients with respect to the serum calcitriol levels required to achieve normocalcemia (468 vs. 4000 pg/ml), the stimulation of 24-hydroxylase activity in fibroblasts, and the presence of alopecia suggests that additional factors determine phenotypic expression in affected individuals.

Patient 4 showed a decrease in sterol-binding capacity without demonstrable change in receptor protein levels. Although this combination of findings could result from an unstable sterol-binding site that is unable to bind sterol after a brief residency in the nucleus or from production of both a normal and a variant receptor protein, it could simply represent an artifact of the extraction procedures. However, this patient had limited clinical response to calciferol during relapse for no apparent cause.

Only 1 of the 5 patients without demonstrable calcitriol binding achieved normocalcemia after treatment with extraordinary doses of vitamin D (4–7 $\times$ 10^6 IU daily), reaching a serum calcitriol level of 4280 pg/ml. Another (RK) had only intermittent calcemic responses to calcitriol but nonetheless experienced spontaneous improvement in rickets. The remaining patients had persistent hypocalcemia and severe rickets.

2. ABNORMAL CHROMATIN BINDING

Two kindreds (1A, 1B, and 2B) have absent chromatin binding by the calcitriol-receptor complex but response to calciferol treatment with healing of rachitic bone lesions. Receptor content (RLIA) was normal, and fibroblasts from both kindreds responded to very high concentrations of calcitriol (10^{-7} to 10^{-6} M) with an elevation of 24-hydroxylase activity. This response is consistent with the clinical response to high-dose calciferols that has been observed.

3. ABNORMAL 24-HYDROXYLASE RESPONSE TO CALCITRIOL

In other patients (*, 3, and D$_2$), calcitriol-receptor binding and chromatin association are normal, but cultured fibroblasts fail to demonstrate 24-hydroxylase activity in response to calcitriol treatment. In addition, lectin-stimulated peripheral mononuclear cells from these patients continued to proliferate after exposure to calcitriol (146). Since nuclear chromatin binding by the receptor-ligand complex and initiation of transcription may be distinct events, these findings may reflect a subtle binding defect.

Evidence for such a subtle anomaly was elicited in cells from patients D$_1$ and D$_2$ by subjecting the fibroblast extracts to the chromatographic techniques that have been employed for purification of the avian calcitriol receptor. In these samples, the tritiated calcitriol-receptor complex eluted from DNA cellulose at a lower KCl concentration (107 vs. 171 mM) than that required for elution of the normal human receptor. Additionally, sedimentation through sucrose gradients with a low KCl concentration revealed that the receptor failed to form the expected 6S receptor aggregate. However, sedimentation in 300 mM KCl demonstrated the expected 3S monomeric receptor. Examination of the interaction of calcitriol-receptor complexes with the genome has been attempted with two techniques: direct assay of the association of tritiated calcitriol with nuclear chromatin, and stimulation of 24-hydroxylase activity. Although there are discrepancies between the ability of the receptor to bind chromatin and the 24-hydroxylase activity of fibroblasts (patients 1A and 2B), Marx has suggested that there is a good correlation between the in vitro synthesis of 24-hydroxylase and the patient response to calciferols (137). However, there is not consistent support for this suggestion (129,132–134).

4. Correlation with Aberrant Steroid Receptors

The cDNAs of the estrogen and glucocorticoid receptors have recently been purified, and the primary structure of the correspondent mRNA has been obtained (147–151). These receptors share several structural features that are probably representative of other intranuclear steroid receptors. Their cDNAs ($\approx$ 6 kb) are far larger than is necessary for coding for the receptor proteins ($\approx$ 2 kb). The function of the extensive 3' noncoding regions is unknown. The coding regions for the proteins contain at least three distinct functional domains. The 3' region codes for the carboxy-terminal portion of the receptor, which contains the steroid-binding regions. The existence of two RNA splicing sites for the glucocorticoid receptor results in the expression of a functional receptor, as well as a nonfunctional receptor that does not bind cortisol. The midportion of the receptor contains a region homologous to the v-erb-A oncogene, with nine conserved cysteine residues. It has been postulated that this region forms a coordination complex with metallic cations that tightly binds double-stranded DNA and initiates DNA transcription. A mutant glucocorticoid receptor, which binds cortisol but not DNA, has a single amino acid substitution in this region (152). The estrogen receptor forms a stable complex with a 90-kD protein in the absence of steroid binding. This protein, a member of a group a heat shock proteins secreted in increased quantities during cellular stress, stabilizes the ligandless receptor and disassociates with steroid binding. Many of the identified glucocorticoid receptor mutations have clinical equivalents in VDDR II (Table 5.5).

The cloning of the calcitriol receptor gene is in progress (153), and the gene mutations in VDDR II are likely to be identified in the near future. Some of the cells displaying these mutations will be valuable in experiments designed to evaluate the mechanism of calcitriol action and regulation of the calcitriol receptor gene.

TABLE 5.5. Possible molecular bases of receptor anomalies identified in patients with VDDR type II.

Receptor abnormality	Protein	DNA
Absent sterol binding	Absent receptor protein	Gene deletion
		Mutation preventing RNA processing
		Absent or altered 5′ promoter region
		Absent enhancer element
	Nonfunctional sterol-binding domain	Gene mutation
Decreased binding affinity for calcitriol	Unstable or altered sterol-binding domain	Gene mutation
Decreased binding capacity for calcitriol	Unstable binding domain or protein	Gene mutation
	Production of nonfunctional receptor variant	Alternative splicing of mRNA of the coding region for binding domain
Decreased chromatin binding	Altered DNA-binding domain	Gene mutation
	Absent or dysfunctional RNP or receptor associated protein	Gene mutation in other gene or gene associated elements
	Phosphorylation defect	Gene mutation in receptor gene or phosphorylase gene
Absent or decreased 24-hydroxylase activity	Abnormal Receptor	Gene mutation
	Absent or dysfunctional RNP or receptor associated protein	See above
	Phosphorylation defect	See above
	Abnormal 24-OH-lase enzyme	Mutation in phosphorylase gene
		Defective calcitriol-receptor binding to 5′ region of the 24-hydroxylase gene

Abbreviation: RNP, ribonucleoprotein.

VII. Summary

Calcitriol, the most biologically active vitamin D metabolite, is synthesized in a precisely regulated manner in response to PTH, phosphate, and calcitriol. The receptor for calcitriol is found in a wide variety of tissues and cell types. Binding of calcitriol to its intranuclear receptor results in tissue-specific responses and, in the majority of tissues, in the synthesis of a class of vitamin D–dependent CaBPs. In the intestine, the appearance of CaBP is associated with increased calcium transport. However, in the majority of tissues, the mechanism of action and the effects of calcitriol are unknown.

Rickets, the clinical manifestation of vitamin D deficiency, remains a major disease in agrarian and newly industrialized nations. In Western Europe and North America, hereditary rachitic syndromes and extreme

infant prematurity are now the most difficult health problems associated with rickets. Although all rachitic patients have similar signs, symptoms, and radiographic features, there are some differences, and distinct etiologies are evident in the biochemical profiles of affected individuals.

In spite of the limited number of patients with the hereditary vitamin D–resistant form of rickets, a wide spectrum of clinical and biochemical features has emerged. The variable features of VDDR Type II include the degree of hypocalcemia (absent to severe), the presence of alopecia totalis, the severity and type of the radiographic evidence of hypomineralization, the age of onset (6 months to 45 years), and the therapeutic requirement and success in attaining normocalcemia and healing of the bony lesion. Utilizing cultured fibroblasts or monocytes from VDDR II patients, several calcitriol-receptor abnormalities have been identified. These abnormalities include altered sterol binding, decreased chromatin binding by the calcitriol-receptor complex, and absent or impaired activation of the calcitriol-dependent genes. The characterized receptor defects are almost as numerous as the kindreds studied; each kindred may represent a distinct mutation. Precise identification of the molecular mechanisms for true resistance to calcitriol will lead to a better understanding of the ways calcitriol influences cellular function.

References

1. Mellanby E (1919) An experimental investigation on rickets. Lancet 2:407–412.
2. Albright F, Buttler AM, Bloomberg E (1937) Rickets resistant to vitamin D therapy. Am J Dis Child 54:531–547.
3. Prader VA, Illig R, Heidi E (1961) Eine besondere from der primären vitamin D–resistenten rachitis mit hypocalcämie und autosomal-dominanten erbgang: Die hereditäre pseudo-mangelrachitis. Helv Paediatr Acta 16:452–468.
4. Holick MF, MacLaughlin JA, Clark MB, Holick SA, Potts JT Jr, Anderson RR, Blank IH, Parish JA (1980) Photosynthesis of previtamin D_3 in human skin and the physiologic consequences. Science 210:203–205.
5. Holick MF, Kleiner-Bossaller A, Schnoes HK, Kasten PM, Boyle IT, DeLuca HF (1975) 1,24,25-Trihydroxyvitamin D_3, a metabolite of vitamin D_3 effective on intestine. J Biol Chem 248:6691–6696.
6. Fraser DR (1980) Regulation of the metabolism of vitamin D. Physiol Rev 60:551–613.
7. Weisner RH, Kumar R, Seemans E, Go VLW (1980) Enterohēpatic physiology of 1,25-dihydroxyvitamin D_3 metabolites in normal man. J Lab Clin Med 96:1094–1100.
8. Kumar R, Nagubandi S, Mattox VR, Londowski JM (1980) Enterohepatic physiology of 1,25-dihydroxyvitamin D_3. J Clin Invest 65:277–284.
9. Clements MR, Chalmers TM, Fraser DR (1984) Enterohepatic circulation of vitamin D: A reappraisal of the hypothesis. Lancet 1:1376–1379.
10. Holick MF, MacLaughlin JA, Doppelt SH (1981) Regulation of cutaneous previtamin D_3 photosynthesis in man: Skin pigment is not an essential regulator. Science 211:590–593.

11. Bhattacharyya MH, DeLuca HF (1974) The regulation of calciferol-25-hydroxylase in chick. Biochem Biophys Res Commun 59:734–741.
12. Mawer EB, Reeve A (1977) The use of an isolated perfused liver to study the control of cholecalciferol-25-hydroxylase activity in the rat. Calcif Tissue Res 22 (Suppl):24–28.
13. Milne ML, Baran DT (1985) End product inhibition of hepatic 25-hydroxyvitamin D production in the rat: Specificity and kinetics. Arch Biochem Biophys 242:488–492.
14. Baran DT, Milne ML (1983) 1,25-Dihydroxyvitamin D–induced inhibition of ^{3}H-25 hydroxyvitamin D production by the rachitic rat liver in vitro. Calcif Tissue Int 35:461–464.
15. Halloran BP, Bikle DD, Levens MJ, Castro ME, Globus RK, Holton E (1986) Chronic 1,25-dihydroxyvitamin D_3 administration reduces the serum concentration of 25-hydroxyvitamin D by the increasing metabolic conversion rate. J Bone Mineral Res 1 (Suppl):16 (abstract).
16. Stanbury SW, Mawer EB (1983) Vitamin D metabolism in man: Contribution of clinical studies. In Frame B, Potts JT (eds): Clinical Disorders of Bone and Mineral Metabolism. Excerpta Medica, Princeton, NJ, pp 72–77.
17. Whyte MP, Haddad JG, Walters D, Stamp TCB (1979) Vitamin D bioavailability: Serum 25-hydroxyvitamin D levels in man after oral, subcutaneous, intramuscular, and intravenous vitamin D administration. J Clin Endocrinol Metab 48:906–911.
18. Bell NH, Shaw S, Turner RT (1984) Evidence that 1,25-dihydroxyvitamin D_3 inhibits the hepatic production of 25-hydroxyvitamin D in man. J Clin Invest 74:1540–1544.
19. Lore F, DiCairano G, Periti P, Caniggia A (1982) Effect of the administration of 1,25-dihydroxyvitamin D_3 on the serum levels of 25-hydroxyvitamin D in postmenopausal osteoporosis. Calcif Tissue Int 34:539–541.
20. Bell NH (1984) Vitamin D–endocrine system. J Clin Invest 76:1–6.
21. LoCascio V, Adami S, Galvanini G, Ferrari M, Cominacini L, Tartarotti D (1985) Substrate-product relation of 1-hydroxylase activity in primary hyperparathyroidism. N Engl J Med 313:1123–1125.
22. Henry HL (1979) Regulation of the hydroxylation of 25-hydroxyvitamin D_3 in vivo and in primary cultures of chick kidney cells. J Biol Chem 254:2722–2729.
23. Baksi SN, Kenny AD (1979) Acute effects of parathyroid extract on renal vitamin D hydroxylases in Japanese quail. Pharmacology 18:169–174.
24. Trechsel U, Bonjour J-P, Fleisch H (1979) Regulation of the metabolism of 25-hydroxyvitamin D_3 in primary cultures of chick kidney cells. J Clin Invest 64:206–217.
25. Stern PH, Taylor AB, Bell NH, Epstein S (1981) Demonstration that circulating 1alpha,25-dihydroxyvitamin D is loosely regulated in normal children. J Clin Invest 68:1374–1377.
26. Booth BE, Tsai HC, Morris RC (1985) Vitamin D status regulates 25-hydroxyvitamin D_3-1α-hydroxylase and its response to parathyroid hormone in the chick. J Clin Invest 75:155–161.
27. Baxter LA, DeLuca HF (1976) Stimulation of 25-hydroxyvitamin D_3-1α-hydroxylase by phosphate depletion. J Biol Chem 251:3158–3161.
28. Friedlander EJ, Henry HL, Norman AW (1977) Studies on the mode of action of calciferol. XII. Effects of dietary calcium and phosphorus on the relationship between the 25-hydroxyvitamin D-1-hydroxylase and production of chick intestinal calcium binding protein. J Biol Chem 252:8677–8683.
29. Booth BE, Tsai HC, Morris RC (1977) Parathyroidectomy reduces 25-hydroxyvitamin D_3-1α-hydroxylase activity in the hypocalcemic vitamin D–deficient chick. J Clin Invest 60:1314–1320.

30. Bikle DD, Rasmussen H (1975) The ionic control of 1,25-dihydroxyvita-min D_3 production in isolated chick renal tubules. J Clin Invest 55:127–138.
31. Portale AA, Halloran BP, Murphy MM, Morris RC (1986) Oral intake of phosphorus can determine the serum concentration of 1,25-dihydroxy-vitamin D by determining its production rate in humans. J Clin Invest 77:7–12.
32. Gray RW, Wilz DR, Caldas AE, et al. (1977) The importance of phosphate in regulating plasma 1,25-$(OH)_2$-vitamin D levels in humans: Studies in healthy subjects in calcium-stone formers and in patients with primary hyperparathyroidism. J Clin Endocrinol Metab 451:299–306.
33. Tanaka Y, Halloran B, Schnoes HK, DeLuca HF (1979) In vitro produc-tion of 1,25-dihydroxyvitamin D_3 by rat placental tissue. Proc Natl Acad Sci USA 76:5033–5035.
34. Howard GA, Turner RT, Sherrard DJ, Baylink DJ (1981) Human bone cells in culture metabolize 25(OH) D_3 to 1,25$(OH)_2D_3$ and 24,25-$(OH)_2D_3$. J Biol Chem 256:7738–7740.
35. Gray TK, Maddux FW, Lester GE, Williams ME (1982) Rodent macro-phages metabolize 25-hydroxyvitamin D_3 in vitro. Biochem Biophys Res Commun 109:723–729.
36. Adams JS, Sharma OP, Gacad MA, Singer FR (1983) Metabolism of 25-hydroxyvitamin D_3 by cultures pulmonary alveolar macrophages in sarcoidosis. J Clin Invest 72:1856–1860.
37. Teitelbaum SL, Bar-Shavitz Z, Perry HM, Welgris HG, Kahn AJ, Reitsma P, Gray R, Horst R (1985) In Norman AW (ed): Proceedings of the Sixth Workshop on Vitamin D. DeGruyter, New York, pp A64 (abstract).
38. Turner RT, Howard GA, Puzas E, Baylink DJ, Knapp DR (1983) Calvarial cells synthesize 1α,25-dihydroxyvitamin D_3 from 25-hydroxyvitamin D_3. Biochemistry 22:1073–1076.
39. Adams JS, Singer FR, Gacad MA, Sharma OP, Hayes MJ, Vouros P, Holick MF (1985) Isolation and structural identification of 1,25-dihydrox-yvitamin D_3 produced by cultured alveolar macrophages in sarcoidosis. J Clin Endocrinol Metab 60:960–966.
40. Puzas EJ, Farley JR, Turner RT, Baylink DJ (1984) 1,25-Dihydroxyvita-min D: Response to and production by bone cells. In Kumar R (ed): Vitamin D. Basic and Clinical Aspects. Martin Nijhoff, Boston, pp 125–150.
41. Fetchick DA, Bertolini DR, Sarin PS, Weintraub ST, Mundy GR, Dunn J (1986) Production of 1,25-dihydroxyvitamin D_3 by human T cell lympho-trophic virus-1-transformed lymphocytes. J Clin Invest 78:592–595.
42. Gray TK, Lester GE, Lorenc RS (1979) Evidence for extra-renal 1α-hydroxylation of 25-hydroxyvitamin D_3 in pregnancy. Science 204:1311–1313.
43. Reeve L, Tanaka Y, DeLuca HF (1983) Studies on the site of 1,25-dihydroxyvitamin D_3 synthesis in vivo. J Biol Chem 258:3615–3617.
44. Papapoulos SE, Clemens TL, Fraher LJ, Lewin IG, Sandler LM, O'Rior-dan JL (1979) 1,25-Dihydroxycholecalciferol in the pathogenesis of the hypercalcemia of sarcoidosis. Lancet 1:627–630.
45. Bell NH, Stern PH, Pantzer E, Sinha TK, DeLuca HF (1979) Evidence that the increased circulating 1α,25-dihydroxyvitamin D is the probable cause for abnormal calcium metabolism in sarcoidosis. J Clin Invest 64:218–225.
46. Mason RS. Frankel T, Chan YL, Lissner D, Posen S (1984) Vitamin D conversion by sarcoid lymph node homogenate. Ann Intern Med 100:59–61.

47. Gkonos PJ, London R, Hendler ED (1984) Hypercalcemia and elevated 1,25-dihydroxyvitamin D levels in a patient with end-stage renal disease and active tuberculosis. N Engl J Med 311:1683–1685.
48. Kozeny GA, Barbato AL, Bansal VK, Vertuno LL, Hano JE (1984) Hypercalcemia associated with silicone-induced granulomas. N Engl J Med 311:1103–1105.
49. Breslau NA, McGuire JL, Zerwekh JE, Frenkel EP, Pak CY (1984) Hypercalcemia associated with increased serum calcitriol levels in three patients with lymphoma. Ann Intern Med 100:1–6.
50. Haussler MR, Norman AW (1967) The subcellular distribution of physiological doses of vitamin D_3. Arch Biochem Biophys 118:145–153.
51. Zull JE, Czarnowska-Misztal E, DeLuca HF (1966) On the relationship between vitamin D action and actinomycin-sensitive processes. Proc Natl Acad Sci USA 55:177–184.
52. Norman AW, Roth J, Orci L (1982) The vitamin D endocrine system: Steriod metabolism, hormone receptors, and biological response (calcium binding proteins). Endocr Rev 3:331–366.
53. Lawson DEM, Wilson PW, Barker DC, Kodicek E (1969) Isolation of chick intestinal nuclei: Effect of vitamin D_3 on nuclear metabolism. Biochem J 115:263–268.
54. Stumpf WE, Sar M, Narbaitz R, Reid FA, DeLuca HF, Tanaka Y (1980) Cellular and subcellular localization of 1,25-$(OH)_2$-vitamin D_3 in rat kidney: Comparison with the localization of parathyroid hormone and estradiol. Proc Natl Acad Sci USA 77:1149–1153.
55. Colston K, Hirt M, Feldman D (1980) Organ distribution of the cytoplasmic 1,25-dihydroxycholecalciferol receptor in various mouse tissues. Endocrinology 107:1916–1922.
56. Halloran BP, DeLuca HF (1981) Appearance of the intestinal cytosolic receptor for 1,25-dihydroxyvitamin D_3 during neonatal development in the rat. J Biol Chem 256:7338–7342.
57. Bishop JE, Hunziker W, Norman AW (1982) Evidence for multiple molecular weight forms of the chick intestinal 1,25-dihydroxyvitamin D_3 receptor. Biochem Biophys Res Commun 108:140–145.
58. Pike JW, Haussler MR (1979) Purification of chicken intestinal receptor for 1,25-dihydroxyvitamin D. Proc Natl Acad Sci USA 76:5485–5489.
59. Simpson RU, DeLuca HF (1982) Purification of the chicken intestinal receptor for 1alpha,25-dihydroxyvitamin D_3 to apparent homogeneity. Proc Natl Acad Sci USA 79:16–20.
60. Pike JW, Donaldson CA, Marion SL, Haussler MR (1982) Development of hybridomas secreting monoclonal antibodies to the chicken intestinal 1α,25-dihydroxyvitamin D_3 receptor. Proc Natl Acad Sci USA 79:7719–7723.
61. Pike JW, Marion SL, Donaldson CA, Haussler MR (1983) Serum and monoclonal antibodies against the chick intestinal receptor for 1,25-dihydroxyvitamin D_3. Generation by a preparation enriched in a 64,000-dalton protein. J Biol Chem. 258:1289–1296.
62. Dame MC, Pierce EA, DeLuca HF (1985) Identification of the porcine intestinal 1,25-dihydroxyvitamin D_3 receptor on sodium dodecyl sulfate/polyacrylamide gels by renaturation and immunoblotting. Proc Natl Acad Sci USA 82:7825–7829.
63. Hunziker W, Walters MR, Bishop JE, Norman AW (1982) Effect of vitamin D status on the equilibrium between occupied and unoccupied 1,25-dihydroxyvitamin D intestinal receptor in the chick. J Clin Invest 69:826–834.

64. Pike JW, Sleator NM (1985) Hormone-dependent phosphorylation of the 1,25-dihydroxyvitamin D_3 receptor in mouse fibroblasts. Biochem Biophys Res Commun 131:378–385.
65. Wasserman RH, Brindale ME, Fullmer CS (1981) Calcium-binding protein (CaBP) and other vitamin D responsive proteins. In Bronner F, Peterlik M (eds): Calcium and Phosphate Transport across Biomembranes. Academic Press, New York, pp 279–295.
66. Hunziker W, Schrickel S (1986) Chick and rat vitamin D dependent calcium binding protein both have a six domain structure and shared extensive amino acid sequence homologies. J Bone Mineral Res 1 (Suppl):22 (abstract).
67. Kendrick NC, Barr CR, Moriarity D, DeLuca HF (1981) Effects of vitamin D deficiency on in vitro labeling of chick intestinal proteins: Analysis by 2-dimensional electrophoresis. Biochemistry 20:5288–5294.
68. Wasserman RH, Brindak ME, Meyer SA, Fullmer CS (1982) Evidence for multiple effects of vitamin D_3 on calcium absorption: Response of rachitic chicks, with or without partial vitamin D_3 repletion, to 1,25-dihydroxyvitamin D_3. Proc Natl Acad Sci USA 79:7939–7943.
69. Nemere I, Yoshimoto Y, Norman AW (1984) Calcium transport in perfused duodena from normal chicks: Enhancement within fourteen minutes of exposure to 1,25-dihydroxyvitamin D_3. Endocrinology 115:1476–1483.
70. O'Doherty PJA (1979) 1,25-Dihydroxyvitamin D_3 increases the activity of the intestinal phosphatidylcholine deacylation-reacylation cycle. Lipids 14:75.
71. Matsumoto T, Fontaine O, Rasmussen H (1981) Effect of 1,25-dihydroxyvitamin D_3 on phospholipid metabolism in the chick duodenal mucosal cells. j Biol Chem 256:3354.
72. Hall AK, Norman AW (1986) Use of a sensitive oligonucleotide probe to study the 1,25-dihydroxyvitamin D_3 regulation of calbindin-D_{9K}-mRNA in the rat and chick intestine. J Bone Mineral Res 1 (Suppl):21 (abstract).
73. Dominguez JH, Gray RW, Lemann J (1976) Dietary phosphate deprivation in men and women: Effects of mineral and acid balances, parathyroid hormone and the metabolism of 25-OH vitamin D. J Clin Endocrinol Metab 43:1056–1068.
74. Cheng L, Dersch C, Kraus E, Spector D, Sacktor B (1984) Renal adaptation to phosphate load in the acutely thyroparathyroidectomized rat: Rapid alteration in the brush border membrane phosphate transport. Am J Physiol 246:F488–F494.
75. Dousa TP, Kempson SA (1982) Regulation of renal brush border membrane transport of phosphate. Mineral Electrolyte Metab 7:113–121.
76. Kurnik BRC, Hruska KA (1984) Effects of 1,25-dihydroxycholecalciferol on phosphate transport in the vitamin D–deprived rat. Am J Physiol 247:F177–F182.
77. Roodman GD, Ibbotson KJ, MacDonald BR, Kuehl TJ, Mundy GR (1985) 1,25-Dihydroxyvitamin D_3 causes formation of multinucleated cells with several osteoclast characteristics in cultures of primate marrow. Proc Natl Acad Sci USA 82:8213–8217.
78. Canalis E (1983) The hormonal and local regulation of bone formation. Endocr Rev 4:62–77.
79. Kurihara N, Ishizuka S, Kiyoki M, Haketa Y, Ikeda K, Kuegawa M (1986) Effects of 1,25-dihydroxyvitamin D_3 on osteoblastic MC3T3-E1 cells. Endocrinology 118:940–947.
80. Price PA (1985) Vitamin K–dependent formation of the bone gla protein (osteocalcin) and its function. Vitam Horm 42:65–108.
81. Haussler MR, Norman AW (1969) Chromosomal receptor for a vitamin D metabolite. Proc Natl Acad Sci USA 62:155.

152   J. F. McLeod and J. G. Haddad Jr.

82. Hughes MR, Haussler MR (1978) 1,25-Dihydroxyvitamin D$_3$ receptors in parathyroid glands: Preliminary characterization of the cytoplasmic and nuclear binding components. J Biol Chem 253:1065–1073.
83. Au WY (1984) Inhibition by 1,25-dihydroxycholecalciferol of hormone secretion of rat parathyroid gland in organ culture. Calcif Tissue Int 36:384–391.
84. Silver J, Russel J, Sherwood LM (1985) Regulation by vitamin D metabolites of messenger ribonucleic acid for preproparathyroid hormone in isolated bovine parathyroid cells. Proc Natl Acad Sci USA 82:4270–4273.
85. Silver J, Naveh T, Mayer H, Schmelzer H, Popovtzer MM (1986) Regulation by vitamin D metabolites of preproparathyroid hormone mRNA in vivo. J Bone Mineral Res 1 (Suppl):433 (abstract).
86. Slatopolsky E, Weerts C, Theilan J, Horst R, Harter H, Martin TJ (1984) Marked suppression of secondary hyperparathyroidism by administration of 1,25-dihydroxycholecalciferol in uremic patients. J Clin Invest 74:2136–2143.
87. Monolagas SC, Provvedini DM, Tsoukas CD (1985) Interactions of 1,25-dihydroxyvitamin D$_3$ and the immune system. Mol Cell Endocrinol 43:113–122.
88. Werntz DA, Tsoukas CD, Provvedini DM, Vaughan JH, Deftos LJ, Manolagas SC (1984) Expression of 1,25-dihydroxyvitamin D$_3$ receptors in lymphocytes from patients with rheumatoid arthritis. Calcif Tissue Int 36:528 (abstract).
89. Lammi-Keefe CJ, Jensen RG (1984) Fat-soluble vitamins in human milk. Nutr Rev 42:365–371.
90. Brommage R, DeLuca HF (1985) Evidence that 1,25-dihydroxyvitamin D$_3$ is the physiologically active metabolite of vitamin D$_3$. Endocr Rev 6:491–511.
91. Papapoulos SE, Clemens TL, Fraher LJ, Gleed J, O'Riordan JL (1980) Metabolites of vitamin D in human vitamin D deficiency: Effect of vitamin D$_3$ and dihydroxycholecalciferol. Lancet 2:612–615.
92. Garabedian M, Vainsel M, Mallet E, Huquette G, Guillozo BA, Toppet M, Grimberg R, Nguyen TM, Balsan S (1983) Circulating vitamin D metabolite concentrations in children with nutritional rickets. J Pediatr 103:381–386.
93. Lyles KW, Clark AG, Drezner MK (1982) Serum 1,25-dihydroxyvitamin D levels in subjects with hypophosphatemic rickets and osteomalacia. Calcif Tissue Int 34:125–130.
94. Scriver CR, Reade TM, DeLuca HF, Hamstra AJ (1978) Serum 1,25-dihydroxyvitamin D levels in normal subjects and in patients with hereditary rickets or bone disease. N Engl J Med 299:976–979.
95. Glorieux FH, Morin CL, Travers R, Delvin EE, Poirier R (1976) Intestinal phosphate transport in familial hypophosphatemic rickets. Pediatr Res 10:691–696.
96. Condon JR, Nassim JR, Rutter A (1970) Defective intestinal absorption in familial and non-familial hypophosphatemia. Br J Med 3:138–141.
97. Glorieux FH, Morin CL, Travers R, Delvin EE, Poirier R (1976) Intestinal phosphate transport in familial hypophosphatemic rickets. Pediatr Res 10:691–696.
98. Lyles KW, Drezner MK (1982) Parathyroid hormone effects on serum 1,25-dihydroxyvitamin D levels in patients with X-linked hypophosphatemic rickets: Evidence for abnormal 25-hydroxyvitamin D-1-hydroxylase activity. J Clin Endocrinol Metab 54:638–644.
99. Insogna KL, Broadus AE, Gertner JM (1983) Impaired phosphorus conservation and 1,25-dihydroxyvitamin D generation during phosphate deprivation in familial hypophosphatemic rickets. J Clin Invest 71:1562–1569.

100. Adams JS, Gacad M, Singer FR (1984) Specific internalization and action of 1,25-dihydroxyvitamin D_3 in cultured dermal fibroblasts from patients with X-linked hypophosphatemia. J Clin Endocrinol Metab 59:556–560.
101. Drezner MK (1984) The role of abnormal vitamin D metabolism in X-linked hypophosphatemic rickets and osteomalacia. Adv Exp Med Biol 178:399–404.
102. Meyer RA (1986) Animal models of human diseases: X-linked hypophosphatemia (familial or sex-linked vitamin D-resistance rickets). Am J Pathol 118:340–342.
103. Tieder M, Modai D, Samuel R, Arie R, Halabe A, Bab I, Gabizon D, Liberman UA (1985) Hereditary hypophosphatemic rickets with hypercalcuria. N Engl J Med 312:611–617.
104. Rasmussen H, Pechet M, Anast C, Mazur A, Gertner J (1981) Long-term treatment of familial hypophosphatemic rickets with oral phosphate and 1 alpha-hydroxyvitamin D_3. J Pediatr 99:16–25.
105. Harrel RM, Lyles KW, Harrelson JM, Friedman NE, Drezner MK (1985) Healing of bone disease in X-linked hypophosphatemic rickets/osteomalacia. Induction and maintenance with phosphorus and calcitriol. J Clin Invest 75:1858–1868.
106. Cotton GE, Van Puffelen P (1986) Hypophosphatemic osteomalacia secondary to neoplasia. J Bone Joint Surg 68:129–133.
107. Agus Z (1983) Oncogenic hypophosphatemic osteomalacia. Kidney Int 24:113–123.
108. Weidner N, Bar RS, Weiss D, Strottmann MP (1985) Neoplastic pathology of oncogenic osteomalacia/rickets. Cancer 55:1691–1705.
109. Stein JH, Kunau RT (1979) Miscellaneous disorders of tubular function. In Early LE, Gottschalk CW (eds): Strauss and Welt's Diseases of the Kidney. Little, Brown, Boston, pp 1071–1073.
110. Roth KS, Foreman JW, Segal S (1981) The Fanconi syndrome and the mechanism of tubular transport dysfunction. Kidney Int 20:705–716.
111. Lipsett MB, Chrousos GP, Tomita M, Brandon DD, Loriaux DL (1985) The defective glucocorticoid receptor in man and nonhuman primates. Recent Prog Horm Res 41:199–247.
112. Fraser D, Kooh SV, Kind HP, Holick MF, Tanaka Y, DeLuca HF (1973) Pathogenesis of hereditary vitamin-D-dependent rickets. N Engl J Med 289:817–822.
113. Delvin EE, Glorieux FH, Marie PJ, Pettifor JM (1981) Vitamin D dependency: Replacement therapy with calcitriol. J Pediatr 99:26–34.
114. Brooks MH, Bell NH, Love L, Stern PH, Orfei E, Queener SF, Hamstra AJ, DeLuca HF (1978) Vitamin-D-dependent rickets type II. Resistance of target organs to 1,25-dihydroxyvitamin D. N Engl J Med 298:996–999.
115. Marx SJ, Spiegel AM, Brown EM, Gardner DG, Downs RW, Attie M, Hamstra AJ, DeLuca HF (1978) A familial syndrome of decreased in sensitivity to 1,25-dihydroxyvitamin D. J Clin Endocrinol Metab 47:1303–1310.
116. Zerwekh JE, Glass K, Jowsey J, Pak CY (1979) A unique form of osteomalacia associated with end organ refractoriness to 1,25-dihydroxyvitamin D and apparent defective synthesis of 25-hydroxyvitamin D. J Clin Endocrinol Metab 49:171–175.
117. Adams JS, Wahl TO, Moore WV, Horton WA, Lukert BP (1979) Familial vitamin D dependent rickets: Further evidence for end organ resistance to active vitamin D metabolites. In Program and Abstracts, 61st Annual Meeting of the Endocrine Society, p 767 (abstract).
118. Fujita T, Nomura M, Okajima S, Furuya H (1980) Adult-onset vitamin

D—resistant osteomalacia with unresponsiveness to parathyroid hormone. J Clin Endocrinol Metab 50:927–931.

119. Kudoh T, Kumagai T, Uetsuji N, Tsugawa S, Oyanagi K, Chiba Y, Minami R, Nakao T (1981) Vitamin D dependent rickets: Decreased sensitivity to 1,25-dihydroxyvitamin D. Eur J Pediatr 137:307–311.

120. Rosen JF, Fleischman AR, Finberg L, Hamstra A, DeLuca HF (1979) Rickets with alopecia: An inborn error of vitamin D metabolism. J Pediatr 94:729–735.

121. Liberman UA, Samuel R, Halabe A, Kuali R, Edelstein S, Weisman Y, Papapoulos SE, Clemen T, Fraher LJ, O'Riordan JLH (1980) End organ resistance to 1,25 dihydroxycholecalciferol. Lancet 1:504–506.

122. Sockalosky JJ, Ulstrom RA, DeLuca HF, Brown DM (1980) Vitamin D—dependent rickets in a child: Alopecia and marked end-organ unresponsiveness to 1,25-(OH)$_2$D$_3$. J Pediatr 96:701–703.

123. Tsuchiya Y, Matsuo N, Cho H, Kumgai M, Yasaka A, Suda T, Orima H, Shiraki M (1980) An unusual form of vitamin D—dependent rickets in a child: Alopecia and marked end organ hyposensitivity to biologically active vitamin D. J Clin Endocrinol Metab 51:685–690.

124. Feldman D, Chen T, Cone C, Hirst M, Shani S, Benderli A, Hochberg Z (1982) Vitamin D resistant rickets with alopecia: Cultured skin fibroblasts exhibit defective cytoplasmic receptors and unresponsiveness to 1,25(OH)$_2$D$_3$. J Clin Endocrinol Metab 55:1020–1022.

125. Balsan S, Garabedian M, Liberman UA, Eil C, Bourdeau A, Guillozo H, Grimberg R, Le Deunff MJ, Lieberherr M, Guimbaud P, Broyer M, Marx SJ (1983) Rickets and alopecia with resistance to 1,25-dihydroxyvitamin D: Two different clinical courses with two different cellular defects. J Clin Endocrinol Metab 57:803–811.

126. Beer S, Tieder M, Kohelet D, Liberman OA, Vure E, Bar-Joseph G, Gabizon D, Borochowitz ZU, Varon M, Modai D (1981) Vitamin D resistant rickets with alopecia: A form of end organ resistance to 1,25-dihydroxyvitamin D. Clin Endocrinol (Oxf) 14:395–402.

127. Silver J, Landau H, Bab I, Shvil Y, Friedlaender MM, Rubinger D, Popovtzer MM (1985) Vitamin D—dependent rickets type I and II. Diagnosis and response to therapy. Isr J Med Sci 21:53–56.

128. Hochberg Z, Benderli A, Levy J, Vardi P, Weisman Y, Chen T, Feldman D (1984) 1,25-Dihydroxyvitamin D resistance, rickets, and alopecia. Am J Med 77:805–811.

129. Hirst MA, Hochman HI, Feldman D (1985) Vitamin D resistance and alopecia: A kindred with normal 1,25-dihydroxyvitamin D binding, but decreased receptor affinity for deoxyribonucleic acid. J Clin Endocrinol Metab 60:490–495.

130. Gamblin GT, Liberman UA, Eil C, Down RW, DeGrange DA, Marx SJ (1985) Defective induction of 25-hydroxyvitamin D$_3$-24-hydroxylase by 1,25-dihydroxyvitamin D$_3$ in cultured skin fibroblasts. J Clin Invest 75:954–960.

131. Tietze HU, Burgert A, Schaff A, Hennes U (1981) Familial rickets with alopecia: Inborn and organ unresponsiveness to 1,25-(OH)$_2$D$_3$. Acta Pediatr 240 (Suppl 35).

132. Castells S, Greig F, Fusi M, Finberg L, Yasumura S, Liberman UA, Eil C, Marx S (1986) Severely deficient binding of 1,25-dihydroxyvitamin D to its receptors in a patient responsive to high doses of this hormone. J Clin Endocrinol Metab 63:252–256.

133. Castells S, Greig F, Fusi M, Finberg L, Yasumura S, Liberman UA, Eil C, Marx SJ (1985) Manifestations of 1,25 hydroxyvitamin D receptors with abnormally low hormone affinity: Comparison of two kindreds. Pediatr Res 19 (Suppl):310F (abstract).

134. Bliziotes M, Yergey A, Nanes M, Muenzer J, Vieira N, Begley M, Kher K, Marx S (1986) Lack of calcium transport response to $1,25(OH)_2D_3$: Documentation in vivo and successful treatment. J Bone Miner Res 1 (Suppl):14 (abstract).
135. Marx SJ, Liberman UA, Eil C, Gamblin GT, DeGrange DA, Balsan S (1984) Hereditary resistance to 1,25-dihydroxyvitamin D. Recent Prog Horm Res 40:589–620.
136. Marx SJ, Swart EG, Hamstra AJ, DeLuca HF (1980) Normal intrauterine development of the fetus of a woman receiving extraordinarily high dose of 1,25-dihydroxyvitamin D_3. J Clin Endocrinol Metab 51:1138–1142.
137. Marx SJ, Liberman UA, Eil C, DeGrange DA, Bliziotes MM (1985) Resistance to 1,25-dihydroxycholecalciferol in man and other species. In Norman AW (ed): Sixth International Conference on Vitamin D. DeGruyter, New York, pp 107–116.
138. Simpson RU, DeLuca HF (1980) Characterization of a receptor-like protein for 1,25-dihydroxyvitamin D_3 in rat skin. Proc Natl Sci USA 77:5822–5826.
139. Eil C, Liberman UA, Rosen JF, Marx SJ (1981) A cellular defect in hereditary vitamin D–dependent rickets type II: Defective nuclear uptake of the 1,25-dihydroxyvitamin D in cultures fibroblasts. N Engl J Med 304:1588–1591.
140. Griffin JE, Zerwekh JE (1983) Impaired stimulation of 25-hydroxyvitamin D-24-hydroxylase in fibroblasts from a patient with vitamin D–dependent rickets type II. J Clin Invest 72:1190–1199.
141. Liberman UA, Eil C, Marx SJ (1983) Resistance to 1,25-dihydroxyvitamin D. J Clin Invest 71:192–200.
142. Clemens TL, Adams JS, Horiuchi N, Gilchrest BA, Cho H, Tsuchiya Y, Matsuo N, Suda T, Holick MF (1983) Interaction of 1,25-dihydroxyvitamin D_3 with keratinocytes and fibroblasts from the skin of normal subjects and a subject with vitamin D–dependent rickets, type II: A model study of the mode of action of 1,25-dihydroxyvitamin D_3. J Clin Endocrinol Metab 56:824–830.
143. Liberman UA, Eil C, Holst P, Rosen JF, Marx SJ (1983) Hereditary resistance to 1,25-dihydroxyvitamin D: Defective function of receptors for 1,25-dihydroxyvitamin D in cells cultured from bone. J Clin Endocrinol Metab 57:958–962.
144. Pike JW, Dokoh S, Haussler MR, Liberman UA, Marx SJ, Eil C (1984) Vitamin D_3–resistant fibroblast have immunoassayable 1,25-dihydroxyvitamin D_3 receptors. Science 224:879–881.
145. Chen TL, Hirst MA, Cone CM, Hochberg Z, Tietze H-U, Feldman D (1984) 1,25-dihydroxyvitamin D–resistance, rickets and alopecia: Analysis of receptors and the bioresponse in the cultured fibroblasts from patients and parents. J Clin Endocrinol Metab 59:383–388.
146. Koren R, Ravid A, Liberman UA, Hochberg Z, Weisman Y, Novogrodsky A (1985) Defective binding and function of the 1,25-dihydroxyvitamin D_3 receptors in peripheral mononuclear cells of patients with end-organ resistance to 1,25-dihydroxyvitamin D. J Clin Invest 76:2012–2015.
147. Hollenberg SM, Weinberg C, Ong E, Cerelli G, Oro A, Lebo R, Thompson EB, Rosenfeld MG, Evans RM (1985) Primary structure of a functional human glucocorticoid receptor. Nature 318:635–641.
148. Weinberberg C, Hollenberg SM, Rosenfeld MG, Evans RM (1985) Domain structure of human glucocorticoid receptor and its relationship to the v-erb-A oncogene. Nature 318:670–672.
149. Walter P, Green S, Green G, Krust A, Bornert J-M, Staub A, Jensen E, Scrace G, Waterfield M, Chambon P (1985) Cloning of the human estrogen receptor cDNA. Proc Natl Acad Sci USA 82:7889–7893.

150. Green S, Walter P, Kumar V, Krust A, Bornert J-M, Argos P, Chambon P (1986) Human estrogen receptor cDNA: Sequence, expression, and homology to V-erb-A oncogene. Nature 320:134–139.
151. Greene GL, Gilna P, Waterfield M, Baker A, Hort Y, Shine J (1986) Sequence and expression of human estrogen receptor complementary DNA. Science 231:1150–1154.
152. Bishop JM (1986) Oncogenes as steroid receptors. Nature 321:112–113.
153. Mangelsdorf DJ, Pike JW (1986) Characterization of the 1,25-dihydroxyvitamin D receptor and their regulation by in vitro translation. In Program and Abstracts, 68th Annual Meeting of the Endocrine Society, p 88B (abstract).

6

Androgen Insensitivity Syndromes: Paradox of Phenotypic Feminization with Male Genotype and Normal Testicular Androgen Secretion

TERRY R. BROWN and CLAUDE J. MIGEON

The syndromes of androgen insensitivity have attracted a great deal of attention and interest, probably because of the incongruity of the presence of a male genotype in an individual who appears to have a female phenotype and who, in cases of complete hormonal resistance, has adopted a feminine gender identity. Furthermore, the normal physiologic synthesis and secretion of testosterone by the testes coupled to the pathophysiologic absence of biologic responsiveness to androgens presents an apparent paradox of endocrine function which can be explained only at the cellular level of target tissues.

Although the actual incidence of the syndrome is difficult to determine, it is considered a rare abnormality. Despite its low frequency and its consequently limited importance as an endocrinopathy, this syndrome has played a major role in our understanding of the physiology of male sex differentiation on the basis of pathophysiology rather than phenotype.

I. Historical Considerations

A review of the medical literature shows that the earliest reports of cases of androgen insensitivity might be those of a "woman with testes" in 1802 by Tonni et al. (1) and of a normally developed 23-year-old female who at autopsy was found to have inguinal testes and absence of uterus by Steglehner (2) in 1817.

The first report of a syndrome, of course, does not mean that it did not exist prior to that time. Indeed, Goodman (3), in his book on genetic disorders among Jewish people, suggests that the Talmud contains the report of a case of androgen insensitivity. Rabbi Gamliel, the elder, had to deal with a man whose wife was of the family of Dorkagi. In the Dorkagi family, "most women had neither menstrual flow nor vaginal bleeding." The Hebrew name of the family was explained on the basis of Dor meaning generation and Kagi meaning to cut, the Dorkagis being a family where most women were cut off from producing a new generation.

The main clinical features of the syndrome were recognized to include (a) a female habitus with breast development and female external genitalia, and (b) the presence of testes located either in the abdomen or at various levels of the inguinal canal. It is therefore logical that many early authors used the term "women with testes" to designate this disorder. Because a number of the patients also lacked sexual hair, Wilkins (4), in 1950, coined the term "hairless women with testes." In 1953, Morris (5) reviewed the literature and reported 80 cases of what he considered to be the syndrome, adding 2 personal observations. At that time he introduced the term "testicular feminization," which has since been widely used. The terms mentioned above are basically descriptive of the major clinical signs of the syndrome.

In the 1950 edition of his textbook *The Diagnosis and Treatment of Endocrine Disorders in Childhood and Adolescence,* Wilkins (4) described in detail the case of a 30-year-old patient with breasts who appeared entirely feminine but lacked pubic and axillary hair. She had married at 22 years of age, had normal libido and intercourse, but was amenhorreic and infertile. The report includes a detailed description of the pathology of the gonads. Most importantly, Wilkins recognized the etiology of the syndrome. After his patient had been gonadectomized, it was decided to test her response to the administration of androgen. Daily applications of testosterone ointment, followed by administration of methyltestosterone, up to 50 mg daily, produced no growth of sexual hair, enlargement of the clitoris, deepening of the voice, or other androgenic effects. These end organs were "superfeminine" in their resistance to androgen. Wilkins therefore concluded that the syndrome was related to an "insensitivity of end organs to androgens."

We believe that "androgen insensitivity" is the best term to designate the syndrome, since it is based on the etiology of the abnormality (6). This last term is also preferred by psychologists (7) who feel that the designation of testicular feminization tends to place emphasis on the male gonadal differentiation, whereas the important fact for the patients is the insensitivity of their end organs to androgen.

The presence of a normal XY karyotype (8), of a 17-ketosteroid excretion in a normal male range, and of a high level of bioassayable gonadotropins (4) had been established relatively early. In 1964, Southren et al. (9) demonstrated that the concentrations of testosterone in the plasma of 4 patients with androgen insensitivity were in the normal male range. This observation, which was confirmed by numerous other investigators, indicated not only that the patients were genetic males with male gonads but also that they had a hormonal production of the male type. It further supported the earlier evidence of an end-organ unresponsiveness.

In the early 1970s, our group became interested in the use of human skin fibroblasts as a model for the study of the mode of action of steroid hormones at the cellular level. We felt that it was a valid approach, since skin is a target organ for the action of androgens. In addition, multiplying fibroblasts closely simulate an in vivo situation. Eventually Keenan et al. (10) were able to demonstrate the presence of a dihydrotestosterone (DHT) receptor in skin fibroblasts. Its characteristics were studied, and an assay for its measurement was developed. In 1974, our group used this assay to demonstrate that cytoplasmic androgen receptor binding was absent in several subjects with complete androgen insensitivity (11). Two years later, evidence for genetic heterogeneity in the syndrome was reported. Amrhein et al. (12) observed quantitatively normal cytoplasmic and nuclear DHT binding in fibroblasts from patients clinically indistinguishable from those of Keenan et al. (10) and therefore concluded that there were at least two distinct genetic variants within the syndrome of complete androgen insensitivity. In the first variant, deficient cytoplasmic androgen binding results from a mutation of the X-linked gene specifying for the DHT receptor. The androgen insensitivity in the second variant must be due to a defect in steroid hormone action after the binding of DHT to the cytoplasmic receptor and translocation of the steroid receptor complex to the nucleus (13).

The familial incidence of the syndrome was recognized very early. The study of various pedigrees suggested that the pattern of transmission was compatible with either X-linked recessive or sex-limited autosomal-dominant inheritance (14). Unfortunately, numerous linkage studies were not helpful in discriminating among these alternatives. By studying the androgen receptor phenotype of an obligate heterozygote, Meyer et al. (15) were able to demonstrate that the transmission of the receptor deficiency was X-linked. Furthermore, it was shown that this X-linked locus was subject to the inactivation that characterizes other X-linked loci (15). It is of interest that although many genes have been assigned to the X chromosome, the DHT receptor locus is the only one known to be directly involved in determining sex differentiation.

A related condition, descriptively termed "pseudovaginal perineoscrotal hypospadias" was described by Nowakowski and Lenz in 1961 (16). These patients resemble those with other forms of male pseudohermaphroditism; an XY karyotype, normally differentiated testes, male internal ducts, and ambiguous external genitalia. However, at puberty, striking but selective signs of masculinization appear. In 1974, Walsh et al. (17) and Imperato-McGinley et al. (18) reported a defect in the conversion of testosterone to its 5α-reduced metabolite, dihydrotestosterone, in patients with this syndrome. The correlation of the pathophysiologic condition with its biochemical abnormality has resulted in the current designation of this clinical entity as 5α-reductase deficiency.

Most of this chapter will be devoted to the description of the pathophysiology of these conditions of androgen insensitivity and the biochem-

TABLE 6.1. Biochemical heterogeneity of androgen target organ defects.

A. 5α-Reductase deficiency
   1. Severe deficiency of enzyme activity
   2. Unstable enzyme with intermediate activity but decreased affinity for the cofactor, NADPH
   3. Unstable enzyme with decreased affinity for substrate, testosterone, and cofactor NADPH
B. Complete androgen insensitivity
   1. Receptor absent with no or very low steroid binding (R−)
   2. Receptor present with quantitatively normal steroid binding (R+) but qualitatively abnormal receptor structure (thermolability, instability of steroid-receptor complex, no up-regulation)
   3. Receptor present with presumed postreceptor defect
C. Partial androgen insensitivity
   1. Partial deficiency of receptor (R±)
   2. Receptor present with quantitatively normal steroid binding (R+) but qualitatively abnormal receptor
   3. Receptor present with decreased nuclear translocation
   4. Receptor present with presumed postreceptor defect
D. Infertile men
   1. Partial deficiency of receptor (R±)

ical abnormalities that are associated with them. Thus, the phrase androgen insensitivity refers to a heterogeneous group of biochemical defects at the level of androgen target organs (Table 6.1). However, to fully appreciate the pathophysiology of abnormalities of sex differentiation, a complete understanding of normal androgen physiology is essential.

II. Normal Androgen Physiology

A. Overview

Androgens exert their complex actions as determinants of male sex differentiation and development very early during fetal life (19–22). The carefully orchestrated chronology of androgen secretion by the fetal testes during embryonic and neonatal life is responsible for the initial growth and differentiation of many organs of the male reproductive tract, such as the Wolffian ducts, the urogenital sinus, and external genitalia primordia. Androgens also imprint regions of the central nervous system that control the masculine modes of gonadotropin secretion. During puberty, androgens promote the appearance of secondary male sex characteristics, including growth of the external genitalia, development of the prostate and seminal vesicles, male distribution of body hair, and increase in total muscle mass. These hormones are essential for maintenance of spermatogenesis, and they exert a feedback control on the output of gonadotropins by the hypothalamic-pituitary axis. Although the actions of androgens are not confined to the male reproductive tract, it is within these so-called target tissues of the body that the molecular mechanisms of androgen action have been characterized.

Testosterone is the predominant male sex hormone synthesized and secreted by the testes (19). Adequate blood levels of circulating testosterone are necessary for normal androgen biologic activity; however, testosterone itself is not sufficient to evoke the full complement of normal androgenic responses. At least two key factors at the cellular level enter into the scheme of androgen action and determine the target tissue specificity of androgen responses. These factors are the enzymatic conversion of testosterone to an active metabolite, 5α-dihydrotestosterone (DHT) (23,24), and the specific intracellular receptor proteins that form ligand-receptor complexes with the active steroid and direct the androgen to its site of action within the nucleus (25,26). Several human pathologic conditions, by virtue of their aberrant biochemical features, mutually reinforce and illustrate the processes of male sex differentiation and development and the mechanism for androgenic steroid action (27,28). It is within this framework that the discussion in this chapter focuses on defects in androgen action at the cellular level that lead to male pseudohermaphroditism.

B. Normal Male Sex Differentiation in the Fetus

Multiple factors contribute to the complex and integrated sequence of events that result in normal male sex differentiation (Table 6.2). It is usually agreed that four major steps can be distinguished: genetic sex, formation of undifferentiated sexual structures, gonadal sex, and development of phenotypic sex (19,20).

1. GENETIC SEX

Fertilization occurs with the union of two haploid cells, the ovum which contains 23 chromosomes including one X chromosome, and a 23,X or a 23,Y sperm. The sperm is attracted by follicular fluid which accompanies the ovum. Although several sperm can reach the perivitelline space of a single ovum, only one of them fuses with the ovum. The second polar body, including 23 chromosomes and a small amount of cytoplasm, is then extruded into the perivitelline space, close to the first polar body. Thereafter, the chromatin of the sperm head enlarges, forming a male pronucleus while the chromatin of the ovum forms the female pronucleus. Eventually, the two pronuclei come in close contact and fuse, resulting in a fertilized egg whose chromatin complement is either 46,XX or 46,XY. This represents the first step in sexual differentiation.

TABLE 6.2. Stages in sex differentiation.

Differentiating events	Determinants	Genetic influences	Fetal age (weeks)
Genetic sex	Fusion of ovum and acrosome of sperm	X and (X or Y) fertilization	
Undifferentiated structures			
Wolffian ducts	Structural genes	(?)	4
Genital ridges	Structural genes	X(?); others(?)	5
Müllerian ducts	Structural genes	(?)	4–6
Undifferentiated external genitalia	Structural genes	(?)	6
Gonadal sex			
Testicular differentiation	Testicular determinant (TDF)	Y	6–9
	H-Y antigen	Y	Tubules: 7
	Others	Autosomes	Leydig cells: 8
Phenotypic sex			
Müllerian regression	MIF	X(?)	8–10
Wolffian development	Androgen receptor, testosterone	X Autosomes	10–11
Masculinization of external genitalia	Androgen receptor, dihydrotestosterone	X Autosomes	11–14
Penile growth, testicular descent	Androgen receptor, dihydrotestosterone	X Autosomes	23–28

2. UNDIFFERENTIATED SEXUAL STRUCTURES

Both sexes progress through the earliest stages of development in the same manner. An early event (3-week-old embryo) is the formation of primitive germ cells that are located on the walls of the yolk sac near its union with the allantois (29). One week later, the genital ridges (Fig. 6.1) appear in the posterior wall of the coelomic cavity, medial to the anlage for the urinary system (30). By the sixth week of gestation, the germ cells migrate along the allantois through the dorsal mesentery and reach the genital ridges.

At the beginning of the sixth week of fetal life, both male and female embryos have two sets of internal ducts, the Müllerian and Wolffian ducts (30). The Wolffian ducts originate from the excretory tubular system of the kidney. The Müllerian ducts develop as an invagination of the coelomic epithelium near the genital ridge. At their cranial end, the ducts form a funnel-shaped orifice which opens into the coelomic cavity (31).

The external genitalia are still undifferentiated by the fourth week of gestation and consist of two cloacal folds that surround the cloacal

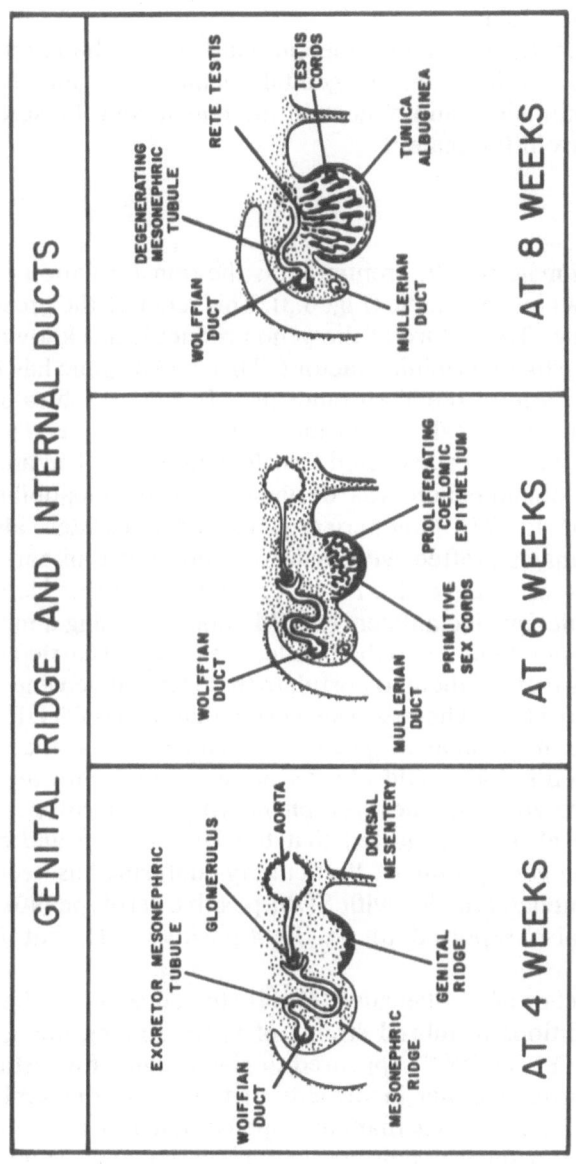

FIGURE 6.1. Formation of the genital ridges and the internal ducts (Müllerian and Wolffian).

membrane laterally, and an anterior gential tubercle (Fig. 6.2). At approximately 6 weeks, the posterior part of the cloacal folds separates to form the anal folds. At that time, the external genitalia include the urogenital membrane which is limited by the genital tubercle and the genital folds. In addition, the genital swellings have become visible on each side of the genital folds.

Later in development, the genital tubercle will become the clitoris of the female or the penis of the male. The genital swellings and genital folds will become the labia majora and minora in the female and the scrotum and the penile urethra in the male.

3. GONADAL SEX

An important step in male sex differentiation is the transformation of the undifferentiated gonad into a testicle (Fig. 6.3). This requires the presence of one Y chromosome. The nature of the gene product is not known but has been called the testis-determining factor (TDF), and its gene has been localized to the short arm of the Y chromosome. In 1955, Eichwald and Silmser (32) showed that females from inbred strains of mice rejected male skin grafts but that males accepted female skin grafts. The antigen responsible for this phenomenon was designated histocompatibility Y antigen (H-Y antigen). In 1971, Goldberg et al. (33) demonstrated, also in inbred mice, that females grafted with male skin produced an antibody that destroyed male sperm, and called the responsible factor the "serologically determined male-specific antigen" (SDM antigen). Using a method to detect SDM, Wachtel, Ohno, and their colleagues (34–36) carried out a series of experiments that led them to postulate that H-Y antigen and TDF were the same gene product. Their data suggested that the H-Y/TDF gene was highly conserved in mammalian species (37). Finally, it was theorized that testicular differentiation would always be associated with an H-Y antigen–positive serotype regardless of phenotype or karyotype (38). However, Silvers et al. (39) suggested that the H-Y antigen and SDM antigen were different gene products. Particularly confusing has been the report of some XO human females with SDM-positive serotype (40). XO female mice were also reported to be SDM-positive (41), but H-Y-negative (42).

More recently, Eicher and colleagues (43,44) studied several inherited "sex reversal" conditions in inbred strains of mice. Among them, two Y-linked conditions (Y^{pos} and Y^{orb}) appeared to give a signal for testicular differentiation which reads either incorrectly or too late during embryogenesis, resulting in XY progenies that developed as females or hermaphrodites. Two other autosomal conditions, one recessive (Tda-1) and the other dominant (mouse chromosome 17), resulted also in XY offspring that developed as females or hermaphrodites. The implications of these studies are that several genes, including some on the autosomes, are

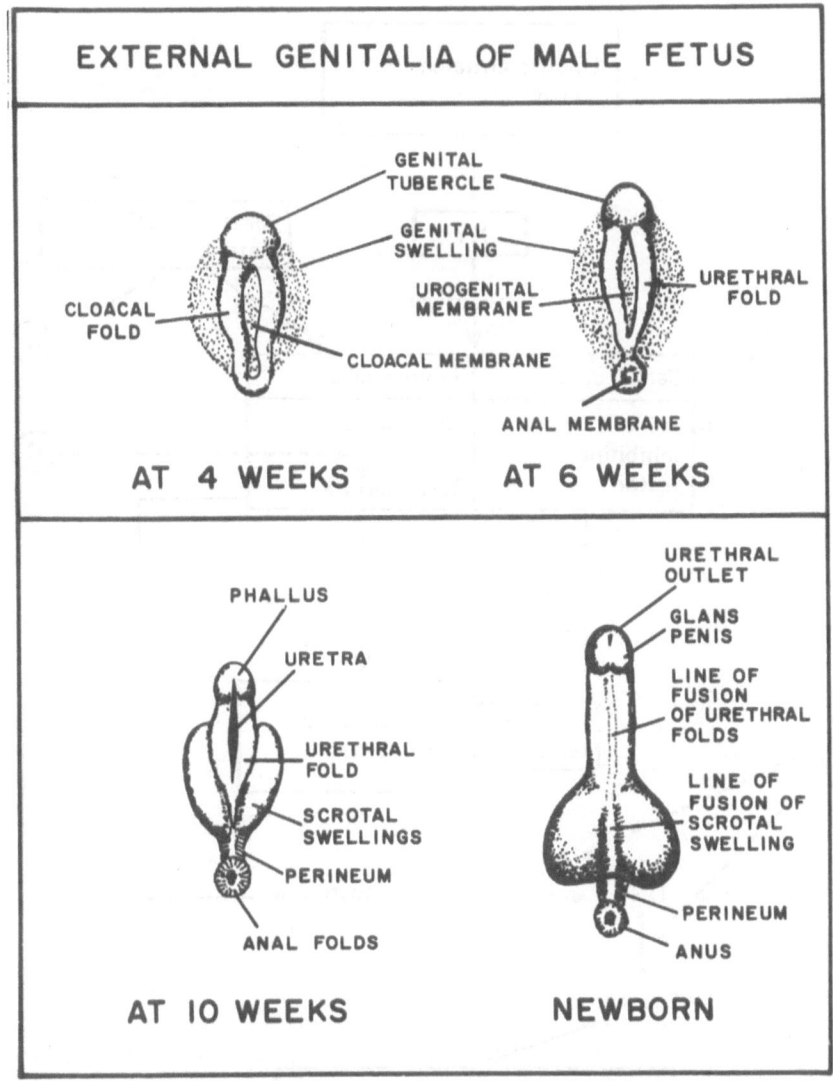

FIGURE 6.2. Normal masculinization of the external genitalia in a male fetus.

involved in testicular differentiation. In the human, the syndrome of 46,XY pure gonadal dysgenesis is X-linked (45), suggesting that there is a gene on the X chromosome needed for the adequate development or maintenance of the testes.

Although the complete mechanism of testicular differentiation is still unknown, the following aspects are now firmly established: the TDF and H-Y antigen genes are located on the Y-chromosome but are most likely two different genes; there is strong evidence that genes other than TDF

166   T. R. Brown and C. J. Migeon

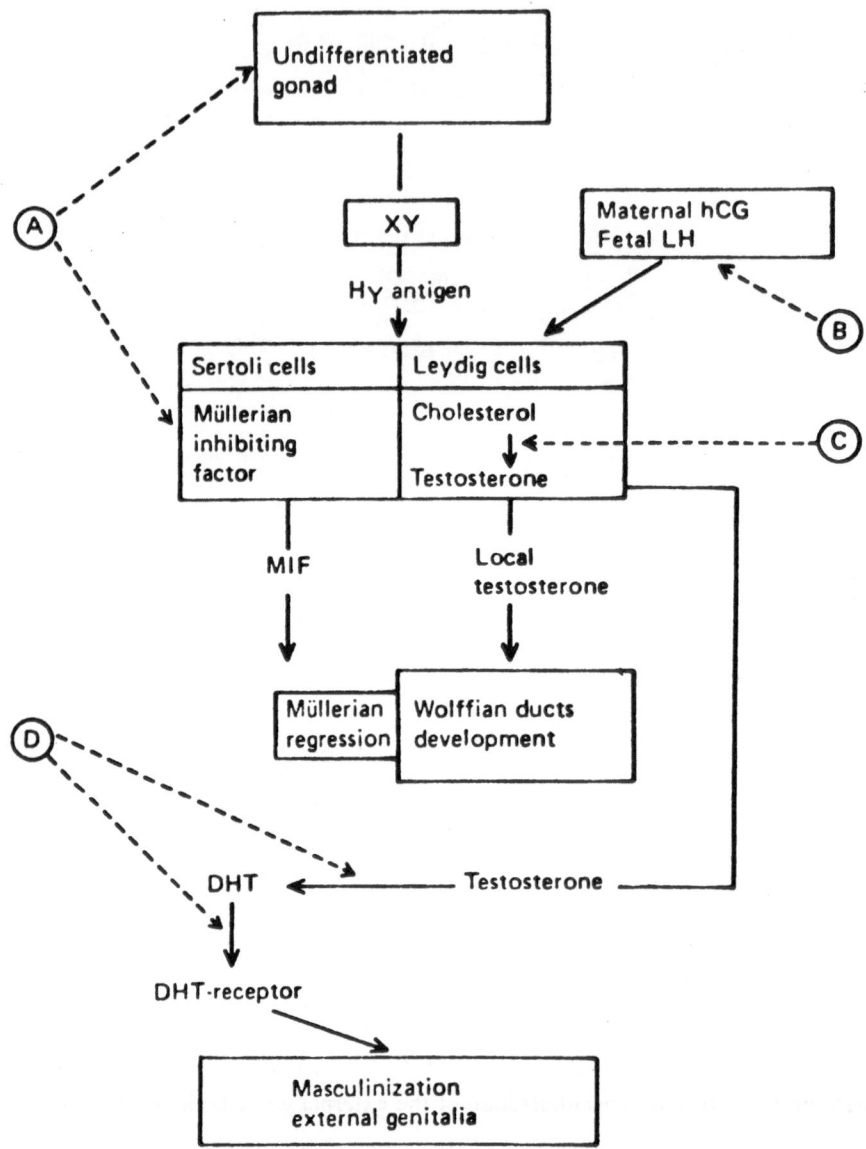

FIGURE 6.3. Abnormalities of male sex differentiation that will result in male pseudohermaphroditism. (A) Gonadal dysgenesis, including pure and partial gonadal dysgenesis, as well as isolated absence of Leydig cells or absence of Müllerian inhibiting factor. (B) Abnormalities of gonadotropins resulting in decreased Leydig cell function. (C) Partial or complete deficiency of any one of the enzymes necessary to transform cholesterol into testosterone. (D) Abnormalities of the target cells, including deficiency of 5α-reductase or abnormality of the androgen receptor.

influence testicular differentiation; and the methods of measurement of H-Y antigen (skin graft) and SDM antigen determine two different gene products.

Whatever the specific influences of the Y chromosome and other chromosomes, the primitive sex cords of the gonads in the male surround the germ cells as they reach the genital ridge (46). The cords then proliferate into a network and penetrate deeply into the medullary part of the gonad. (47). At the same time, the cortical part of the gonad disappears except for a thick layer of connective tissue, the tunica albuginea, which covers the testes (21). By the seventh week of gestation, the sex cords lose their connections to the coelomic surface, and the epithelial portion of the cords develops into the peripheral membrane and the adjacent Sertoli cells (48). The testicular mesenchyme in the space between the seminiferous tubules gives rise to the Leydig cells, which will be responsible for steroid production.

4. Phenotypic Sex

Two events, the differentiation of the internal ducts and the differentiation of the external genitalia, determine the phenotypic sex. Our understanding of the development of the male phenotype is based on the very important experiments of Jost et al. (19,20) in fetal rabbits. They demonstrated that the fetal testes produce testosterone, which is necessary for the development of the Wolffian ducts, and a Müllerian-inhibiting factor (MIF), which is necessary for the regression of the Müllerian ducts (Fig. 6.3). Both androgen and MIF affect the ducts locally. Testicular androgens will also masculinize the external genitalia.

a. Production of MIF

MIF acts locally to cause the regression of the Müllerian ducts (49). The regression begins at 62–65 days of fetal life and initially causes an obliteration of the opening of the duct which would have been the abdominal ostium of the Fallopian tube in the female fetus. There is then a progressive narrowing of the lumen of the ducts until they are completely obliterated by fibrous tissue. Sertoli cells within the testes are the source of MIF (49,50). A protein with MIF activity and molecular weight of 200,000 has been isolated from the incubation medium of cultured Sertoli cells (49). There appears to be a crucial time when the Müllerian ducts are most sensitive to MIF, and studies of human testicular explants from abortuses of various ages have shown that MIF activity is lost during the perinatal period.

b. Testosterone Secretion

Leydig cells in the interstitium of the testes synthesize and secrete testosterone. Studies of testicular tissue incubated with labeled precursors have shown that the onset of androgen production correlates with the appearance of the Leydig cells at about 8–10 weeks. Furthermore, the concentration of testosterone in fetal serum parallels the proliferation of Leydig cells (51–53). The levels of testosterone reach values nearly similar to those of adult males at 14–18 weeks of gestation. Then, the production of testosterone decreases somewhat between 20 and 24 weeks, as the number of Leydig cells appears to diminish. Although the plasma testosterone levels are low at birth, they rise significantly in male newborns during the first 3 months of life (54,55), at a time when the Leydig cells are virtually undetectable by histologic examination.

As in the adult Leydig cells, testosterone is synthesized in the fetal cells from cholesterol. Fetal Leydig cells are capable of synthesizing cholesterol from acetate (56), but they can also internalize cholesterol carried by the low-density lipoproteins via specific receptors on the cell membrane. Cholesterol is then rapidly esterified and stored in droplets within the cytosol of the cell. When the Leydig cell is activated by gonadotropins, an esterase removes the fatty-acid side chain, and cholesterol is then available for testosterone biosynthesis. Three enzymatic steps—20-hydroxylase, 22-hydroxylase, and 20,22-desmolase, commonly referred to as side-chain cleavage enzyme (SCC)—transform cholesterol into pregnenolone, the precursor of all steroid hormones. Three additional enzyme systems—3β-hydroxysteroid dehydrogenase, 17α-hydroxylase, and 17,20-desmolase—are needed for the conversion of pregnenolone to androstenedione. This last steroid has 19 carbons and is considered to be an androgen. However, to acquire biologic activity, androstenedione must be metabolized to testosterone by a 17-ketosteroid reductase.

c. Gonadotropin Regulation

Fetal Leydig cells, like adult Leydig cells, must be stimulated by a tropic hormone to secrete testosterone. Initially, androgen production is induced by human chorionic gonadotropin (hCG). hCG can be detected in the cells of the blastula 10 days after fertilization and, therefore, prior to implantation (57). hCG is formed by the placenta, from α and β subunits. Free α subunits are present in the placenta, whereas the β subunits are incorporated entirely into hCG (58), the ratio of α subunits to hCG being about 5. In fetal serum, hCG levels parallel those of the mother but are about 10 times lower (59). The peak concentration of hCG in fetal blood occurs at 10 weeks' gestation. The levels decline after 20 weeks of gestation (58).

Concomitant with the decline in hCG, there is an increase in luteinizing hormone (LH) secreted by the fetal pituitary gland. LH is first detected in fetal pituitaries at about 10 weeks and reaches a maximum level at 20–24 weeks (60). The secretion of LH is itself under the control of a small peptide hormone of hypothalamic origin, the LH-releasing hormone (LHRH). The pituitaries of female fetuses contain more LH activity than those of the males; the lower levels in males are attributed to the negative feedback regulation resulting from the high levels of circulating testosterone. LH, like hCG, is composed of an α and β subunit, the α subunits of the two peptides being nearly identical in their amino acid composition, whereas the β subunits confer the biologic and immunologic specificity of these hormones.

Follicle-stimulating hormone (FSH) is the other gonadotropin produced by the anterior pituitary gland. It regulates the activity of Sertoli cells and may play a role in the regulation of MIF production. Its concentration in the fetal pituitary parallels that of LH. It is also under the regulation of LHRH from the hypothalamus (61).

d. Androgen Target Cells

Cells that are stimulated by androgens to carry out functions that do not occur otherwise are called target cells (62). Many types of cells responsive to androgens have been identified. Some of them carry out functions normally associated with male differentiation (external genitalia, internal ducts) whereas others do not (bone marrow, liver, kidney). It must be noted that target cells are present in both sexes, the potential response to androgen being limited only by the level of secretion of testosterone (63). How target cells respond to androgen is not thoroughly understood, but a great deal of information has been gained in the recent past (see Section IV).

III. Classification of Male Pseudohermaphroditism

Male pseudohermaphroditism is defined as incomplete masculinization of the external genitalia in an individual with a normal 46,XY karyotype (30,64–66). It is a heterogeneous condition in which the gonads are exclusively testes but the genital ducts or external genitalia are incompletely masculinized and exhibit, to varying degrees, the phenotypic characteristics of a female. The clinical spectrum varies from individuals in whom the configuration of the external genitalia is completely feminized to milder forms in phenotypic male subjects characterized by hypospadias, cryptorchidism, and minimal ambiguity of the external genitalia. Indeed, some patients present a seemingly isolated "adult-onset" form of infertility with an otherwise normal male phenotype.

In general, the defective male development in these patients can be ascribed to a specific failure of the fetal testes to overcome the inherent tendency to feminize the somatic sex structures. This failure may stem either from a secretory failure of the testes themselves during the critical period of sex differentiation or from a failure of target tissues to respond normally to androgen stimulation. With the elucidation of etiologic mechanisms (67), systems of nomenclature based on phenotype have become less important. A detailed classification of male pseudohermaph-roditism is presented in Table 6.3.

Complete dysgenesis of the fetal gonads results in the absence of Sertoli cells which produce MIF and of Leydig cells which produce testosterone. These subjects have female external genitalia, normal Müllerian ducts, and streak gonads (68). In cases of partial dysgenesis, the phenotype depends on the degree of testicular function. Whereas some patients have a deficient production of both MIF and testosterone, others have either an isolated deficiency of MIF (69) or an agenesis of the Leydig cells (70).

As discussed earlier, testosterone plays a major role in the development of the male internal ducts and in the masculinization of the external genitalia. Its biosynthesis from cholesterol requires a series of enzymes, any one of which can be deficient. These enzymatic deficiencies form another group of causes for male pseudohermaphroditism (71). Since the secretion of testosterone by the Leydig cells is controlled by the gonadotropins, a deficiency of gonadotropin secretion will also result in male pseudohermaphroditism. Gonadotropin deficiency can be due either to a structural abnormality of the LH molecule (72) or a deficient secretion by the fetal pituitary. It can also be due to an abnormality of the LH receptor on the surface of the Leydig cell (73) or a defect in the timing of LH secretion (74).

The effect of androgens on target cells requires the presence of a specific protein receptor (25,26,75). As any protein, these receptors are

TABLE 6.3. Classification of male pseudohermaphroditism.

Dysgenesis of fetal gonads associated with normal karyotype
  XY Gonadal dysgenesis
  Congenital anorchia
  Leydig cell agenesis
  Deficiency of MIF
Abnormalities of gonadotropins
  Hypopituitarism
  Kallman's syndrome
  Abnormal endogenous LH
  Deficiency of gonadotropin receptor
  Probable delayed development of the HCG/LH Receptor
Deficiencies of enzymes needed for the biosynthesis of testosterone
  Defects involving both the adrenal glands and the testes
    Deficiency of 20-hydroxylase, 22-hydroxylase or 20,22-desmolase
    Deficiency of 3-hydroxysteroid dehydrogenase
    Deficiency of 17-hydroxylase
  Deficiencies of enzymes of testicular origin
    Deficiency of 17,20-desmolase
    Deficiency of 17-ketosteroid reductase
Abnormalities of androgen target cells
  Deficiency of 5α-reductase
  Syndromes of androgen insensitivity

Reprinted with permission from reference 71.

encoded by a gene. Absence of this gene or presence of an abnormal gene will result in lack of androgen receptor or a structurally abnormal receptor in target cells. Errors in transcription of the gene or in translation of the messenger RNA will also result in structurally abnormal receptors. Since the androgen receptor has a specific acceptor within the chromatin, structural abnormalities of the receptor can involve either the androgen-binding site or the acceptor-binding site. Male pseudohermaphroditism due to abnormal androgen receptors will be discussed in greater detail in Section VI of this chapter.

Finally, in some target cells testosterone must be metabolized to DHT in order to exert the appropriate biologic activity (23,24). This transformation is carried out by the 5α-reductase enzyme. A deficiency of this enzyme during fetal life will result in incomplete androgen effect and male pseudohermaphroditism (18).

Except for dysgenetic testicular male pseudohermaphroditism and the persistent Müllerian duct syndrome, male pseudohermaphroditism is characterized by the absence of Müllerian duct derivatives. Many forms of male pseudohermaphroditism, except for some variants of dysgenetic gonadal formation, are familial and characterized by genetic heterogeneity. No doubt many additional subtypes will be defined and characterized in the future by refined biochemical techniques.

Furthermore, the ability of the testes to virilize the patient at adolescence is frequently a recapitulation of their capacity to masculinize the external genitalia in utero. The greater the development of the phallus, the greater the likelihood that secondary sex characteristics will emerge. Individuals with ambiguous genitalia may remain eunuchoid, exhibit mild virilism, or develop breast enlargement and other feminine secondary sex characteristics. Those with an external female phenotype will usually either feminize or remain sexually infantile. However, the development of male sex characteristics at adolescence is encountered in incomplete androgen insensitivity and in patients with 5α-reductase deficiency.

IV. Molecular Mechanisms of Androgen Action in Target Cells

A. Androgen Receptors (Fig. 6.4)

The first step is the transfer of testosterone from extracellular fluid into the cytosol of the cell (25,26,75). Testosterone circulates in blood mainly bound to testosterone-estradiol binding globulin (TeBG). Only 0.5% of the total circulating testosterone is available for transfer out of the blood vessels and into the extracellular fluid. The passage of the steroid from the

Androgen action in a target cell

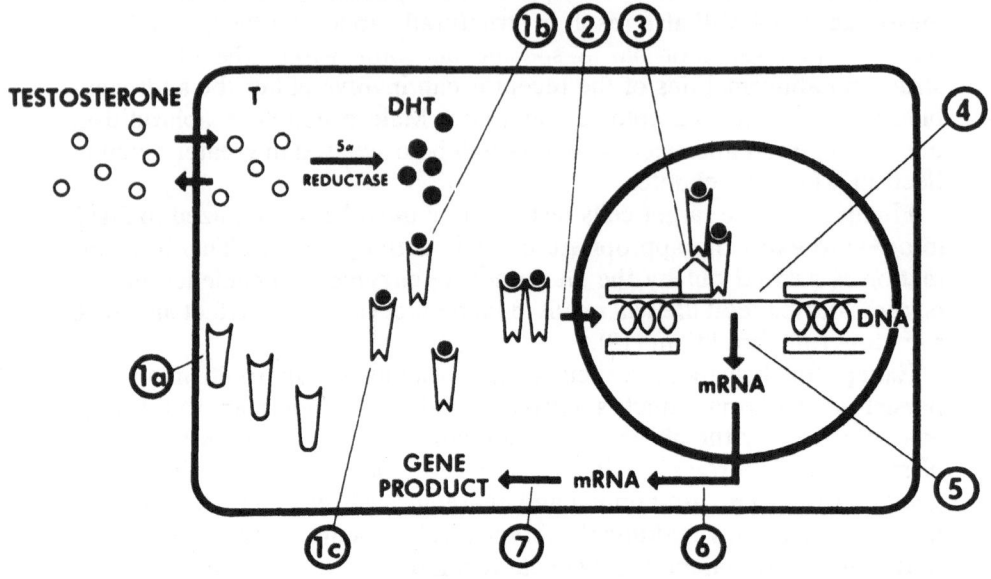

FIGURE 6.4. This conceptual diagram of normal target cell activity indicates processes potentially affected by mutations that might cause androgen insensitivity. First are mutations affecting the conversion of testosterone to dihydrotestosterone (DHT) and the cytosolic receptor: absent receptor (1a); abnormal binding site for the steroid (1b); and abnormal receptor site or recognition of the nuclear acceptor (1c). Site 2 has the potential for a defect in the nuclear translocation of the steroid-receptor complex. Site 3, the nuclear acceptor site, might be abnormal, as might a regulatory molecule (4), mRNA transcription (5), mRNA processing (6), or translation of androgen-dependent proteins (7).

exterior to the interior of the cell is thought to be passive as steroids are small nonpolar molecules that can readily go through the cell membrane. Upon entry into the cell, testosterone can be converted to DHT or estradiol. For many years, it has been agreed that testosterone and DHT bind to a single, high-affinity cytoplasmic receptor protein specific for androgens. The androgen receptor has a greater binding affinity for DHT than for testosterone (76), is encoded by a gene located near the centromere on the X chromosome (77), and is present in androgen-sensitive target tissues of males and females (78). The hormone-receptor complex undergoes a transformation step, following which the complex is translocated to the nucleus. Transformation consists of a conformational change in the receptor steroid complex from the larger cytosolic form ("8S," or larger sedimentation coefficient) to the smaller ("4S") nuclear complex. However, several investigators have recently suggested that steroid receptors are localized exclusively in the nucleus of intact cells, whether or not they are occupied by steroid (79,80). On the basis of these studies, it was proposed that the so-called cytosolic receptors were biochemical artifacts due to the physical disruption of the cell and the subsequent redistribution of receptors, particularly those unoccupied receptors and steroid-receptor complexes that have low affinity for acceptor sites within the nucleus. Regardless of the steps leading to the intranuclear formation of the receptor-steroid complex, it is thought that this complex binds to specific, high-affinity acceptor sites of the chromatin, where this interaction leads to the transcription of specific messenger RNA(s). These RNA transcripts are processed, and translation of the steroid-specific mRNA(s) on the polyribosomes results in the synthesis of gene products controlled by androgens. These products in turn alter the function of target cells. In some but not all androgen-responsive tissues, hormone action results in enhanced DNA replication and cell division. A defect in any of the sequential steps in the action of androgens in a male fetus could cause impaired masculinization of the internal and external genitalia.

Factors that influence the response to androgens in specific tissues include the intracellular concentration of androgen available to the receptor and the relative binding affinity of these steroids for the receptor, the cellular content of androgen receptors, and the structure of chromatin as determined by cellular differentiation, which will influence the quality and/or quantity of acceptor sites.

B. 5α-Reductase Activity

Androgen target cells, both postnatally and during embryonic life, contain a very active steroid 5α-reductase. This single-step reaction rapidly transforms testosterone to DHT. As a consequence, more than 90% of the androgen bound to receptors is DHT.

During fetal life, 5α-reductase activity appears after 12 weeks (24). For this reason, it is testosterone itself, rather than DHT, that promotes the growth of the Wolffian ducts. However, DHT formation is necessary for the masculinization of the external genitalia. Because the biologic activity of DHT is greater than that of testosterone, it has been postulated that the external genitalia are somewhat insensitive to androgens, hence the need for the potentiating effect of 5α-reductase. Studies of patients with 5α-reductase deficiency have also suggested that DHT is necessary for some aspects of maturation at puberty.

C. Cultured Human Genital Skin Fibroblasts as a Model for the Study of Androgen Action

1. Studies of 5α-Reductase Activity

In recent studies from our laboratory with tissues from human fetuses 8–22 weeks old, low but detectable 5α-reductase activity was observed at 8 weeks' gestation in nongenital skin fibroblasts and was present in fibroblasts propagated from a variety of tissues from older fetuses, including testis, kidney, lung, and clitoris (81). Earlier, Wilson (82) had observed that the rate of DHT formation from testosterone was greater in fibroblast monolayers grown from foreskin and scrotal skin than in fibroblasts from nongenital skin sites. The finding of this specialized metabolic function in skin fibroblasts of different anatomic origin was also reported by Pinsky et al. (83), Mulay et al. (84), and by our laboratory (85) (Fig. 6.5). However, the decrease in DHT formation with age that Wilson and Walker (86) had observed in human foreskin homogenates was not apparent with fibroblasts grown from foreskins of subjects at various ages (82).

Recent studies with cultured human skin fibroblasts and pubic skin homogenates have suggested that 5α-reductase activity is androgen-dependent (87–90). Enzyme activity in male foreskin fibroblasts was higher than that in female pubic skin fibroblasts, with intermediate levels observed for pubic skin fibroblasts from hirsute women (87). The difference between males and females in this and other studies might be related to the site of the skin specimen from which the fibroblasts originated, since genital skin fibroblasts appear to have higher 5α-reductase activity than those from pubic skin (82–85,91). However, cloned cell lines of fibroblasts derived from the same skin region display different patterns of steroid metabolism (94), and individual clones may vary markedly in enzyme activity and display a shift from high to low activity upon subcloning (95). In two recent reports (87,96), 5α-reductase activities in skin fibroblast strains increased with serial subculture, perhaps because of selective outgrowth of particular cell types or alterations of the original characteristics of the cells. We (92) and Pinsky et al. (93) have not observed significant differences in 5α-reductase activities of pubic skin fibroblasts from men and women, nor have we observed any significant pattern or change in 5α-reductase activity in four separate strains of foreskin fibroblasts examined in each of 22 consecutive cell passages (T. Brown, unpublished data). Nevertheless, other investigators have recently demonstrated increased 5α-reductase activity in skin from women with hirsutism (97).

Since human genital skin fibroblasts contain an active 5α-reductase (Fig. 6.6), the potential for various compounds to inhibit this enzyme activity can be readily tested. Two such compounds are the secosteroids

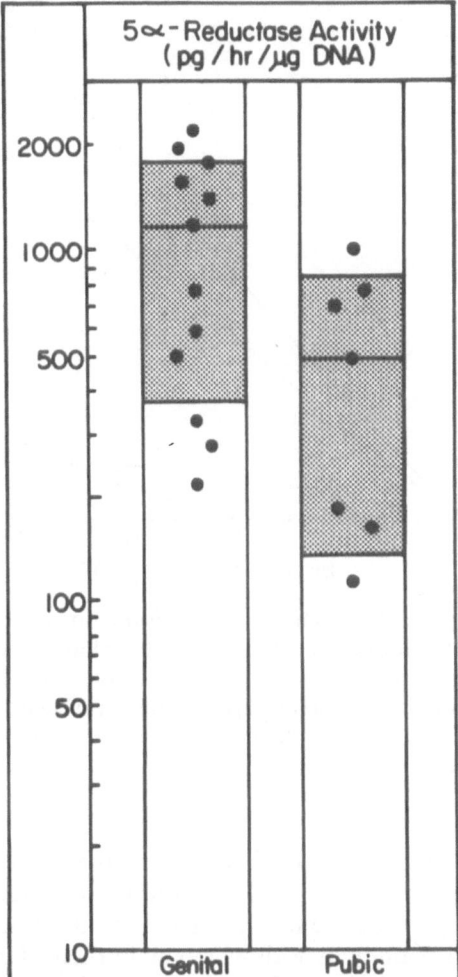

FIGURE 6.5. 5α-Reductase activity in cultured skin fibroblasts propagated from genital or pubic skin specimens of normal subjects. Shaded areas represent the mean ± SD for genital (1063 ± 697) and pubic skin (487 ± 353).

5,10-secoestra-4,5-diene-3,10,17-trione (I) and 5,10-seco-19-norpregna-4,5-diene-3,10,20-trione (II), which we found to be noncompetitive inhibitors of 5α-reductase activity in foreskin fibroblasts (98). Compound I (K_i = 1.6 × 10^{-6} M) was less potent than compound II (K_i = 0.53 × 10^{-6} M) in fibroblast cultures. Neither compound was an effective inhibitor of DHT receptor binding, since their relative affinities were about 10 times less than that of DHT. However, the compound 17β-N,N-diethylcarbamoyl-4-methyl-4-aza-5α-androstan-3-one is also a potent inhibitor of 5α-reductase activity (K_i = 15 nM), and it caused a significant alteration of the

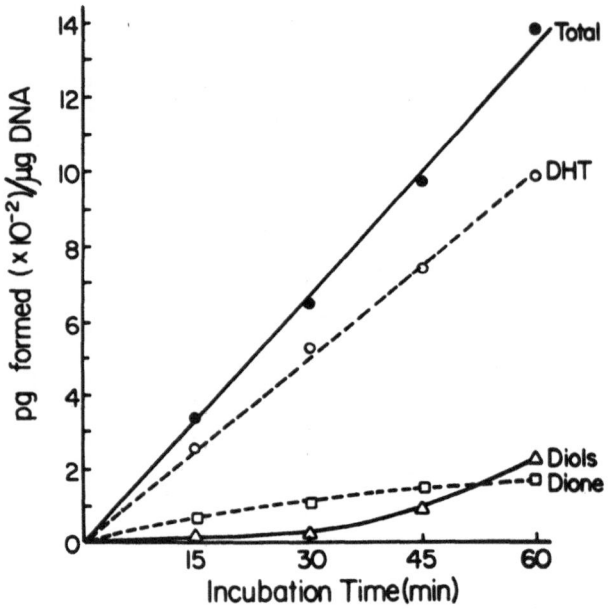

FIGURE 6.6. An example of the time course for formation of 5α-reduced products from testosterone by cultured human newborn foreskin fibroblasts incubated at 37°C. The total (●) 5α-reduced products represent the sum of 5α-dihydrotestosterone (DHT, ○), 5α-androstan-3α and 3β diols (diols, △) and 5α-androstanedione (dione, □). Each point represents the means of four independent determinations on the same cell strain.

nuclear testosterone:DHT ratio in cells incubated in the presence of testosterone and the inhibitor (99). Therefore, the· human genital skin fibroblast system can help identify the mechanism of action of potential antiandrogenic compounds as well as evaluate their efficacy.

The enzyme from genital skin fibroblasts exhibits maximal activity in a narrow pH range, with an optimum of 5.5 in cell-free assays (100). The activity is NADPH-dependent and is localized to the 100,000g particulate (microsomal) fraction of cell sonicates. Nongenital skin fibroblasts do not exhibit a sharp peak of maximal activity at pH 5.5 but rather exhibit lower total activity spread over a wide (pH 6–9) pH range (100). Although testosterone is presumed to be the principal physiologic substrate of the enzyme, 5α-reductase activity is greater with the substrate 20α-hydroxy-pregn-4-en-3-one (101). According to Moore and Wilson (101), the metabolism of this latter pregnene derivative suggests the existence of two separate enzyme activities in genital skin fibroblasts: a normal one with an optimum at pH 5.5, which may be deficient owing to a genetic abnormality in the 5α-reductase activity, and the other with an optimum at pH between 7 and 9, that is lower but similar in normal and mutant cells. The investigators have suggested that these two different enzyme activities may derive from different genes or from different posttranslational processing of a common precursor. The overall significance of this finding is unclear and awaits purification of the enzyme. However, it is clear that the physiologic consequence of 5α-reductase enzyme deficiency is reflected in the absence of measureable enzyme activity at pH 5.5 in cell-free preparations or at physiologic pH in whole cells.

2. STUDIES OF ANDROGEN RECEPTORS

In 1974, our laboratory first reported the use of human skin fibroblasts as a model for the study of androgen action at the cellular level (11). Receptors for DHT (Table 6.4) in cultured human skin fibroblasts exhibit high affinity ($K_d = 10^{-10}$ M) and a high degree of specificity for androgens (78). Scatchard analysis of DHT binding revealed the presence of 2500–15,000 binding sites per cell with an equal distribution of binding activity between the nuclear and cytosolic compartments when assayed following incubation of cells at 37°C. A higher binding capacity was observed for DHT in fibroblasts derived from genital skin (foreskin, labia) than in fibroblasts derived from pubic or nongenital skin (abdomen, forearm), without a significant difference in the binding affinity. We also observed that total androgen receptor-binding capacity did not change with age in human foreskin (102). In fact, low but detectable levels of androgen receptors were demonstrated in nongenital skin fibroblasts from human fetuses as early as 8 weeks of gestation (81). The protein nature of the receptor was confirmed by its temperature and protease lability and by an observed decrease in DHT binding following exposure of cells to protein synthesis inhibitors (11). The receptor-steroid complex sedimented at 3–4S following sucrose density gradient centrifugation in the presence of 0.4 M KCl and appeared as a high-molecular-weight macromolecule upon elution following gel filtration column chromatography (78).

TABLE 6.4. Characteristics of androgen receptor from human skin fibroblasts.

Low capacity (2500–15,000 binding sites per cell)
High specificity for DHT
High affinity ($K_d = 10^{-10}$ M)
Found in cytosol and nuclei upon homogenization
Binds to DNA and oligonucleotides
Heat-labile
Protease sensitive
Not affected by DNAse or RNAse
Sedimentation coefficient: low salt = 7–8S
                            high salt = 3–4S
Binds to nuclear matrix

Additional studies have examined other properties associated with the intracellular binding of androgen to its receptor. Thus, the binding affinity of testosterone to androgen receptors was lower than that of DHT when studied in fibroblasts from a patient with 5α-reductase deficiency (76). In cells incubated in the presence of testosterone and the 5α-reductase inhibitor 17β-N,N-diethylcarbamoyl-4-methyl-4-aza-5α-androstan-3-one, testosterone rather than DHT becomes the predominant intracellular androgen that is bound to androgen receptors in the cytoplasm and nucleus (99). The results of these studies help to explain the lack of sex differentiation in the male fetus which requires the action of the more potent androgen (DHT) and the occurrence of virilization at puberty due to increased levels of testosterone in patients with 5α-reductase deficiency (103,104).

Within the nucleus the androgen-receptor steroid complex is bound to chromatin. Studies from our laboratory have demonstrated that androgen receptor–DHT complexes bind to DNA and exhibit preferential binding to synthetic oligonucleotides composed of the deoxyribonucleotides deoxythymidine and deoxyguanine compared to deoxycytosine, deoxyadenosine, and deoxyinosine (105). Approximately 50–60% of the total nuclear binding of DHT or the synthetic androgen methyltrienolone (R1881) is extracted from nuclei by buffers containing salt concentrations of up to 1 M KCl or NaCl (106). The DHT-receptor complex from cells incubated with DHT at 37°C binds to DNA and can be eluted from a chromatographic column containing DNA-agarose at an ionic strength of approximately 100 mM (T. Brown, unpublished data). Ion exchange column chromatography on DEAE-Biogel of the DHT-receptor complex results in the elution of a peak of specifically bound DHT at an ionic strength of 50 mM. In recent studies we have also demonstrated the saturable, high-affinity binding of the androgen receptor complex within the nuclear matrix prepared from human genital skin fibroblasts (107). This residual nuclear structure, which remains following the extraction of much of the lipid, protein, DNA, and RNA from nuclei, has been implicated as a site for DNA replication (108,109) and steroid hormone–specific gene transcription (110,111). Further studies of this nature should lead to a better understanding of the physicochemical and intranuclear binding properties of the human androgen receptor.

V. Clinical Syndrome of 5α-Reductase Deficiency (Table 6.5)

In 1961, Nowakowski and Lenz (16) described a familial form of male pseudohermaphroditism, which they termed "pseudovaginal perineoscrotal hypospadias" (Table 6.5). Its genetic transmission was as an autosomal-recessive trait (112). These patients resemble those with other forms of male pseudohermaphroditism: XY karyotype, normally differentiated testes, male internal ducts, and ambiguous external genitalia. At puberty, striking but selective signs of masculinization appear.

A. Pathophysiology

Walsh et al. (17) and Imperato-McGinley and Peterson et al. (18,113,114) reported a defect in the conversion of testosterone to its 5α-reduced metabolite, dihydrotestosterone, in patients with this syndrome. The latter investigators described studies in a highly inbred population from an isolated village of the Dominican Republic. In 24 families, 38 subjects with male pseudohermaphroditism were identified, 24 of whom were postpubertal (18). In infancy, the typical presentation included a small, clitorus-like phallus bound in chordae of variable degrees, perineal hypospadias, a bifid scrotum, and a urogenital sinus that opened on the perineum. The blind vaginal pouch opened either into the urogenital sinus or into the urethra. The testes were well differentiated and were located in either the inguinal canal or the labioscrotal folds. No Müllerian structures were present. The Wolffian structures (epididymis, vas deferens, and seminal vesicle) terminated in the vaginal pouch. At puberty, plasma testosterone levels reached the normal adult male range, but DHT levels remained low. Affected subjects virilized, but gynecomastia did not occur; the voice deepened, muscle mass increased, and the phallus, although bound in chordae, enlarged to 4–8 cm in length. The bifid scrotum became rugated and pigmented, and the testes enlarged and descended into the labioscrotal folds. However, the postpubertal affected subjects did not develop acne, temporal hair recession, or enlargement of the prostate, and they had scant facial and body hair. Semen analysis in 1 patient revealed 40×10^6 sperm per milliliter, and a testicular biopsy from another affected subject showed complete spermatogenesis (103,104,114,115).

TABLE 6.5. 5α-Reductase deficiency.

Karyotype: 46,XY

Inheritance: Autosomal-recessive

Genitalia: Usually ambiguous with perineal hypospadias, small phallus, and blind vaginal pouch

Wolffian duct derivatives: Normal

Müllerian duct derivatives: Absent

Gonads: Normal testes

Habitus: Virilization at puberty without gynecomastia; decreased facial and body hair; no temporal hair recession; nonpalpable prostate

Hormone profile: Decreased ratio of $5\alpha : 5\beta$ C_{19} and C_{21} steroids in urine; increased plasma T : DHT ratio before and after hCG stimulation; moderate increase in LH; decreased conversion of T to DHT in vivo

Some of the more intriguing observations reported by Imperato-McGinley and co-workers dealt with the gender identity and apparent pubertal gender transformation of patients living in isolated villages in the Dominican Republic (116). Among the 38 affected individuals, 18 had been reared unambiguously as females. However, following pubertal development of male secondary sex characteristics such as phallic enlargement and increased muscle mass, 16 of the 18 subjects changed their gender identity from female to male and behaved as men. In fact, 17 of 18 subjects were psychologically assessed as having a male gender identity. Despite the uniqueness of this isolated genetic population and its social setting, this study raises important questions about the relative influence and interaction of male sex hormones, sex of rearing, social conditions, and learning on psychosexual development. These observations argue against the hypothesis of an early critical period of gender identity imprinting during infancy and early childhood, proposed by Masica et al. (117).

B. Hormonal Profile

The hormonal profile in these patients is consistent with defective 5α-reduction of testosterone to DHT in androgen target tissues (17). In affected postpubertal subjects, plasma testosterone levels are within or above the range for adult men, but the DHT levels are significantly decreased (103,114). The testosterone/DHT ratio in peripheral blood is often greater than 36 in affected men, in contrast to the normal ratio of 8–16 (103,114,118). In prepubertal individuals, hCG stimulation of testosterone secretion is required to provoke the abnormal ratio of testosterone/DHT in plasma samples (118,119). In postpubertal subjects, plasma LH concentrations are often elevated, and plasma FSH levels tend to be higher than in age-matched controls (120). Therefore, affected postpubertal men exhibit the biochemical hallmark of peripheral androgen insensitivity—elevated plasma LH and testosterone concentrations. In patients with 5α-reductase deficiency, the ratios of urinary $5\alpha:5\beta$-reduced C_{19} and C_{21} steroids are decreased (103,104). As a typical example, the normal ratio of androsterone:etiocholanolone becomes less than 1:3 in subjects with this condition. A deficiency of 5α-reductase activity in cultured fibroblasts from genital skin of affected subjects can be demonstrated in vitro (121). The level of androgen receptor in cultured fibroblasts is normal. Adult women homozygous for the defect have no clinical manifestations (104). Heterozygotes for 5α-reductase deficiency have intermediate ratios of urinary $5\alpha:5\beta$-reduced C_{19} steroids (104).

5α-reductase deficiency is inherited as an autosomal-recessive trait and causes abnormal sex differentiation and other clinical manifestations only in male subjects homozygous for the trait. The enzyme defect exhibits heterogeneity (Table 6.6). Three types of biochemical abnormalities have

TABLE 6.6. Heterogeneous properties of 5α-reductase enzyme from cultured skin fibroblasts of families exhibiting 5α-reductase deficiency.

Family	Activity at pH 5.5 (pmol/mg protein/h)	K_m for testosterone (μM)	K_m for NADPH (μM)	Stability after exposure to cycloheximide (%)
Normal (n = 12)	33	0.08 ± 0.01	40 ± 8	>95
Dallas	0.2	1.80	250	>95
Dominican Republic	0.2	3.4	97	>95
Los Angeles	4.5	0.16	1760	< 5
New York	0.6	2.2	4.25	75

Data taken from references 27, 115, 123.

been described. In the first reports, which included the families from the Dominican Republic, enzyme studies on genital skin biopsy specimens and fibroblasts cultured from similar specimens indicated a very low level of 5α-reductase activity (17,121). A second type of defect involving an alteration in the stability of the enzyme was also described (122,123). In one kinship, 5α-reductase activity was low in biopsy specimens of genital skin but in the low normal range in cultured fibroblasts. Testosterone binding to the enzyme was normal, but the cofactor, nicotinamide adenine dinucleotide phosphate, reduced (NADPH), had a decreased affinity. As a consequence, the enzyme was very unstable and had a very rapid rate of turnover within cultured fibroblasts (123). A third variant of the abnormal 5α-reductase occurred in another family. The enzyme was characterized as having a decreased affinity for both the substrate (testosterone) and the cofactor (NADPH), which resulted in a reduced enzyme activity and lowered stability of the protein in cultured fibroblasts (115).

The phenotype of patients with 5α-reductase deficiency supports the hypothesis that testosterone induces differentiation of the urogenital sinus, prostate, and external genitalia (124). Peterson et al. have postulated that the increased muscle mass, deepening of the voice, spermatogenesis, and male libido seen at puberty in these patients are testosterone-mediated, whereas acne, temporal hair recession, facial hair, and prostatic enlargement are DHT-dependent (103,104). The androgen receptor binds both testosterone and DHT, the latter steroid with a higher affinity (76). Therefore, the sustained high levels of circulating testosterone that occur following puberty may be a factor in the growth of the phallus despite the incomplete masculinization of the external genitalia attained during fetal life. Furthermore, the enzyme defect is incomplete, and at puberty the plasma concentration of DHT becomes detectable owing to the increased availability of testosterone as a substrate. In addition, the hormonal environment at puberty is devoid of competing steroids, such as estrogens and progestins, to which the fetus was exposed during differentiation in utero (125). Gynecomastia does not occur at puberty in these patients, since estrogen production is not increased above that of normal men (126).

C. Clinical Diagnosis

The diagnosis of 5α-reductase deficiency should be suspected in all prepubertal subjects with male pseudohermaphroditism as evidenced by perineal hypospadias and a blind vaginal pouch and in those who virilize at puberty without development of gynecomastia. However, virilization at puberty and the absence of gynecomastia in individuals with male pseudohermaphroditism are not unique to 5α-reductase deficiency. For example, patients with 17-ketosteroid reductase deficiency or incomplete androgen insensitivity may present with these features, but they can be distinguished biochemically from 5α-reductase deficiency. The diagnosis of 5α-reductase deficiency can be confirmed prepubertally as well as postpubertally by demonstrating an abnormally high testosterone:DHT ratio in peripheral blood before or following hCG administration (103,104,118,119). As stated earlier, the T:DHT ratio in postpubertal affected male subjects is 35–84, whereas the ratio in normal men is 8–16 (103,120). In male infants, the T:DHT ratio ranges from 1.7–17 during the immediate postnatal period, when these androgens are detectable (118). In prepubertal male subjects, the administration of hCG (1000–2000 units, IM) for each of 3 days is necessary to increase testosterone and DHT levels sufficiently for the detection of the abnormal ratio due to the enzyme defect (118). Patients with 5α-reductase deficiency manifest high T:DHT ratios after hCG administration (normal = 5.2 ± 1.5 SD for 17 days to 6 months and 11 ± 4.4 SD for 6 months to 14 years). Similarly, the ratio of 5α:5β metabolites of testosterone in urine is a marker of the enzyme defect in both prepubertal and postpubertal subjects (103). Less readily available but more direct studies that can be used to confirm the diagnosis include assessment of the in vitro conversion of testosterone to DHT by genital skin fibroblasts (93,104,121) and measurement of the blood production rate of DHT (64).

D. Clinical Management

The early diagnosis of 5α-reductase deficiency is important because of its bearing on the assignment of sex in the affected child. In cases of early detection, infants may be reared as males or females and undergo appropriate plastic repair of the external genitalia. This decision should be based on the extent of sexual ambiguity and on the potential for normal sexual function in adult life. If the patient is raised as a male, testosterone or DHT therapy can be given to augment phallic size and to facilitate surgical repair. In patients raised as females, early surgical repair is advised. Alternatively, individuals who are diagnosed with this condition after infancy and in whom gender identity is unequivocally female should have prophylactic orchidectomy prior to puberty to prevent virilization and clitoroplasty and estrogen therapy at the age of puberty.

VI. Androgen Receptor and Postreceptor Defects

Several forms of androgen insensitivity have been identified, predominantly through studies similar to those described above with human genital skin fibroblasts (27,28,62). The spectrum of phenotypes in 46,XY individuals is highly variable. Some subjects have normal female external genitalia, others have genital ambiguity, and still others have a normal male phenotype with infertility. A correlation between the severity of a defect in androgen receptor activity and the degree of the abnormal phenotype has not been established. Quantitative defects in the androgen receptor are known, as are qualitative abnormalities in receptor-positive forms of androgen insensitivity.

A. Complete Androgen Insensitivity (Table 6.7)

Complete androgen insensitivity is often referred to as "testicular feminization," even though the former term is more descriptive of the pathogenesis of the syndrome and more acceptable to patients who have this disorder (28). The clinical features of the complete form of androgen insensitivity are uniform. The affected subjects are genetic males with a 46,XY karyotype; testes are present and H-Y antigen is detectable at normal male levels (127). Anatomically, these patients have female external genitalia; a blind vaginal pouch with absence of Müllerian

TABLE 6.7. Complete androgen insensitivity.

Karyotype: 46,XY
Inheritance: X-linked recessive
Genitalia: Female with blind vaginal pouch
Wolffian duct derivatives: Absent
Müllerian duct derivatives: Absent
Gonads: Testes; azoospermia, hyperplastic Leydig cells
Habitus: Scant or absent pubic and axillary hair; normal female breast development and female habitus at puberty; primary amenorrhea
Hormone profile: Elevated plasma LH and testosterone; elevated estradiol (for men); normal or elevated FSH
End-organ insensitivity to androgenic and metabolic effects of testosterone; androgen receptor or postreceptor defect

structures; testes located in the labial folds, inguinal canal, or intraab-dominally; and absent or vestigial Wolffian derivatives. Prior to puberty, the gonads are histologically difficult to distinguish from normal pre-pubertal testes (128–130). Postpubertally, small seminiferous tubules with few spermatogonia and absent spermatogenesis are present (130). The Leydig cells are often hyperplastic and tend to form adenomatous clumps. The testes are predisposed to malignant transformation, although the risk of neoplasia is low before the third decade of life (130,131). The risk of testicular malignancy in patients with androgen insensitivity is similar to that for other men with isolated cryptorchidism (132).

At birth and in childhood, the diagnosis should be suspected in phenotypic females with an inguinal hernia and palpation of a testislike mass in the inguinal region or in the labia. At adolescence, female secondary sexual characteristics appear, including well-developed breasts and female body habitus, but no menses (Fig. 6.7). Pubic hair and axillary hair are sparse and frequently absent. The clitoris is normal or small, the vagina is shallow and ends in a blind pouch, and the labia

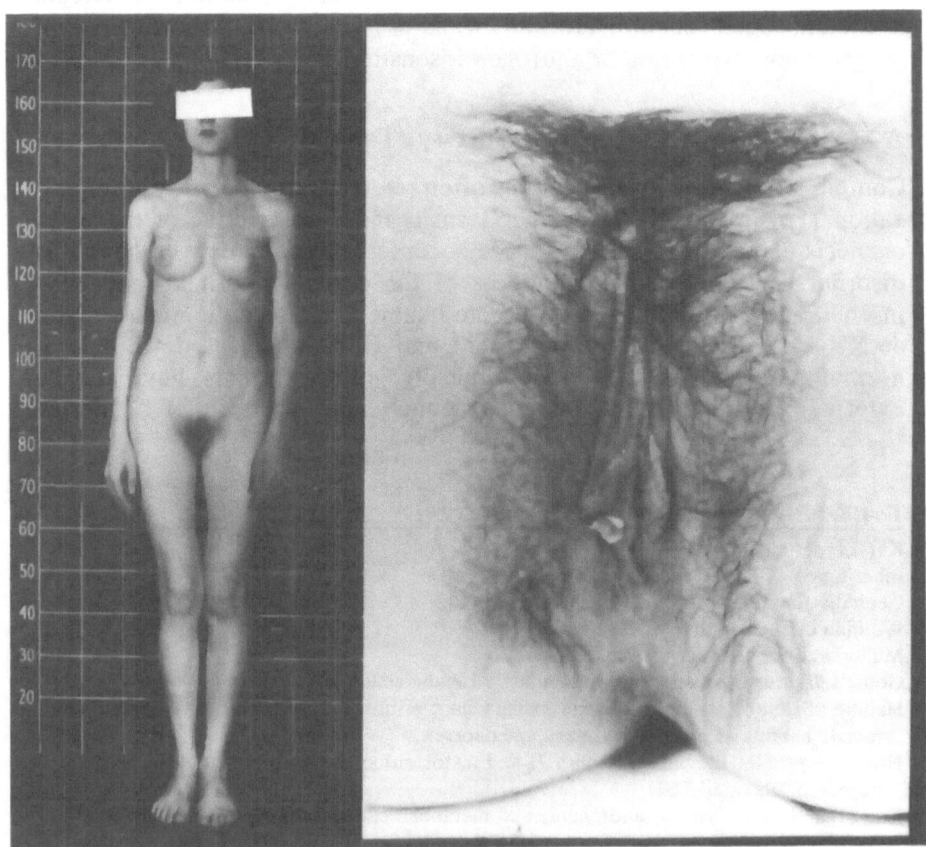

FIGURE 6.7. A patient with complete androgen insensitivity, receptor (+).

minora tend to be underdeveloped. Wolffian duct derivatives are absent, vestigial, or hypoplastic, and no Müllerian structures are found, because secretion of Müllerian-inhibiting factor by the fetal Sertoli cells is normal.

Gender identity is that of a normal female (133,134). Intelligence among affected subjects is normal, and there are no other clinical anomalies. Estimates of the prevalence of this condition vary from 1 in 20,000 to 1 in 64,000 male births (135). Pedigree analysis, cell-cloning studies, and somatic-cell hybridization experiments with human cells that contain balanced X autosome translocations indicate that the androgen receptor is encoded by a gene located on the X chromosome between Xq13 and Xp11 (15,77).

1. PATHOPHYSIOLOGY

In 1950, Wilkins (4) suggested that failure of androgenization of the male fetus and the development of female rather than male secondary sex characteristics at puberty could be explained by end-organ unresponsiveness to androgen. This contention was supported by further studies in which a lack of clinical or metabolic response to testosterone administration was demonstrated (136). Similar X-linked disorders have been described in other mammalian species, including mouse, rat, bull, and chimpanzee (137,138).

Studies in two animal models, the tfm/Y mouse and the pseudohermaphroditic rat, suggested that the primary defect was a deficient number of androgen receptors for testosterone and dihydrotestosterone (138–142). Subsequently, Keenan et al. (10), in our laboratory, reported the absence of androgen receptor-binding activity in cultured fibroblasts from the genital skin of karyotypic males with evidence of complete androgen insensitivity (Fig. 6.8). These observations were soon confirmed by other investigators (143,144). The finding of lack of androgen binding in genital skin fibroblasts from patients with this disorder provided an explanation for the observed lack of androgen action.

However, other patients with the identical clinical phenotype have subsequently been shown to have normal levels of androgen receptor in fibroblasts cultured from their genital skin (12,13,145). In these cases, the process of nuclear translocation of the receptor-DHT complex is also quantitatively normal (13,145). We have suggested that either a subtle qualitative abnormality exists in the receptor protein structure and/or an undefined postreceptor defect in the mechanism of androgen action occurs. Biochemical studies of receptor-steroid binding indicated that qualitative abnormalities do exist in the receptor protein of many receptor (+) patients with complete androgen insensitivity (13,146–151). For instance, receptor affinity and specificity for DHT may be lower than normal (13). This correlates with an increased dissociation rate for the receptor-steroid complex (13). Other qualitative abnormalities relate to the thermolability of the receptor (13,146), its instability in the presence of

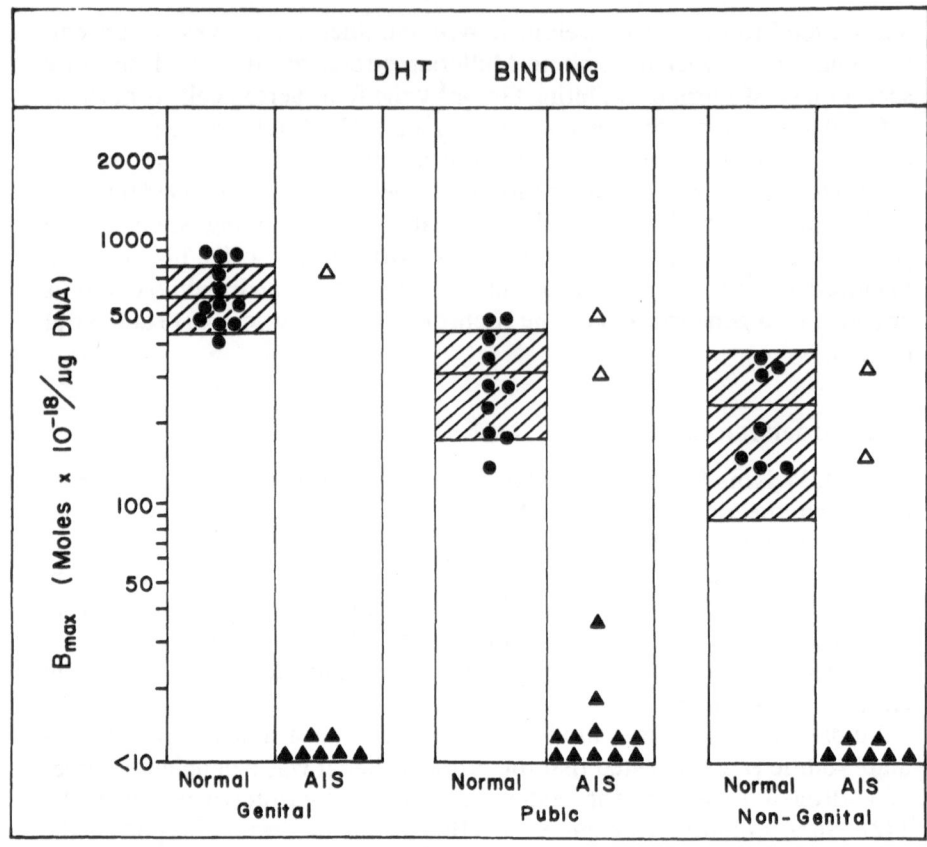

FIGURE 6.8. Androgen receptor-binding capacity (B_{max}) in skin fibroblasts from normal male subjects and subjects with complete androgen insensitivity. Heterogeneity is demonstrated among the patients; 15 had no receptor binding (R−), whereas 5 had normal receptor binding (R+). Hatched areas represent the mean ± SD for normal males.

molybdate (147), and defective "up-regulation" of the receptor number per cell (151). Whereas quantitative nuclear translocation of the receptor-DHT complex may be normal, an increase in the dissociation rate for the complex from the nuclear matrix subfraction has also been observed (107). These data suggest that although structural abnormalities exist within the receptor protein, none are sufficient to pinpoint the molecular mechanism whereby a complete absence of androgen responsiveness occurs within the target cells of these patients.

2. HORMONAL PROFILE

In complete androgen insensitivity, the lack of biologic activity of testosterone and DHT during embryogenesis fails to maintain the Wolffian structures and prevents masculinization of the external genitalia. Normal secretion of Müllerian-inhibiting factor by the fetal Sertoli cells leads to regression of the Müllerian ducts. During infancy, plasma LH and testosterone levels are elevated over those found in age-matched normal males, and LHRH stimulation leads to LH levels that are higher than normal. From the limited studies carried out in prepubertal children with this syndrome, gonadotropins appear to be normal after infancy (152). At puberty, androgen insensitivity at the hypothalamic-pituitary level leads to an increase in pulse frequency and amplitude of LH spikes compared to normal (154). This results in augmented LH secretion, which in turn stimulates an increase in testosterone production by the testicular Leydig cells (28,154,155) (Fig. 6.9). The net result is that plasma testosterone levels and testosterone production rates are higher than those of normal men (28).

The increased testicular secretion of estradiol and the peripheral conversion of androstenedione and testosterone to estradiol result in elevated plasma estradiol concentrations compared to normal men, leading to feminization in the presence of end-organ resistance to androgens (155–157). The elevated estradiol level causes an increase in

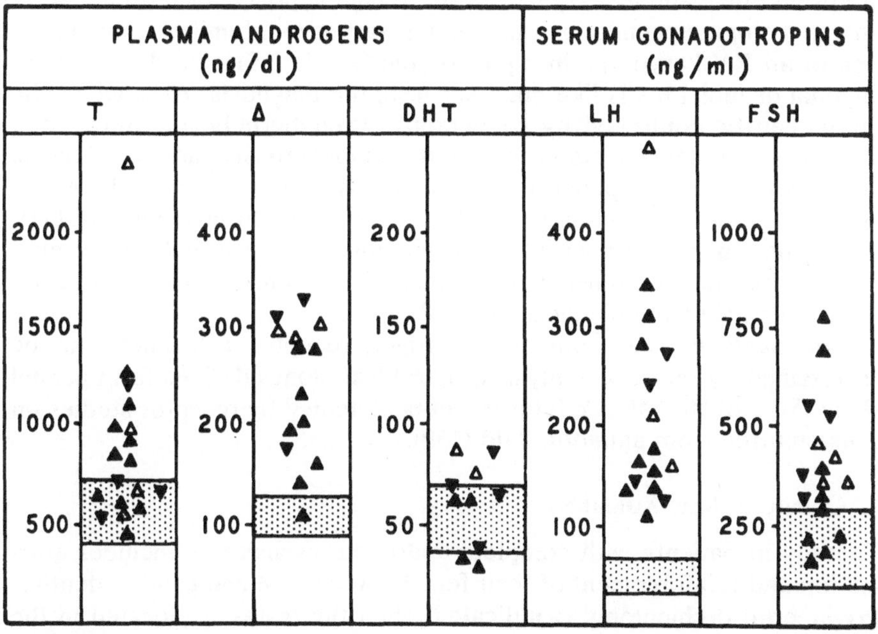

FIGURE 6.9. Concentrations of plasma androgens and serum gonadotropins in subjects with complete androgen insensitivity. Plasma androgens: testosterone (T); androstenedione (Δ); dihydrotestosterone (DHT). Stippled areas represent the mean concentration ± SD for normal male adults.

the concentration of testosterone-estradiol binding globulin (TeBG), which in turn relates to the increase in total plasma testosterone levels (28,126,156). Since the growth of sexual hair is normally mediated by androgens, its presence or absence depends upon the degree of androgen insensitivity. FSH concentrations are more variable and may be either normal or slightly elevated (27,28). As long as the testes remain in situ, LH concentrations remain below levels found after castration, since the hypothalamic-pituitary axis is sensitive to the estrogen feedback (157). Androgen insensitivity coupled with the elevated estradiol production induces the development of female secondary sex characteristics at the expected age of puberty.

3. Clinical Diagnosis

The diagnosis of complete androgen insensitivity can be established by clinical criteria alone in the postpubertal patient and can be strongly suspected in the prepubertal subject. Typically, patients may present with an inguinal hernia or labial mass, primary amenorrhea despite female secondary sex characteristics, and history of an affected sister, aunt, or cousin. The complete form of androgen insensitivity is represented by a phenotypic female with primary amenorrhea, breast development, scant or absent pubic and axillary hair, a short vaginal pouch and absent cervix on gynecologic examination, and an XY karyotype. Similarly, the detection of an XY karyotype in a phenotypic female infant or child with an inguinal or labial testis-like mass suggests the diagnosis. Absence of the uterus can be confirmed by sonography. Prepubertally, the differential diagnosis includes defects in testosterone biosynthesis and 5α-reductase deficiency. In the prepubertal patient, family history, phenotype, endocrine evaluation (including the androgen response to hCG and ACTH), determination of androgen receptor-binding activity in cultured genital skin fibroblasts, and, if necessary, metabolic response to testosterone are used to establish the diagnosis (66,158).

Individuals heterozygous for androgen receptor deficiency can be ascertained by receptor analysis of fibroblast clones derived from genital skin (15). Affected 46,XY fetuses can be detected by receptor studies on cells cultured from amniotic fluid (159).

4. Clinical Management

Therapy in patients with complete androgen insensitivity includes affirmation and reinforcement of their female phenotype and gender identity. Prepubertal orchiectomy is indicated when the testes are located in the inguinal region or the labia majora and are associated with a hernia. Otherwise, the affected subjects are allowed to undergo spontaneous feminization at puberty prior to gonadectomy in late adolescence. It may be advisable to follow the size of intraabdominal testes by periodic sonography because of their potential for malignancy. If the testes are removed, estrogen substitution is necessary to promote and maintain secondary sex characteristics. The vagina may be adequate in length for sexual intercourse, but in patients with a short vaginal pouch, manual dilatation with a prosthesis is effective in adolescence.

B. Partial Androgen Insensitivity

Partial or incomplete androgen insensitivity (Table 6.8) encompasses a heterogeneous group of syndromes in affected subjects with a 46,XY karyotype (27,28,62). The external genitalia are predominantly male or ambiguous. The pedigree analysis, like that for complete androgen insensitivity, is consistent with an X-linked recessive mode of inheritance. The various patients previously described by Lubs, Gilbert-Dreyfus, Reifenstein, Rosewater, Walker, and their collaborators most likely represent variant forms of partial androgen insensitivity (160–163). It should be noted that phenotypic expression (Fig. 6.10) is highly variable in partial androgen insensitivity, even within members of the same family (164,165). In one family, the propositus had marked ambiguity of the external genitalia, whereas his maternal cousin was completely masculinized but presented with a micropenis (165). In another study of 11 affected subjects, 2 had a relatively mild defect in masculinization represented by a small penis and bifid scrotum, 8 had perineal hypospadias, and 1 had hypospadias, a urogenital sinus with a blind vaginal pouch, and absent vas deferens (165). All lacked Müllerian structures. By contrast, the complete form of androgen insensitivity has a uniform appearance within families as an expression of the mutant gene. The most common presentation in infancy is an apparently male child with a third-degree hypospadias (with the urethral opening at the base of the phallus), a small penis, and, frequently, cryptorchidism. Müllerian duct derivatives are absent; in some subjects, Wolffian duct derivatives are present but usually hypo-

TABLE 6.8. Partial androgen insensitivity.

Karyotype: 46,XY

Inheritance: X-linked recessive

Genitalia: Variable; ambiguous with blind vaginal pouch, hypoplastic male, normal male with infertility

Wolffian duct derivatives: Rudimentary, hypoplastic, or normal

Müllerian duct derivatives: Absent

Gonads: Testes, azoospermia, hyperplastic Leydig cells

Habitus: Decreased or normal axillary and pubic hair, beard growth, and body hair; gynecomastia common at puberty

Hormone profile: Elevated plasma LH and testosterone; increased estradiol (for men); normal or elevated FSH

Partial end-organ insensitivity to androgenic and metabolic effects of testosterone; absence of spermatogenesis; androgen receptor or postreceptor defect

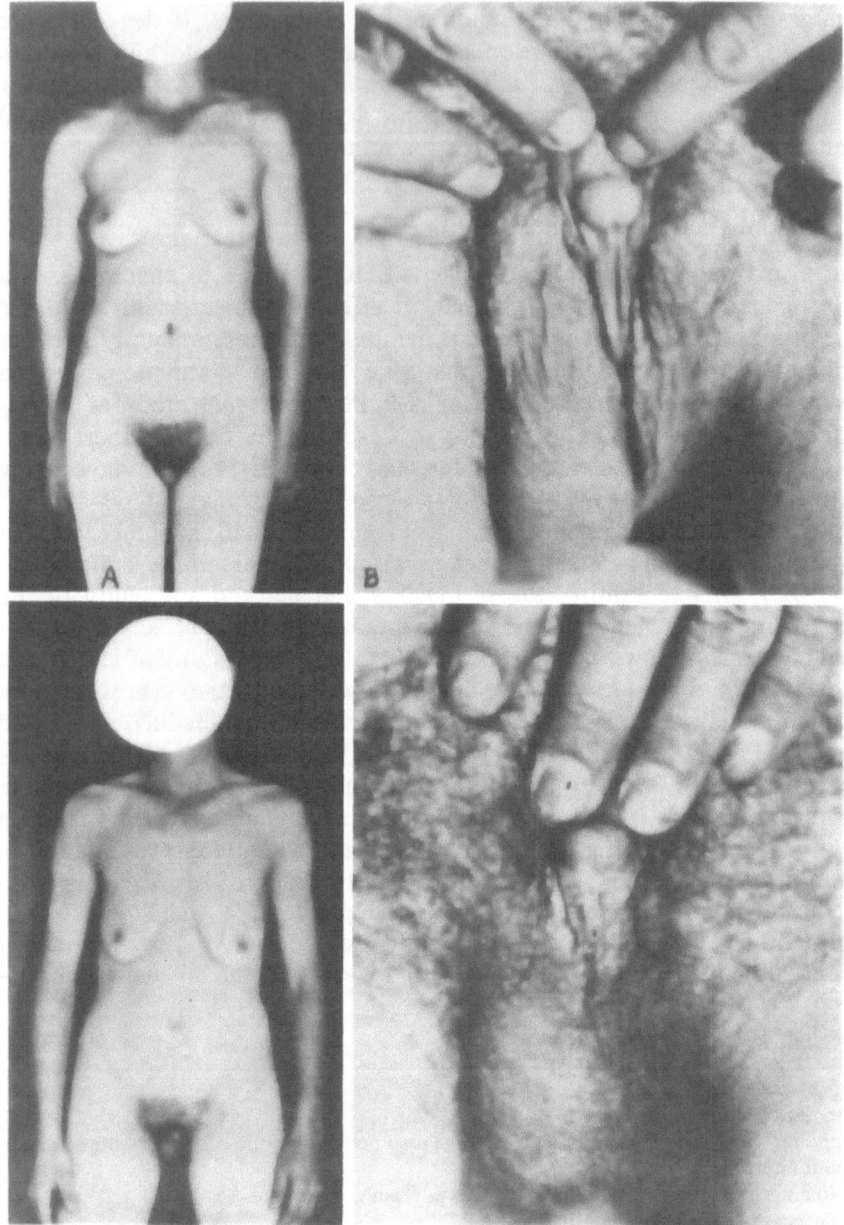

FIGURE 6.10. Partial androgen insensitivity in 2 siblings with variable phenotype.

plastic. At puberty, pubic and axillary hair and gynecomastia usually appear, male secondary sex characteristics are poorly developed, and the testes remain small and exhibit azoospermia owing to germinal cell arrest beyond the primary stage (130).

1. PATHOPHYSIOLOGY

Although a single biochemical defect was suspected as a unifying cause for the clinical findings in these disorders of varying severity, the pathogenesis was clarified only after reports from our laboratory (85) and subsequent confirmation by others (144,166). Studies of DHT binding by cultured fibroblasts from genital skin have demonstrated two types of defect: some patients have reduced high-affinity binding of DHT to the intracellular androgen receptor, a quantitative or partial deficiency of the androgen receptor (AR±); other affected subjects have receptor-positive (AR+) androgen insensitivity, and skin fibroblasts from these patients possess normal levels of high-affinity cytosol and nuclear binding of DHT. Only one of the two patterns, AR± or AR+, occurs within an affected family. As in the AR+ form of complete androgen insensitivity, the molecular defect in the receptor-positive patients with partial androgen insensitivity is uncertain. A decrease in nuclear retention of DHT-receptor complexes has been reported (167,168). The AR+ cases may represent a subtle qualitative abnormality in the androgen receptor in which the activated receptor-steroid complexes bind to nuclear chromatin "acceptor sites" but fail to elicit the events subsequent to nuclear binding that mediate the biologic response to androgen. Qualitative defects in the receptor have been demonstrated in cells from some patients (169–173). Alternatively, the defect could result in a postreceptor abnormality such as in transcription of DNA into RNA, in the levels of steroid-regulated specific messenger RNAs, in the processing of mRNAs, or in the translation of mRNA into protein. Future studies, no doubt, will elucidate new genetic defects in the action of androgen on target cells.

2. HORMONAL PROFILE

Similar to other patients with androgen insensitivity, the concentrations of plasma LH and testosterone are elevated, whereas in most cases the FSH level is normal; the high LH concentrations are resistant to suppression by exogenous testosterone (174,175). Estradiol and testosterone secretion rates are increased. However, the degree of feminization at puberty, despite the elevated estradiol secretion, is less than in the complete form of androgen insensitivity.

3. CLINICAL DIAGNOSIS

The diagnosis of partial androgen insensitivity cannot be made from the phenotype alone (66). Errors in testosterone biosynthesis can also cause subjects with a 46,XY karyotype to present with an abnormally small penis, hypospadias, incomplete fusion of the labioscrotal folds, a blind vaginal pouch, and gynecomastia at puberty. However, the pattern of inheritance and the measurement of plasma LH, and of testosterone and its precursors before and after the administration of hCG, will serve to distinguish patients with androgen insensitivity (66,158). Abnormally elevated plasma LH and testosterone levels during the immediate postnatal period may be indicative of androgen insensitivity (158). Studies of intracellular binding of DHT in cultured fibroblasts from genital skin may show either a partial deficiency or normal number of androgen receptors. The demonstration of a poor or absent metabolic and clinical response to testosterone can serve as a useful adjuvant in the diagnosis of partial androgen insensitivity (170,176).

4. CLINICAL MANAGEMENT

There is no specific therapy for partial androgen insensitivity. Sex of rearing depends on the age at diagnosis and the degree of genital ambiguity. In view of the limited response to testosterone in patients with this condition and the appearance of gynecomastia at puberty, it may be prudent to raise patients with this syndrome who have ambiguous genitalia as females (170,176). In patients assigned a female gender identity, plastic repair of the genitalia and gonadectomy are indicated before 6 months of age. Estrogen substitution therapy at puberty is required.

C. Androgen Insensitivity in Infertile Men

Aiman et al. (177) recently described a new and specific type of infertility due to partial deficiency of the androgen receptor in 3 unrelated men with uninformative family histories. Two of the men had a normal adult male phenotype; 1 had slight gynecomastia, decreased body hair, and a modest reduction in testicular size. All were infertile and had severe oligospermia or azoospermia. The striking hormonal findings were elevated serum concentrations of testosterone in the presence of normal to elevated LH levels. Two of the 3 men had increased plasma production rates for testosterone, androstenedione, and estradiol. A decreased androgen receptor binding capacity for DHT in genital skin fibroblasts was consistent with a partial deficiency of the androgen receptor. We subsequently reported similar findings in 4 subjects with infertility (178).

These findings strongly suggest that infertility in otherwise normal men may be the only clinical manifestation of partial androgen insensitivity and represents one end of a highly variable phenotypic expression of androgen insensitivity in patients with a comparable deficiency in androgen receptors as measured in vitro. High intratesticular concentrations of testosterone are required for normal spermatogenesis; thus, the

TABLE 6.9. Summary of the abnormalities of androgen target cells resulting in male
pseudohermaphroditism.

Disorder	External genitalia	Internal ducts	Gonadal tissue	At puberty	
				Phenotype	Testicular androgens
5α-Reductase deficiency	Feminine or ambiguous	Masculine	Testes	Nearly normal masculine	Testosterone, N DHT,−
Complete androgen insensitivity, R(−)	Feminine	Masculine	Testes	Infantile gynecomastia	++
Complete androgen insensitivity, R(+)	Feminine	Masculine	Testes	Infantile gynecomastia	++
Partial androgen insensitivity, R(±)	Ambiguous	Masculine	Testes	Partial masculine gynecomastia	++
Partial androgen insensitivity, R(+)	Ambiguous	Masculine	Testes	Partial masculine gynecomastia	++
Some cases of infertility, R(±)	Masculine	Masculine	Testes	Normal masculine	++

Key: R(−), no receptor activity; R(+), normal receptor activity; R(±), decreased receptor activity; N, normal
secretion of testicular androgens; −, decreased secretion of testicular androgens; ++, increased secretion of
testicular androgens.
Reprinted with permission from Brown TR, Berkovitz GD, Migeon CJ (1982) Androgen receptors in man. Diagnostic
Medicine, pp 23–37.

infertility in these patients is quite likely a consequence of the androgen
insensitivity, since testosterone levels were not decreased in these patients.
In corollary studies (179), our laboratory has reported that androgen
receptor levels, in cultured testicular cells from normal subjects and patients
with androgen insensitivity, parallel receptor levels found in genital skin
fibroblasts from the same individuals. The pattern of inheritance of androgen
receptor–dependent infertility is uncertain. The prevalence of this new
cause of male infertility is not known, but screening infertile men for
increased plasma concentrations of testosterone and LH or a high testosterone × LH product promises to shed light on this question. A recent study by
Aiman and Griffin (180), although not confirmed by Eil et al. (181), suggested
that as many as 40% of azoospermic men with an otherwise unknown
etiology for their infertility might have quantitatively reduced numbers of
androgen receptors in genital skin fibroblasts.

VII. Concluding Remarks (Table 6.9)

There are fundamental gaps in our knowledge of the mechanism of
androgen action. As yet undefined cellular factors, apart from receptors,

may influence the response of target tissues to androgen. This is well illustrated by the wide variation in phenotype of patients with partial androgen insensitivity, by the lack of correlation of the severity of the defect in masculinization of the external genitalia with the magnitude of the receptor abnormality in vitro, and by the seemingly apparent difference in androgen insensitivity of target tissues in the same patient (the hypothalamic-pituitary-gonadotropin complex vs. the external genitalia). Future efforts by investigators will be directed toward the purification of the androgen receptor and the production of monoclonal antibodies against the protein, cloning and mapping of the nucleotide sequence of the gene for the androgen receptor, and detailed definition of nuclear regulatory mechanisms of androgen action via the receptor.

References

1. Tonni P, Tinnelli J, Paganini J, Ballardi L (1802) Jacqueline Foroni rendue a son veritable sexe; ou rapport, refexions et jugements presentes a l'Academie de Mantoue par la classe de medecine, sur le sexe d'un individu vivant, connu sous le nom de Jacqueline Foroni, No. 534 (Imprimerie francaise et italienne a St. Zeno, Milan).
2. Steglehner (1875) De Hermaphroditorum Natura (Bamberg 1817); cited by Leopold; Arch Gynaek 8:487.
3. Goodman RM (1979) Genetic Disorders among Jewish People. Johns Hopkins University Press, Baltimore.
4. Wilkins L (1950) In The Diagnosis and Treatment of Endocrine Disorders in Childhood and Adolescence. Thomas, Springfield, IL.
5. Morris JM (1953) The syndrome of testicular feminization in male pseudohermaphrodites. Am J Obstet Gynecol 65:1192–1211.
6. Migeon CJ, Amrhein JA, Keenan BS, Meyer WJ, Migeon BR (1979) The syndrome of androgen insensitivity in man: Its relation to our understanding of male sex differentiation. In Vallet LH, Porter IH (eds): Genetic Mechanisms of Sexual Development. Academic Press, New York, pp 93–128.
7. Money J, Ehrhardt A (1972) Man and Woman, Boy and Girl. Johns Hopkins University Press, Baltimore.
8. Jacobs PA, Baikie AG, Court-Brown WM, Forrest H, Roy JR, Stewart JSS, Lennox B (1959) Chromosomal sex in the syndrome of testicular feminization. Lancet 2:591–592.
9. Southren AL, Sharma DC, Ross H, Sherman DH, Gordon G (1964) Plasma concentration and biosynthesis of testosterone in the syndrome of feminizing testes. Bull NY Acad Med 40:86.
10. Keenan BS, Meyer WJ, Hadjian AJ, Jones HW, Migeon CJ (1974) Syndrome of androgen insensitivity in man: Absence of 5-alpha-dihydrotestosterone binding protein in skin fibroblasts. J Clin Endocrinol Metab 38:1143–1146.
11. Keenan BS, Meyer WJ, Hadjian AJ, Migeon CJ (1975) Androgen receptor in human skin fibroblasts: Characterization of a specific 17-beta-hydroxy-5-alpha-androstan-3-one protein complex in cell sonicates and nuclei. Steroids 25:535–552.
12. Amrhein JA, Meyer WJ III, Jones HW Jr, Migeon CJ (1976) Androgen insensitivity in man: Evidence for genetic heterogeneity. Proc Natl Acad Sci USA 73:891–894.
13. Brown TR, Maes M, Rothwell SW, Migeon CJ (1982) Human complete

androgen insensitivity with normal dihydrotestosterone receptor binding capacity in cultured genital skin fibroblasts: Evidence for a qualitative abnormality of the receptor. J Clin Endocrinol Metab 55:61–69.

14. Morris JM, Mahesh VB (1963) Further observations on the syndrome "testicular feminization." Am J Obstet Gynecol 87:731–748.

15. Meyer WJ III, Migeon BR, Migeon CJ (1975) Locus on human X chromosome for dihydrotestosterone receptor and androgen insensitivity. Proc Natl Acad Sci USA 72:1469–1472.

16. Nowakowski H, Lenz W (1961) Genetic aspects of male hypogonadism. Recent Prog Horm Res 17:53–95.

17. Walsh PC, Madden JD, Harrod MJ, Goldstein J, MacDonald PC, Wilson JD (1974) Familial incomplete male pseudohermaphroditism, type 2. Decreased dihydrotestosterone formation in pseudovaginal perineoscrotal hypospadias. N Engl J Med 291:944–949.

18. Imperato-McGinley JL, Guerrero L, Gautier T, Peterson RE (1974) Steroid 5α-reductase deficiency in man: An inherited form of male pseudohermaphroditism. Science 186:1213–1215.

19. Jost A (1953) Problems of fetal endocrinology: The gonadal and hypophyseal hormones. Recent Prog Horm Res 8:379–418.

20. Jost A, Vigier B, Prepin J, Perchellet JP (1973) Studies on sex differentiation in mammals. Recent Prog Horm Res 29:1–41.

21. Wilson JD (1978) Sexual differentiation. Annu Rev Physiol 40:279–306.

22. Josso N (1981) Physiology of sex differentiation. In Laron Z (ed): Pediatric and Adolescent Endocrinology: The Intersex Child, Vol 8. S. Karger, Paris, pp 1–13.

23. Bruchovsky N, Wilson J (1968) The conversion of testosterone to 5α-androstran-17β-ol-3-one by rat prostate in vivo and in vitro. J Biol Chem 243:2012–2021.

24. Siiteri PK, Wilson JD (1974) Testosterone formation and metabolism during male sexual differentiation in the human embryo. J Clin Endocrinol Metab 38:113–125.

25. Chan L, O'Malley BW (1976) Mechanism of action of the sex steroid hormones (3 parts). N Engl J Med 294:1322–1328, 1372–1381, 1430–1437.

26. Liao S (1977) Molecular actions of androgens. In Litwack G (ed): Biochemical Actions of Hormones. Vol 4. Academic Press, New York, pp 351–406.

27. Griffin JE, Wilson JD (1980) The syndromes of androgen resistance. N Engl J Med 302:198–209.

28. Migeon CJ, Brown TR, Fichman KR (1981) Androgen insensitivity syndrome. In Josso N (ed): Pediatric and Adolescent Endocrinology: The Intersex Child, vol 8. S. Karger, Paris, pp 171–202.

29. Witschi E (1948) Migrations of germ cells of human embryos from the yolk sac to the primitive gonadal folds. Contrib Embryol 32:67–80.

30. Jirasek JE (1971) Development of the Genital System and Male Pseudohermaphroditism. Johns Hopkins University Press, Baltimore, p 43.

31. Stretter GL (1945) Developmental horizons in human embryos. Age groups XI to XXIII. Contrib Embryol 32:133–203.

32. Eichwald EJ, Silmser CR (1955) Skin. Transplant Bull 2:148–149.

33. Goldberg EH, Boyse EA, Bennett D, Scheid M, Carswell EA (1971) Serologic demonstration of H-Y (male) antigen on mouse sperm. Nature 232:478–480.

34. Wachtel SS, Ohno S, Koo GC, Boyse EA (1975) Possible role for H-Y antigen in the primary determination of sex. Nature 257:235–236.

35. Ohno S (1976) Major regulatory genes for mammalian sexual development. Cell 7:315–321.

36. Ohno S, Nagai Y, Ciccarese S, Iwata H (1979) Testis organizing H-Y antigen and the primary sex determining mechanism of mammals. Recent Prog Horm Res 35:449–476.
37. Wachtel SS (1981) Conservatism of the H-Y/H-W receptor. Hum Genet 58:54–58.
38. Wachtel SS, Koo GC (1981) H-Y antigen in gonad differentiation. In Austin CR, Edwards RG (eds): Mechanism of Sex Differentiation in Animals and Man. Academic Press, New York, pp 255–299.
39. Silvers WK, Gasser DL, Eicher EM (1982) H-Y antigen, serologically detectable male antigen, and sex determination. Cell 28:439–440.
40. Wolf U, Fraccaro M, Mayerova A, Hecht T, Zuffardi O, Hameister H (1980) Turner patients are H-Y positive. Hum Genet 54:315–318.
41. Engel W, Klemme B, Ebrecht A (1981) Serologic evidence for H-Y antigen in XO-female mice. Hum Genet 57:68–70.
42. Celada F, Welshons WJ (1963) An immunogenetic analysis of the male antigen in mice utilizing animals with an exceptional chromosome constitution. Genetics 48:139–151.
43. Eicher EM, Washburn LL, Whitney JB III, Morrow KE (1982) *Mus psochiavinus* Y chromosome in the C57BL/6J murine genome causes sex reversal. Science 217:535–537.
44. Eicher EM (1982) Primary sex determining genes in mice. In Amann RP, Siedel EG Jr (eds): Prospects for Sexing Mammalian Sperm. Colorado Associated University Press, Boulder, p 121.
45. Simpson J (1979) Gonadal dysgenesis and sex chromosome abnormalities, phenotype-karyotypic correlations. In Vallet LH, Porter IH (eds): Genetic Mechanisms of Sexual Development. Academic Press, New York, pp 365–406.
46. Heller CG, Clermont Y (1964) Kinetics of the germinal epithelium in man. Recent Prog Horm Res 20:545–575.
47. Gruenwald P (1942) The development of the sex cords in the gonads of man and animals. Am J Anat 70:359–398.
48. Mancini R, Narbaitz R, LaVieri J (1960) Origin and development of the germinal epithelium and sertoli cells in the human testis: Cytological, cytochemical and quantitative study. Anat Rec 136:477–489.
49. Josso N, Picard JY, Tran D (1977) The anti-Müllerian hormone. Recent Prog Horm Res 33:117–167.
50. Blanchard MG, Josso N (1974) Source of the anti-Müllerian hormone synthesized by the fetal testis: Müllerian-inhibiting activity of fetal bovine Sertoli cells in tissue culture. Pediatr Res 8:968–971.
51. Reyes FI, Winter JSD, Faiman C (1973) Studies on human sexual development. I. Fetal gonadal and adrenal sex steroids. J Clin Endocrinol Metab 37:74–78.
52. Abramovich DR, Rowe P (1973) Fetal plasma testosterone levels at midpregnancy and at term: Relationship to fetal sex. J Endocrinol 56:621–622.
53. Diez d'Aux RC, Murphy BEP (1974) Androgens in the human fetus. J Steroid Biochem 5:207–210.
54. Forest MG, Sizonenko PC, Cathiard AM, Bertrand J (1974) Hypophyso-gonadal function in infants during the first year of life. I. Evidence for testicular activity in early infancy. J Clin Invest 53:819–828.
55. Forest MG, Cathiard AM (1975) Pattern of plasma testosterone and Δ^4-androstenedione in normal newborns: Evidence for testicular activity at birth. J Clin Endocrinol Metab 41:977–980.
56. Rice BR, Johanson CA, Sternberg WH (1966) Formation of steroid hormones from acetate-1-^{14}C by a human fetal testis preparation grown in organ culture. Steroids 7:79–90.
57. Jaffe RB, Lee PA, Midgely AR Jr (1969) Serum gonadotropins before, at the

inception of, and following human pregnancy. J Clin Endocrinol Metab 29:1281–1283.
58. Vaitukaitis JL (1974) Changing placental concentrations of hCG and its subunits during gestation. J Clin Endocrinol Metab 38:755–760.
59. Hagen C, McNeilly AS (1977) The gonadotropins and their subunits in fetal pituitary glands and circulation. J Steroid Biochem 8:537–543.
60. Kaplan SL, Grumbach MM, Aubert ML (1976) The ontogenesis of pituitary hormones and hypothalamic factors in the human fetus: Maturation of the central nervous system, regulation of the anterior pituitary function. Recent Prog Horm Res 32:161–243.
61. Silber-Khodr TM, Khodr GS (1978) Studies in human fetal endocrinology. I. Luteinizing hormone releasing factor content of the hypothalamus. Am J Obstet Gynecol 130:795–800.
62. Brown TR, Migeon CJ (1986) Androgen receptors in normal and abnormal male sexual differentiation. In Chrousos GP, Loriaux DL, Lipsett MB (eds): Steroid Hormone Resistance: Mechanisms and Clinical Aspects. Plenum, New York, pp 227–255.
63. Bardin CW, Catterall JF (1981) Testosterone: A major determinant of extragenital dimorphism. Science 211:1285–1294.
64. Simpson JL (1978) Male pseudohermaphroditism: Genetics and clinical delineation. Hum Genet 44:1–49.
65. Griffin JE, Wilson JD (1978) Hereditary male pseudohermaphroditism. Clin Obstet Gynaecol 5:457–479.
66. Migeon CJ (1980) Male pseudohermaphroditism. Ann Endocrinol 41:311–343.
67. Berkovitz GD, Lee PA, Brown TR, Migeon CJ (1984) Etiologic evaluation of male pseudohermaphroditism in infancy and childhood. Am J Dis Child 138:755–759.
68. Cohen M, Shaw M (1965) Two XY siblings with gonadal dysgenesis and a female phenotype. N Engl J Med 272:1083–1086.
69. Armendares S, Buentello L, Frenk S (1973) Two male sibs with uterus and Fallopian tubes, a rare, probably inherited disorder. Clin Genet 4:291–296.
70. Berthezene F, Forest M, Grimaud J, Clasutrat B, Mornex R (1976) Leydig cell agenesis: A cause of male pseudohermaphroditism. N Engl J Med 295:969–972.
71. Fichman KR, Migeon BR, Migeon CJ (1980) Genetic disorders of male sexual differentiation. In Harris H, Hirschhorn K (eds): Advances in Human Genetics, Vol 10. Plenum, New York, pp 333–377.
72. Park J, Burnett L, Jones H, Migeon C, Blizzard R (1976) A case of male pseudohermaphroditism associated with elevated LH, normal FSH and low testosterone possibly due to the secretion of an abnormal LH molecule. Acta Endocrinol (Copenh) 83:173–181.
73. Perez-Palacios G, Scaglia H, Kofman S, Saavedra D, Ochoa S, Laroza O, Perez A (1975) Inherited deficiency of gonadotropin receptor in Leydig cells: A new form of male pseudohermaphroditism. Am J Hum Genet 27:71A.
74. Meyer W, Keenan B, DeLacerda L, Park I, Jones H, Migeon C (1978) Familial male pseudohermaphroditism with normal Leydig cell function at puberty. J Clin Endocrinol Metab 46:593–603.
75. Mainwaring WIP (1977) The Mechanism of Action of Androgens. Springer-Verlag, New York, pp 1–178.
76. Maes M, Sultan C, Zerhouni N, Rothwell SW, Migeon CJ (1979) Role of testosterone binding to the androgen receptor in male sexual differentiation of patients with 5α-reductase deficiency. J Steroid Biochem 11:1385–1390.
77. Migeon BR, Brown TR, Axelman J, Rothwell SW, Migeon CJ (1981) Studies

on the locus for androgen receptor localization on the human X chromosome and evidence for homology with the Tfm locus in the mouse. Proc Natl Acad Sci USA 78:6339–6343.

78. Brown TR, Migeon CJ (1981) Cultured human skin fibroblasts: A model for the study of androgen action. Mol Cell Biochem 36:3–22.

79. King WJ, Greene GL (1984) Monoclonal antibodies localize oestrogen receptor in the nuclei of target cells. Nature 307:745–747.

80. Welschons WV, Krummel BM, Gorski J (1984) Nuclear localization of unoccupied receptors for glucocorticoids, estrogens and progesterone in GH$_3$ cells. Endocrinology 117:2140–2147.

81. Sultan C, Migeon BR, Rothwell SW, Maes M, Zerhouni N, Migeon CJ (1980) Androgen receptors and metabolism in cultured human fetal fibroblasts. Pediatr Res 14:67–69.

82. Wilson JD (1975) Dihydrotestosterone formation in cultured human fibroblasts. Comparison of cells from normal subjects and patients with familial incomplete male pseudohermaphroditism, type 2. J Biol Chem 250:3498–3504.

83. Pinsky L, Kaufman M, Straisfeld C, Shanfield B (1974) Lack of difference in testosterone metabolism between cultured skin fibroblasts of human adult males and females. J Clin Endocrinol Metab 39:395–398.

84. Mulay S, Finkelberg R, Pinsky L, Solomon S (1972) Metabolism of 4-^{14}C-testosterone by serially subcultured human skin fibroblasts. J Clin Endocrinol Metab 34:133–143.

85. Amrhein JA, Klingensmith G, Walsh PC, McKusick VA, Migeon CJ (1977) Partial androgen insensitivity: The Reifenstein syndrome revisited. N Engl J Med 297:350–356.

86. Wilson JD, Walker JD (1969) The conversion of testosterone to 5α-androstan-17β-ol-3-one (dihydrotestosterone) by skin slices. J Clin Invest 48:371–379.

87. Mowszowicz I, Kirchhoffer M-O, Kuttenn F, Mauvais-Jarvis P (1980) Testosterone 5α-reductase activity of skin fibroblasts increase with serial subcultures. Mol Cell Endocrinol 17:41–50.

88. Mowszowicz I, Malanitor E, Kirchhoffer M-O, Mauvais-Jarvis P (1983) Dihydrotestosterone stimulates 5α-reductase activity in pubic skin fibroblasts. J Clin Endocrinol Metab 56:320–325.

89. Kuttenn F, Mauvais-Jarvis P (1975) Testosterone 5α-reduction in the skin of normal subjects and of patients with abnormal sex development. Acta Endocrinol (Copenh) 79:164–176.

90. Mauvais-Jarvis P (1977) Androgen metabolism in human skin: Mechanism of control. In Martini L, Motta M (eds): Androgens and Antiandrogens. Raven Press, New York, pp 229–245.

91. Pinsky L, Finkelberg R, Straisfeld C, Zilahi B, Kaufman M, Hall G (1972) Testosterone metabolism by serially subcultured fibroblasts from genital and nongenital skin of individual human donors. Biochem Biophys Res Commun 46:346–349.

92. Amrhein JA, Meyer WJ, Danish RK, Migeon CJ (1977) Studies of androgen production and binding in 13 males with micropenis. J Clin Endocrinol Metab 45:732–738.

93. Pinsky L, Kaufman M, Straisfeld C, Zilahi B, St-G Hall C (1978) 5α-Reductase activity of genital and nongenital skin fibroblasts from patients with 5α-reductase deficiency, androgen insensitivity, or unknown forms of male pseudohermaphroditism. Am J Med Genet 1:407–416.

94. Kaufman M, Pinsky L, Straisfeld C, Stanfield B, Zilahi B (1975) Qualitative differences in testosterone metabolism as an indication of cellular heterogeneity in fibroblast monolayers derived from human preputial skin. Exp Cell Res 96:31–36.

95. Griffin JE, Allman DR, Durrant JL, Wilson JD (1981) Variation in steroid 5α-reductase activity in cloned human skin fibroblasts. J Biol Chem 256:3662–3666.
96. Lamberigts G, Dierick P, DeMoor P, Verhoeven G (1979) Comparison of the metabolism and receptor binding of testosterone and 17β-hydroxy-5α-androstan-3-one in normal skin fibroblast cultures: Influence of origin and passage number. J Clin Endocrinol Metab 48:924–930.
97. Serafini P, Ablan F, Lobo RA (1985) 5α-Reductase activity in the genital skin of hirsute women. J Clin Endocrinol Metab 60:349–355.
98. Zerhouni N, Maes M, Sultan C, Rothwell SW, Migeon CJ (1979) Selective inhibition by secosteroids of 5α-reductase activity in human sex skin fibroblasts. Steroids 33:277–285.
99. Berkovitz GD, Brown TR, Migeon CJ (1984) Inhibition of 5α-reductase activity and alteration of nuclear testosterone:dihydrotestosterone ratio in human genital skin fibroblasts. J Androl 5:171–175.
100. Moore RJ, Griffin JE, Wilson JD (1975) Diminished 5α-reductase activity in extracts of fibroblasts cultured from patients with familial incomplete male pseudohermaphroditism, type 2. J Biol Chem 250:7168–7172.
101. Moore RJ, Wilson JD (1976) Steroid 5α-reductase in cultured human fibroblasts: Biochemical and genetic evidence for two distinct enzyme activities. J Biol Chem 251:5895–5900.
102. Fichman KR, Nyberg LM, Bujnovszky P, Brown TR, Walsh PC (1981) The ontogeny of the androgen receptor in human foreskin. J Clin Endocrinol Metab 52:919–923.
103. Peterson RE, Imperato-McGinley J, Gautier T, Sturla E (1979) Hereditary steroid 5α-reductase deficiency: A newly recognized cause of male pseudo-hermaphroditism. In Vallet HL, Porter IH (eds): Genetic Mechanisms of Sexual Development. Academic Press, New York, pp 149–173.
104. Imperato-McGinley J, Peterson RE, Gautier T (1984) Primary and secondary 5α-reductase deficiency. In Serio M, Motta M, Zanisi M, Martini L (eds): Sexual Differentiation: Basic and Clinical Aspects. Raven Press, New York, pp 233–245.
105. Brown TR, Rothwell SW, Migeon CJ (1982) Human androgen insensitivity mutation does not alter oligonucleotide recognition by the androgen receptor–DHT complex. Mol Cell Endocrinol 32:215–231.
106. Brown TR, Rothwell SW, Migeon CJ (1983) Comparison of methyltrienolone and dihydrotestosterone binding and metabolism in human genital skin fibroblasts. J Steroid Biochem 14:1013–1022.
107. Brown TR, Migeon CJ (1985) Androgen binding in nuclear matrix of human genital skin fibroblasts from patients with androgen insensitivity syndrome. J Clin Endocrinol Metab 62:542–550.
108. Pardoll DM, Vogelstein B, Coffey DS (1980) A fixed site of DNA replication in eukaryotic cells. Cell 19:527–536.
109. Robinson SI, Nelkin BD, Vogelstein B (1982) The ovalbumin gene is associated with the nuclear matrix of chicken oviduct cells. Cell 28:99–106.
110. Barrack ER, Coffey DS (1982) Biological properties of the nuclear matrix: Steroid hormone binding. Recent Prog Horm Res 38:133–195.
111. Barrack ER (1983) The nuclear matrix of the prostate contains acceptor sites for androgen receptors. Endocrinology 113:430–432.
112. Opitz JM, Simpson JL, Sarto GE, Sumitt RL, New M, German J (1972) Pseudovaginal perineoscrotal hypospadias. Clin Genet 3:1–26.
113. Imperato-McGinley JL, Peterson RE (1976) Male pseudohermaphroditism: The complexities of male phenotypic development. Am J Med 61:251–272.
114. Peterson RE, Imperato-McGinley J, Gautier T, Sturla E (1977) Male

pseudohermaphroditism due to steroid 5α-reductase deficiency. Am J Med 62:170–191.

115. Imperato-McGinley JL, Peterson RE, Leshin M, Griffin JE, Cooper G, Draghi S, Berenyi M, Wilson JD (1980) Steroid 5α-reductase deficiency in a 65-year-old male pseudohermaphrodite: The natural history, ultrastructure of the testes and evidence for inherited enzyme heterogeneity. J Clin Endocrinol Metab 50:15–22.

116. Imperato-McGinley JL, Peterson RD, Gautier T, Sturla E (1979) Androgens and the evolution of male-gender identity among male pseudohermaphrodites with 5α-reductase deficiency. N Engl J Med 300:1233–1237.

117. Masica DN, Money J, Ehrhardt AA (1971) Fetal feminization and female gender identity in the testicular feminizing syndrome of androgen insensitivity. Arch Sex Behav 1:131–142.

118. Pang S, Levine LS, Chow D, Sagiani F, Saenger P, New MI (1979) Dihydrotestosterone and its relationship to testosterone in infancy and childhood. J Clin Endocrinol Metab 48:821–826.

119. Saenger P, Goldman AS, Levine LS, Korth-Schutz S, Muecke EC, Katsumata M, Doberne Y, New MI (1978) Prepubertal diagnosis of steroid 5α-reductase deficiency. J Clin Endocrinol Metab 46:627–634.

120. Imperato-McGinley J, Peterson RE, Gautier T, et al. (1982) Hormonal evaluation of a large kindred with complete androgen insensitivity; evidence for secondary 5α-reductase deficiency. J Clin Endocrinol Metab 54:931–941.

121. Moore RJ, Griffin JE, Wilson JD (1975) Diminished 5α-reductase activity in extracts of fibroblasts cultured from patients with familial incomplete male pseudohermaphroditism, type 2. J Biol Chem 250:7168–7172.

122. Fisher KL, Kogut MD, Moore RJ, Goebelsman U, Weitzman J, Hart I Jr, Griffin JE, Wilson JD (1978) Clinical, endocrinological, and enzymatic characterization of two patients with 5α-reductase deficiency: Evidence that a single enzyme is responsible for the 5α-reduction of cortisol and testosterone. J Clin Endocrinol Metab 47:653—664.

123. Leshin M, Griffin JE, Wilson JD (1978) Hereditary male pseudohermaphroditism associated with an unstable form of 5α-reductase. J Clin Invest 62:685–691.

124. Wilson JD (1972) Recent studies on the mechanism of action of testosterone. N Engl J Med 287:1284–1291.

125. Hodgins MB (1983) Possible mechanisms of androgen resistance in 5α-reductase deficiency: Implications for the physiological role of 5α-reductase. J Steroid Biochem 19:555–559.

126. Wilson JD, Aiman J, MacDonald PC (1980) The pathogenesis of gynecomastia. Adv Intern Med 25:1–32.

127. Wachtel SS (1983) H-Y Antigen and the Biology of Sex Determination. Grune and Stratton, New York.

128. O'Leary JA (1965) Comparative studies of the gonad in testicular feminization and cryptorchidism. Fertil Steril 16:813–819.

129. Ferenczy A, Richart RM (1972) The fine structures of the gonads in the complete from of testicular feminization syndrome. Am J Obstet Gynecol 113:399–409.

130. Muller J (1984) Morphometry and histology of gonads from twelve children and adolescents with the androgen insensitivity (testicular feminization) syndrome. J Clin Endocrinol Metab 59:785–789.

131. Manuel M, Katayma K, Jones HW Jr (1976) The age of occurrence of gonadal tumors in intersex patients with a Y chromosome. Am J Obstet Gynecol 124:293–300.

132. Simpson JL, Photopulos G (1976) The relationship of neoplasia to disorders of abnormal sexual differentiation. Birth Defects 12:15–50.

133. Money J, Ehrhardt AA, Masica DN (1968) Fetal feminization induced by androgen insensitivity in the testicular feminizing syndrome: Effect on marriage and maternalism. Johns Hopkins Med J 123:105–114.
134. Masica DN, Money J, Ehrhardt AA, Lewis VG (1969) IQ, fetal sex hormones and cognitive patterns: Studies in the testicular feminizing syndrome of androgen insensitivity. Johns Hopkins Med J 124:34–43.
135. Jagiello G, Atwell JD (1962) Prevalence of testicular feminization. Lancet 2:329.
136. French FS, Van Wyk JJ, Baggett B, Easterling, WE, Talbert LM, Johnston FR, Forchielli E, Dey AC (1966) Further evidence of a target organ defect in the syndrome of testicular feminization. J Clin Endocrinol Metab 26:493–503.
137. Lyon MF, Hawkes SG (1970) X-linked gene for testicular feminization in the mouse. Nature 227:1217–1219.
138. Bardin CW, Bullock LP, Sherins RJ, Mowszowicz I, Blackburn WR (1973) Androgen metabolism and mechanism of action in male pseudohermaphroditism: A study of testicular feminization. Recent Prog Horm Res 29:65–109.
139. Goldstein JL, Wilson JD (1972) Studies on the pathogenesis of the pseudohermaphroditism in the mouse with testicular feminization. J Clin Invest 51:1647–1658.
140. Gehring U, Tomkins GM, Ohno S (1971) Effect of the androgen insensitivity mutation on a cytoplasmic receptor for dihydrotestosterone. Nature 232:106–107.
141. Attardi B, Ohno S (1974) Cytosol androgen receptors from kidney of normal and testicular feminized (tfm) mice. Cell 2:205–212.
142. Gehring U, Tomkins GM (1974) Characterization of a hormone receptor defect in the androgen insensitivity mutant. Cell 3:59–64.
143. Griffin JE, Punyashthiti K, Wilson JD (1976) Dihydrotestosterone binding by cultured fibroblasts: Comparison of cells from control subjects and from patients with hereditary male pseudohermaphroditism due to androgen resistance. J Clin Invest 57:1342–1351.
144. Kaufman M, Straisfeld C, Pinsky L (1976) Male pseudohermaphroditism presumably due to target organ unresponsiveness to androgens: Deficient 5α-dihydrotestosterone binding in cultured skin fibroblasts. J Clin Invest 58:345–350.
145. Kaufman M, Pinsky L, Baird PH, McGillivray BC (1979) Complete androgen insensitivity with a normal amount of 5α-dihydrotestosterone-binding activity in labium majus skin fibroblasts. Am J Med Genet 4:401–411.
146. Griffin JE (1979) Testicular feminization associated with a thermolabile androgen receptor in cultured fibroblasts. J Clin Invest 64:1624–1631.
147. Griffin JE, Durrant JL (1982) Qualitative receptor defects in families with androgen resistance: Failure of stabilization of the fibroblast cytosol androgen receptor. J Clin Endocrinol Metab 55:465–474.
148. Pinsky L, Kaufman M, Summitt RL (1981) Congenital androgen insensitivity due to a qualitatively abnormal androgen receptor. Am J Med Genet 10:91–99.
149. Kaufman M, Pinsky L, Simard L, Wong SC (1982) Defective activation of androgen receptor complexes: A marker of androgen insensitivity. Mol Cell Endocrinol 25:151–162.
150. Kovacs WJ, Griffin JE, Weaver DD, Carlson BR, Wilson JD (1984) A mutation that causes lability of the androgen receptor under conditions that normally promote transformation to the DNA-binding state. J Clin Invest 73:1095–1104.
151. Kaufman M, Pinsky L, Feder-Hollander R (1981) Defective up-regulation of the androgen receptor in human androgen insensitivity. Nature 293:735–737.
152. Faiman C, Winter JSD (1974) The control of gonadotropin secretion in complete testicular feminization. J Clin Endocrinol Metab 39:631–638.
153. Boyar RM, Moore RJ, Rosner W, Aiman J, Chipman J, Madden JD, Marks JF,

Griffin JE (1978) Studies on gonadotropin-gonadal dynamics in patients with androgen insensitivity. J Clin Endocrinol Metab 47:1116–1117.

154. Tremblay RR, Foley TP Jr, Corvol P, Park IJ, Kowarski A, Blizzard RM, Jones HW, Migeon CJ (1972) Plasma concentration of testosterone, dihydrotestosterone, testosterone-oestradiol binding globulin, and pituitary gonadotropins in the syndrome of male pseudohermaphroditism with testicular feminization. Acta Endocrinol (Copenh) 70:331–341.

155. MacDonald PC, Madden JD, Brenner PF, Wilson JD, Siiteri PK (1979) Origin of estrogen in normal men and in women with testicular feminization. J Clin Endocrinol Metab 49:905–916.

156. Kelch RP, Jenner MR, Weinstein R, Kaplan SL, Grumbach MM (1972) Estradiol and testosterone secretion by human, simian, and canine testes, in males with hypogonadism and in male pseudohermaphrodites with the feminizing testes syndrome. J Clin Invest 51:824–830.

157. Conte FA, Grumbach MM (1984) Bearing of abnormalities of sex differentiation on the hypothalamic-pituitary-gonadal axis at puberty. In Serio M, Motta M, Zanisi M, Martini M (eds): Sexual Differentiation: Basic and Clinical Aspects. Raven Press, New York, pp 275–285.

158. Lee PA, Brown TR, LaTorre HA (1986) Diagnosis of the partial androgen insensitivity syndrome during infancy. JAMA 255:2207–2209.

159. Sultan C, Emberger JM, Devillier C, Chavis C, Terraza A, Descomps B, Jean R (1984) Specific 5α-dihydrotestosterone receptor and 5α-reductase activity in human amniotic fluid cells. Am J Obstet Gynecol 150:956–960.

160. Lubs HA Jr, Vilar O, Bergenstal DM (1959) Familial male pseudohermaphroditism with labial testes and partial feminization: Endocrine studies and genetic aspects. J Clin Endocrinol Metab 19:1110–1120.

161. Gilbert-Dreyfus S, Sebaoun CA, Belaisch J (1975) Etude d'un cas familial d'androgynoidisme avec hypospadias grave, gynecomastie et hyperoestrognie. Ann Endocrinol 18:93–101.

162. Bowen P, Lee CSN, Migeon CJ, Kaplan NM, Whalley PJ, McKusick VA, Reifenstein EC (1965) Hereditary male pseudohermaphroditism with hypogonadism, hypospadias, and gynecomastia (Reifenstein's syndrome). Ann Intern Med 62:252–270.

163. Rosewater S, Gwinup G, Hamwi GJ (1965) Familial gynecomastia. Ann Intern Med 63:377–385.

164. Wilson JD, Harrod MJ, Goldstein JL, Hemsell DL, MacDonald PC (1974) Familial incomplete male pseudohermaphroditism, type I: Evidence for androgen resistance and variable clinical manifestations in a family with the Reifenstein syndrome. N Engl J Med 290:1097–1103.

165. Maes M, Lee PA, Jeffs RD, Sultan C, Migeon CJ (1980) Phenotypic variation in a family with partial androgen insensitivity syndrome. Am J Dis Child 134:470–473.

166. Griffin JE, Wilson JD (1977) Studies on the pathogenesis of the incomplete forms of androgen resistance in man. J Clin Endocrinol Metab 45:1137–1143.

167. Eil C (1982) Familial incomplete male pseudohermaphroditism associated with impaired nuclear androgen retention. J Clin Invest 71:850–858.

168. Gyorki S, Warne GL, Khalid BAK, Funder JW (1983) Defective nuclear accumulation of androgen receptors in disorders of sexual differentiation. J Clin Invest 72:819–825.

169. Griffin JE, Wilson JD (1984) Disorders of androgen receptor function. Ann NY Acad Sci 438:61–71.

170. Pinsky L, Kaufman M, Killinger DW, Burko B, Shatz D, Volpe R (1984) Human minimal androgen insensitivity with normal dihydrotestosterone-binding capacity in cultured genital skin fibroblasts: Evidence for an

androgen-selective qualitative abnormality of the receptor. Am J Hum Genet 36:965–978.

171. Jukier L, Kaufman M, Pinsky L, Peterson RE (1984) Partial androgen resistance associated with secondary 5α-reductase deficiency: Identification of a novel qualitative androgen receptor defect and clinical implications. J Clin Endocrinol Metab 59:679–688.

172. Kaufman M, Pinsky L, Bowin A, Au MWS (1984) Familial external genital ambiguity due to a transformation defect of androgen-receptor complexes that is expressed with 5α-dihydrotestosterone and the synthetic androgen methyltrienolone. Am J Med Genet 18:493–507.

173. Pinsky L, Kaufman M, Chudley AE (1985) Reduced affinity of the androgen receptor for 5α-dihydrotestosterone but not methyltrienolone in a form of partial androgen resistance. J Clin Invest 75:1291–1296.

174. Madden JD, Walsh PC, MacDonald PC, Wilson JD (1973) Clinical and endocrinologic characterization of a patient with the syndrome of incomplete testicular feminization. J Clin Endocrinol Metab 41:751–760.

175. Keenan BS, Kirkland JL, Kirkland RT, Clayton GW (1977) Male pseudohermaphroditism with partial androgen insensitivity. Pediatrics 59:224–231.

176. Price P, Wass JAH, Griffin JE, Leshin M, Savage MO, Large DM, Bu'Lock DE, Anderson DC, Wilson JD, Besser GM (1984) High dose androgen therapy in male pseudohermaphroditism due to 5α-reductase deficiency and disorders of the androgen receptor. J Clin Invest 74:1496–1508.

177. Aiman J, Griffin JE, Gazak JM, Wilson JD, MacDonald PC (1979) Androgen insensitivity as a cause of infertility in otherwise normal men. N Engl J Med 300:223–227.

178. Migeon CJ, Brown TR, Lanes R, Palacios A, Amrhein JA, Schoen EJ (1984) A clinical syndrome of mild androgen insensitivity. J Clin Endocrinol Metab 59:672–678.

179. Brown TR, Spinola-Castro A, Berkovitz GD, Migeon CJ (1985) Androgen receptor in cultured human testicular fibroblasts. J Clin Endocrinol Metab 61:134–141.

180. Aiman J, Griffin JE (1982) The frequency of androgen receptor deficiency in infertile men. J Clin Endocrinol Metab 54:725–732.

181. Eil C, Gamblin GT, Hodge JW, Clark RV, Sherins RJ (1985) Whole cell and nuclear androgen uptake in skin fibroblasts from infertile men. J Androl 6:365–371.

7
States of Aldosterone Deficiency or Pseudodeficiency

MORRIS SCHAMBELAN and ANTHONY SEBASTIAN

Sodium, chloride, potassium, and acid-base homeostasis in humans is dependent in part on the "mineralocorticoid activity" of steroid hormones secreted by the adrenal cortex and is reflected in the electrolyte and acid-base composition of the extracellular fluid and in the volume of the arterial blood. Aldosterone is the only adrenal steroid known to participate in the physiologic feedback regulation of electrolyte and volume homeostasis. Thus, primary or secondary disorders of the adrenal cortex that result in subnormal rates of aldosterone production are characterized by abnormalities in extracellular fluid volume, blood pressure, and electrolyte and acid-base composition. In addition, rare disorders in which signs and symptoms of aldosterone deficiency are present despite normal or elevated levels of aldosterone (so-called pseudohypoaldosteronism) also result in major alterations in mineral homeostasis. This chapter will review the pathophysiology and treatment of disorders due to aldosterone deficiency and pseudodeficiency in the context of our current understanding of the physiologic effects of mineralocorticoid hormones, summarized in Section I.

I. Physiologic Effects of Mineralocorticoid Hormones

By direct or indirect mechanisms, mineralocorticoid hormones modulate transport of sodium, potassium, hydrogen, bicarbonate, chloride, calcium, magnesium, and water by target epithelia in kidney, large bowel, sweat glands, and salivary ducts. The renal effects of mineralocorticoid hormones are critical for the maintenance of mineral homeostasis in humans who are not excreting copious amounts of sweat or liquid stool, and reflect specific actions on the tubules that can occur independently of measurable changes in glomerular filtration rate or renal hemodynamics (1).

Renal functional responses attributed to mineralocorticoid hormones have been observed in the "distal nephron" of dogs (2), the proximal and distal convoluted tubules of rats (3), the connecting tubules of rabbits (4), the cortical collecting tubules of rabbits (4,5), and the medullary collecting ducts of rabbits (6) and rats (7). It is questionable whether the observed proximal tubular effects of aldosterone have homeostatic significance, and it appears that earlier studies that revealed an effect of mineralocorticoid hormones in the distal convoluted tubule (3) may have reflected changes occurring in the initial collecting tubules. Studies of isolated nephron segments in the rabbit indicate that the cortical collecting tubules are responsive to mineralocorticoid hormones (5). At least two cell types ("light" and "dark" cells) have been identified in the collecting tubules (8), but it is not known if both cell types respond to mineralocorticoid hormones.

Studies of hormone receptor binding and nuclear effects at the subcellular level provide an understanding of steroid action as well as a biochemical basis for the distinction between glucocorticoid and mineralocorticoid activity. The steroid crosses the cell membrane, presumably by simple diffusion, where it binds to high-affinity receptor proteins in the cytoplasm of the appropriate target tissue (9). In rat kidney, there appear to be multiple steroid hormone-binding sites (10) with different affinities. Indeed, the "mineralocorticoid" and "glucocorticoid" potency of a particular steroid hormone correlates with its relative affinity for the so-called mineralocorticoid and glucocorticoid receptors (11). Steroid action is dependent on the formation of an active steroid-receptor complex that is transferred to the nucleus for subsequent action on the genome (12).

Studies with inhibitors of ribosomal protein synthesis suggest that, in the case of aldosterone, this action results in the formation of a protein that plays a primary role in mediating the sodium transport effect of the hormone (13,14). On the other hand, steroids that act as competitive antagonists (e.g., progesterone, 17α-hydroxyprogesterone, and spironolactone) bind with high affinity to the mineralocorticoid hormone receptor but fail to initiate the subsequent steps required for mineralocorticoid hormone action (15). Autoradiographic studies in isolated nephron segments from rabbits have led to the conclusion that, in the cortical collecting tubule, aldosterone binds predominantly to mineralocorticoid sites, whereas in the loop of Henle, distal convoluted tubule, and medullary collecting duct, aldosterone binds to both mineralocorticoid and glucocorticoid sites (16).

A. Effects on Sodium Transport

The effects of mineralocorticoid hormones on ion and water transport at the cellular level have been studied in toad and turtle bladders, which have epithelia with morphologic and functional characteristics similar to mammalian collecting ducts, and in isolated segments of mammalian renal tubules. In these tissues, mineralocorticoid hormones increase sodium reabsorption despite opposing transepithelial electrical and chemical gradients; i.e., mineralocorticoid hormones stimulate the active transport of sodium (17). The stimulation of active sodium transport in turn increases the lumen-negative transtubular potential gradient; i.e., the transport process is electrogenic (5). In isolated perfused tubule segments (from rabbits), mineralocorticoids increase the rate of sodium absorption and the transtubular electrical potential, which is normally lumen-negative (5,18), but have no effect on the outer medullary collecting tubules, which are normally at near zero potential or are lumen-positive and which evidence no net sodium transport (18). The increased lumen-negative potential may account in part for the stimulation of hydrogen ion and potassium secretion observed in response to mineralocorticoid hormones, although aldosterone can stimulate the secretion of these ions even when the effect of the hormone on sodium transport is prevented (19,20). Furthermore, aldosterone stimulates hydrogen ion secretion in a segment of the rabbit medullary collecting duct in which sodium transport is not stimulated by aldosterone (6). Mineralocorticoid hormones also stimulate net chloride reabsorption in isolated perfused collecting tubules of rabbits (21), and aldosterone enhances the effect of vasopressin to increase transepithelial water movement in the toad bladder (22).

According to current concepts, the alterations in transport induced by mineralocorticoid hormones are mediated by molecular-ionic events at the luminal and peritubular cell membranes. Transepithelial transport of sodium from lumen to cell along a favorable electrical and chemical (concentration) gradient is followed by active extrusion of sodium into peritubular fluid by a Na^+-K^+ exchange pump on the peritubular membrane. Mineralocorticoid hormones increase both the permeability of the luminal cell membrane to sodium and potassium and the rate of peritubular Na^+-K^+ exchange (23,24). The rate of transport of sodium across the peritubular membrane is increased at least in part by promoting increased cellular entry of sodium from the lumen. The peritubular transport process entails hydrolysis of adenosine triphosphate (ATP) by Na^+-K^+ ATPase that is activated by increasing concentrations of sodium in the cell or of potassium in the peritubular fluids (25). This enzyme is present in highest concentration in the peritubular cell membrane and has been identified as the Na^+-K^+ exchange pump. Studies using discrete nephron segments in rabbits demonstrated that mineralocorticoid hormones increase Na^+-K^+ ATPase activity only in the cortical collecting tubule and in the so-called connecting segment (located between the end of the distal convoluted tubule and the beginning of the cortical collecting tubule) (4). The possibility has been considered that a direct effect of mineralocorticoids on renal Na^+-K^+ ATPase activity may be the initial step in the stimulation of sodium transport, but the evidence for this hypothesis is not conclusive. There are conflicting data from recent studies regarding whether preclusion of cellular uptake of sodium does (26) or does not (27) inhibit stimulation of Na^+-K^+ ATPase activity by mineralocorticoid hormones. From other evidence, it has been proposed that mineralocorticoid hormones increase the supply of ATP and alter the coupling between ATP utilization and the transport process, thereby enhancing the "efficiency" of energy used for transport (28).

B. Effects on Potassium Transport

The specific effects of mineralocorticoid hormones on the transepithelial transport of potassium have not been studied as extensively as those on the transport of sodium. In the lumen-negative post–macula densa segments of the distal nephron, principally the collecting tubules and ducts and "late distal convoluted tubules," the potassium secretory process can be subdivided into several components: (a) active cellular uptake of potassium at the peritubular membrane mediated by Na^+-K^+ ATPase; (b) passive diffusion of potassium from cell to lumen along a favorable electrochemical gradient; (c) active cellular uptake of potassium from the lumen by a pump presumed to be located in the luminal membrane. Indirect evidence suggests that in the collecting duct segments, potassium secretion is mediated by the principal cells whereas potassium reabsorption is mediated by the intercalated cells (29,30). By increasing sodium reabsorption, mineralocorticoid hormones increase lumen-negative transtubular electrical potential gradient in the cortical collecting tubule, where it is known that potassium secretion correlates positively with the magnitude of the transtubular potential difference and rate of sodium reabsorption (31). Mineralocorticoid hormones also appear to increase the permeability of the luminal membrane to potassium (23,24) and the rate of cellular uptake of potassium at the peritubular membrane (secondary to increased Na^+-K^+ ATPase activity) (23), thus facilitating the potassium secretory response to increased transtubular potential difference.

C. Effects on Hydrogen Ion Transport

The cellular mechanism of acidification in the mammalian distal nephron has not been defined. A widely adopted "working model" has been derived from in vitro studies of Colombian toad and freshwater turtle urinary bladders, which contain hydrogen ion–secreting epithelia resembling the mammalian collecting tubules and ducts (in contradistinction to more proximal nephron segments) with respect to embryologic origin and certain characteristics of structure and function (32,33). In these epithelia, the hydrogen ion secretory mechanism is electrogenic (i.e., it is capable of transporting a net positive charge into the lumen), and it is not dependent on sodium resorption, although the rate of hydrogen ion secretion may be influenced by the rate of active sodium resorption, which determines the magnitude of the transepithelial electrical potential difference promoting or opposing hydrogen ion secretion. Secretion of hydrogen ion occurs even under conditions in which sodium transport is prevented and no favorable transepithelial electrical potential difference is permitted to develop. Aldosterone increases the rate of hydrogen ion secretion when the transepithelial hydrogen ion concentration gradient is nonlimiting (i.e., it increases the "conductance" of hydrogen ion in the active transport pathway); it does not influence the magnitude of the transepithelial hydrogen ion concentration gradient required to nullify net hydrogen ion secretion, which is an index of the "driving force" (proton motive force) of the hydrogen ion secretory pump (19). In the mucosal border of the turtle bladder, a reversible proton-translocating ATPase can function as a hydrogen ion pump that can account for urinary acidification (34).

Recent studies of acidification of luminal fluid by the rabbit cortical collecting tubule perfused in vitro indicate that the physiologic characteristics of the hydrogen ion secretory mechanism in the mammalian distal nephron are similar to those defined in the urinary bladder of the toad and turtle (20). The data support the existence of a distinct acidification mechanism that is both active and electrogenic. Electrogenic hydrogen ion secretion is stimulated by pretreatment of the animals with mineralocorticoid hormone (20). Mineralocorticoid hormone stimulation of hydrogen ion secretion in the rabbit cortical collecting tubule is due in part to stimulation of active sodium reabsorption, since, in the presence of luminal sodium, mineralocorticoid hormone administration is associated with a large and significant increase in lumen-negative electrical potential, and due in part to a more direct effect on the proton secretory mechanism, since, in the absence of luminal sodium, mineralocorticoid hormone administration results in a large and significant increase in lumen-positive electrical potential attributable to hydrogen ion secretion (20).

In the outer medullary collecting tubules, unlike the cortical collecting tubules, mineralocorticoid hormones do not stimulate sodium reabsorption, nor do they alter the transtubular lumen-positive electrical potential difference in the collecting tubules that are located in the inner stripe of the outer medulla (18). This lumen-positive potential results from electrogenic hydrogen ion secretion, which occurs despite the absence of sodium reabsorption, but is accompanied by chloride secretion at nearly identical rates (35,36). Aldosterone stimulates hydrogen ion secretion in this nephron segment (6), in which the intrinsic hydrogen ion secretory capacity exceeds that of the cortical collecting tubule and approaches 15% of that of the proximal convoluted tubule (36). Thus, mineralocorticoid hormone-stimulated hydrogen ion secretion in the outer medullary collecting tubule may be the major determinant of distal nephron hydrogen ion secretion rates in the mammalian kidney under ordinary physiologic conditions and in physiologic and pathologic states associated with altered mineralocorticoid levels.

II. Syndromes of Aldosterone Deficiency and Pseudodeficiency

Primary disorders of the adrenal gland that impair significantly the secretion of mineralocorticoid hormones characteristically result in sodium wasting and extracellular fluid volume depletion as well as in abnormalities of plasma electrolyte and acid-base composition. As a consequence of the reduction in the effective intravascular volume, renin secretion is stimulated, and plasma renin activity is markedly elevated. Similar abnormalities of electrolyte and acid-base homeostasis are also found characteristically in those clinical disorders in which hypoaldosteronism is secondary to a deficient stimulation by the renin-angiotensin system, the major known trophic stimulus for aldosterone secretion. Distinction between primary and secondary hypoaldosteronism is therefore facilitated greatly by measurement of plasma renin activity. We have used the terms hyperreninemic and hyporeninemic hypoaldosteronism to distinguish primary from secondary aldosterone deficiency (Table 7.1). Elevated plasma renin levels are also found characteristically in a rare disorder in which the signs and symptoms of aldosterone deficiency are present despite markedly elevated levels of aldosterone, so-called classic (type I) pseudohypoaldosteronism. Mineralocorticoid hormone resistance also appears to be present in another rare disorder in which volume depletion and hyperreninemia are not characteristic, yet in which the electrolyte and acid-base abnormalities suggest aldosterone deficiency. We have designated this disorder type II pseudohypoaldosteronism. In the following sections the pathophysiology and treatment of syndromes of aldosterone deficiency and pseudodeficiency are discussed in detail.

A. Primary (Hyperreninemic) Hypoaldosteronism

Hypoaldosteronism in primary adrenal disorders can occur as an isolated defect of the zona glomerulosa or as a component of a generalized process that also reduces the secretion of steroids that originate in the zona fasciculata. In both instances the renin-angiotensin system is stimulated—hence the term hyperreninemic hyperaldosteronism (Fig. 7.1). In the generalized form, ACTH levels are also increased as consequence of cortisol deficiency.

1. GENERALIZED ADRENOCORTICAL INSUFFICIENCY

Generalized adrenocortical insufficiency (Addison's disease) occurs when destruction of adrenocortical tissue reduces secretory rates of both

TABLE 7.1. Syndromes of aldosterone deficiency and pseudodeficiency.

I. Primary (hyperreninemic) hypoaldosteronism
   A. Combined mineralocorticoid and glucocorticoid deficiency
      1. Addison's disease
         Autoimmune
         Tuberculosis, fungal, sarcoid, hemachromatosis
         Metastatic carcinoma, lymphoma
         Bilateral adrenal hemorrhage
      2. Bilateral adrenalectomy
      3. Adrenal enzyme deficiency states
         21-Hydroxylase deficiency
         3β-ol dehydrogenase deficiency
         Desmolase deficiency
   B. Aldosterone deficiency without glucocorticoid deficiency
      1. Corticosterone methyl oxidase deficiency, types I and II
      2. Isolated zona glomerulosa defect
      3. Critically ill patients
      4. Heparin therapy
      5. Converting enzyme inhibitors
II. Secondary (hyporeninemic) hyperaldosteronism
   A. Hyporeninemic hypoaldosteronism
      1. Diffuse, histologically evident renal disease
         Diabetic nephropathy
         Tubulointerstitial diseases
         Obstructive uropathy
      2. Autonomic neuropathy
      3. Extracellular volume expansion
      4. Impaired conversion of prorenin to active renin
      5. Impaired renal prostaglandin production
         Nonsteroidal antiinflammatory drugs
III. Pseudohypoaldosteronism
   A. Type I (classic) pseudohypoaldosteronism
   B. Type II pseudohypoaldosteronism

glucocorticoid and mineralocorticoid hormones below the physiologic needs of the organism. This combined deficiency is usually fatal if untreated. Although tuberculosis was once the most common cause of Addison's disease, with a decrease in the incidence of this disease in developed countries, autoimmune-mediated adrenal atrophy is now the most common etiology (37). Metastatic carcinoma, lymphoma, adrenal hemorrhage, and fungal infections account for the majority of the remaining cases. A similar combined hormonal deficiency results from surgical removal of both adrenal glands and from certain inherited disorders of steroid biosynthesis that will be described in subsequent sections.

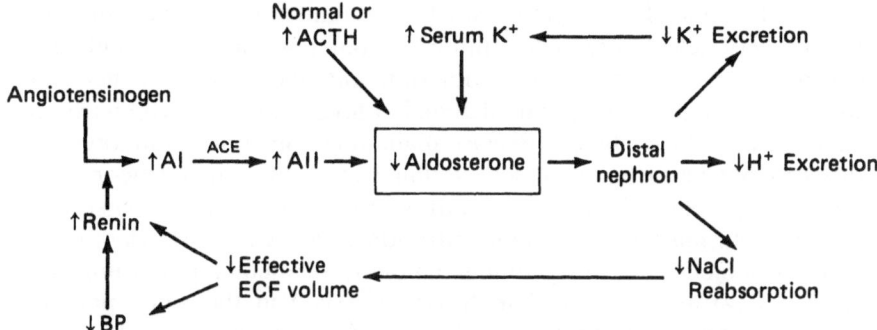

FIGURE 7.1. Pathophysiology of primary (hyperreninemic) hypoaldosteronism. The proximate disorder is a primary abnormality of the adrenal cortex that causes impaired adrenal secretion of aldosterone (rectangle), resulting in reduced sodium chloride reabsorption and diminished potassium and hydrogen ion secretion by the renal tubule. Aldosterone deficiency persists despite increased activity of the renin-angiotensin system and hyperkalemia. Secretion of adrenocorticotropin (ACTH) is normal when the defect is confined to the zona glomerulosa and is increased when the lesion also impairs cortisol production by the zona fasciculata.

a. Pathophysiology

Deficiency of glucocorticoid hormones results in anorexia, weight loss, weakness, apathy, and a general inability to withstand "stress" (38). Glucocorticoids normally inhibit ACTH secretion, so markedly elevated levels of ACTH are characteristic of Addison's disease. Hyperpigmentation results from the increased levels of ACTH or related proopiomelanocortin-derived peptides (lipotropin, melanocortin). Deficiency of mineralocorticoid hormones results in impaired renal sodium conservation as well as impaired potassium and hydrogen ion secretion in the distal nephron. If the sodium intake is sufficiently large, extracellular fluid volume and plasma potassium and bicarbonate levels can be maintained at normal or near normal levels. If the sodium intake is low, however, or if extrarenal losses of sodium occur, the inability to conserve sodium maximally results in marked sodium deficits, hyponatremia, hyperkalemia, acidosis, hypovolemia, and increased plasma renin levels (39). Glucocorticoid deficiency may add to the severity of the hypovolemia by redistributing fluid between vascular and extravascular compartments (40) and to the hyponatremia by impairing renal diluting ability (41).

Studies in patients with adrenal insufficiency provide evidence that normal circulating levels of aldosterone influence renal clearance of potassium in subjects ingesting moderate amounts of dietary potassium, sodium, and chloride. We examined this influence in a group of adrenalectomized patients maintained on physiologic replacement doses of glucocorticoid (dexamethasone) and mineralocorticoid (fludrocortisone) steroids (42). In some of these patients, when administration of the mineralo-corticoid was selectively discontinued, renal potassium and net acid excretion decreased, and plasma potassium and hydrogen ion concentrations increased and were sustained at higher levels (Fig. 7.2). Similarly, in some patients with Addison's disease maintained on constant amounts of hydrocortisone or cortisone, discontinuing treatment with fludrocortisone (or similar potent mineralocorticoid) is associated with a sustained increase in plasma potassium concentration to values greater than those of normal subjects ingesting similar amounts of dietary potassium and sodium (43,44) and greater than those observed in the same patients treated with physiologic replacement doses of fludrocortisone (44). These data suggest that the level of circulating aldosterone is an important determinant of the set point of renal regulation of plasma potassium concentration and body potassium content in humans ingesting moderate to high levels of dietary potassium.

These findings also suggest that the reduced renal potassium and hydrogen ion secretion resulting from aldosterone deficiency can be particularly offset by increasing the delivery of sodium containing fluid to the distal nephron when abundant dietary sodium is provided (Fig. 7.2), but that a limitation of renal potassium clearance persists. Increasing dietary intake of potassium is associated with an exaggerated increase in plasma potassium concentration (45), thus revealing the underlying impairment of renal potassium homeostasis.

b. Diagnosis and Treatment

The diagnosis of generalized adrenocortical insufficiency is confirmed by the finding of subnormal plasma levels of cortisol and aldosterone and by reduced urinary excretion of their major metabolites, the urinary 17-hydroxycorticoids and aldosterone-18-glucuronide, respectively. Evidence that the primary abnormality resides in the adrenal cortex can be provided by the demonstration that plasma levels of ACTH and renin are elevated concomitantly. Inasmuch as measurement of plasma ACTH levels is not available in all laboratories and special precautions are required for handling such samples, the primary nature of the adrenal disorder is frequently demonstrated by measurement of cortisol levels following either acute or chronic stimulation by exogenously administered ACTH. Subnormal levels of cortisol that fail to increase in response to ACTH point to a primary abnormality of cortisol secretion. Aldosterone levels will also fail to increase normally in such patients. Prolonged administration of ACTH may be required in certain circumstances to distinguish primary from secondary syndromes of adrenocortical deficiency.

Generalized adrenocortical insufficiency often presents as a medical emergency; therapy should never be withheld pending laboratory results that establish the diagnosis. When the diagnosis of adrenal insufficiency is

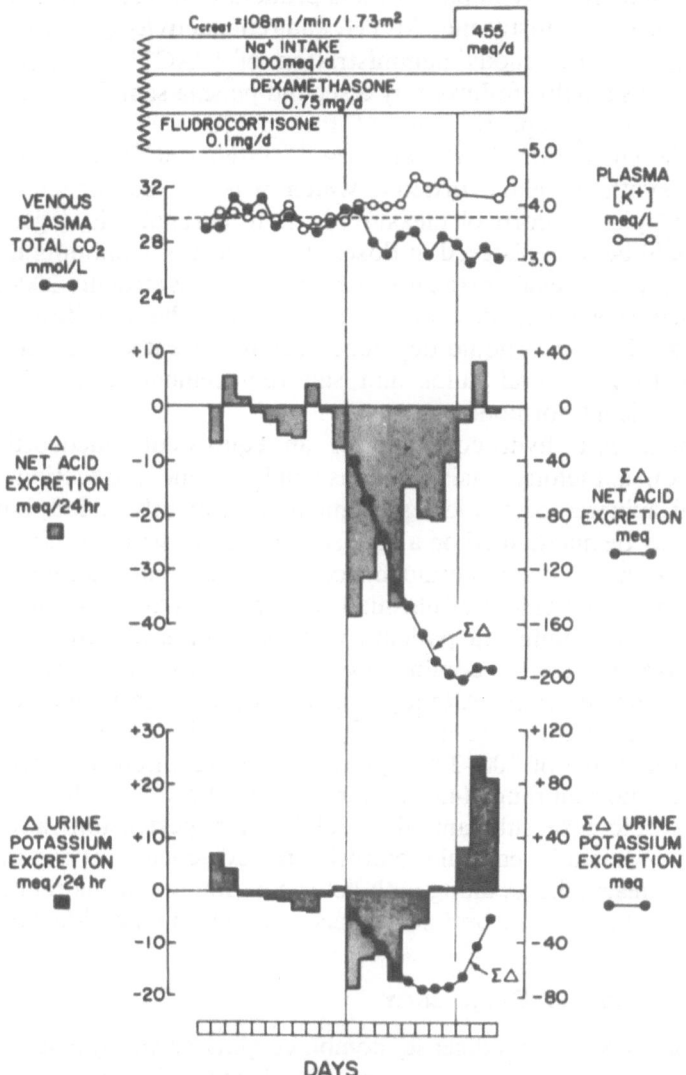

FIGURE 7.2. Effect of selective discontinuation of mineralocorticoid replacement therapy in an adrenalectomized patient who had normal kidneys and who was maintained on a normal intake of sodium chloride. Δ urine potassium and net acid excretion (shaded bars) reflect the differences between the daily excretion rate and the mean excretion rate in the control period (scale at left-hand side of figure). ΣΔ urine potassium and net acid excretion (solid circles) reflect the accumulated daily differences (scale at right-hand side of figure). After mineralocorticoid therapy was discontinued, urine potassium and net acid excretion decreased, plasma potassium concentration increased, and plasma total CO_2 levels decreased below control levels. (Reprinted from *Kidney International* (42) with permission.)

suspected in a critically ill patient, it is prudent to obtain plasma samples in which cortisol, aldosterone, ACTH, and renin activity can be measured subsequently. Intravenous administration of [1-24]ACTH (Cortrosyn, 25 units) given as a bolus followed by a second plasma sample 1 h later is an additional useful step. If delay of therapy for even 1 h is deemed inadvisable, glucocorticoid therapy can be initiated with dexamethasone, a potent synthetic glucocorticoid which will not be reflected in the subsequent plasma cortisol measurement. In adrenal crisis, glucocorticoids should be administered in doses that reflect the amounts normally secreted by the adrenal cortex under maximal stress, usually 200–300 mg of hydrocortisone per day or its equivalent. The manifestations of associated mineralocorticoid deficiency are best treated in such circumstances with parenteral fluids and salt replacement rather than with mineralocorticoid hormones.

Treatment with hydrocortisone or an equivalent glucocorticoid is required on a lifelong basis. Because aldosterone production is also decreased, mineralocorticoid replacement is frequently added, but many patients may be maintained on a high sodium intake without mineralocorticoid. The need for mineralocorticoid replacement therapy can be assessed by measurement of plasma renin activity (43), a sensitive index of extracellular volume in patients with Addison's disease. The usual mineralocorticoid steroid administered therapeutically is fludrocortisone, which is effective orally. Although 50–100 μg fludrocortisone per day is the usual recommended dose (38,46), recent studies have suggested that doses in the range of 200–300 μg/day may be required to reduce renin levels to the normal range (43). At these doses, however, blood pressure levels may increase substantially, and frank hypertension may occur. Thus, there should be careful monitoring for evidence of hypertension or congestive heart failure, along with determination of plasma renin levels, in order to determine the optimal replacement dose in individual patients.

2. 21-HYDROXYLASE DEFICIENCY

In addition to Addison's disease, combined glucocorticoid and mineralocorticoid deficiency also occurs in several congenital disorders of the adrenal gland that are characterized pathologically by diffuse bilateral enlargement of the adrenal cortex and that are collectively referred to as congenital adrenal hyperplasia (47). These disorders result from inherited defects in adrenal steroid biosynthesis assumed to result from adrenal enzyme defects. The reduced secretion of steroids by the adrenal gland leads to supernormal circulating levels of adrenotropic hormones (ACTH and/or angiotensin), leading to adrenal hyperplasia. The most common form of congenital adrenal hyperplasia results from defective 21-hydroxylation in the zona fasciculata and in some but not all patients also in the zona glomerulosa. Secretion by the zona fasciculata of 21-hydroxylated steroids is subnormal, whereas that of progesterone and 17α-hydroxy-progesterone (the normal substrates for 21-hydroxylation) is increased. Signs and symptoms of combined glucocorticoid and mineralocorticoid deficiency are commonly but not invariably present (48,49). The increased levels of 17α-hydroxyprogesterone result in overproduction of adrenal androgens, causing clinical virilization (pseudohermaphroditism in the female, precocious puberty in the male), varying greatly in severity among patients. The increased levels of 17α-hydroxyprogesterone and progesterone in the peripheral circulation contribute to the clinical manifestations of mineralocorticoid insufficiency inasmuch as these steroids are mineralocorticoid receptor antagonists that inhibit the renal action of aldosterone (50,51).

3. CORTICOSTERONE METHYLOXIDASE DEFICIENCY

A deficiency of the mixed-function oxidase (corticosterone methyloxidase) required for the final steps of aldosterone biosynthesis results in a syndrome of isolated mineralocorticoid deficiency characterized by salt wasting, hyperkalemia, and metabolic acidosis (52–54). The terms corticosterone methyloxidase types 1 and 2 have been proposed to describe the aldosterone deficiency states caused by impaired hydroxylation of corticosterone to 18-hydroxycorticosterone and of impaired dehydrogenation of 18-hydroxycorticosterone to aldosterone, respectively (52). In patients with the type 2 defect, plasma levels of aldosterone may occasionally be in the "normal" range (53), but these values should be considered to be inappropriately low for the marked degree of hyperreninemia and hyperkalemia that are present and that in fact reflect the mineralocorticoid-deficient state. The presence of the type 2 defect can be established by the demonstration of an increase in the excretory rate of the major urinary metabolites of 18-hydroxycorticosterone (52). Because corticosterone methyloxidase is not required for the biosynthesis of cortisol, there are no associated abnormalities in the levels of cortisol, ACTH, or adrenal androgens. The severity of the consequences of mineralocorticoid deficiency in affected children can be minimized by the maintenance of a high salt intake or by mineralocorticoid replacement therapy. The requirements for mineralocorticoid therapy and/or a high salt intake may decrease with age (53), but usually such patients can not tolerate dietary salt restriction.

4. SELECTIVE DESTRUCTION OF THE ZONA GLOMERULOSA

Acquired lesions that cause selective destruction of the zona glomerulosa can result in an isolated deficiency of aldosterone and 18-hydroxycorticosterone, steroids that originate solely (aldosterone) or primarily (18-hydroxycorticosterone) in this outermost layer of adrenal cells. Such lesions can occur as a consequence of a chronic autoimmune process in which antibodies are directed only against cells of the zona glomerulosa (55,56) or as the initial phase of a pathologic process that eventually leads to generalized adrenocortical insufficiency (57,58). Mineralocorticoid replacement is given as in other primary adrenal disorders.

Hypoaldosteronism that occurs in critically ill patients (59,60) or in those on heparin therapy (61) rarely results in clinically significant electrolyte disorders, and thus mineralocorticoid therapy may not be indicated. Administration of converting enzyme inhibitors may cause significant hyperkalemia, particularly in patients with underlying renal insufficiency (62). In such patients, hyperreninemic hypoaldosteronism is present as a consequence of reduced angiotensin II levels and does not imply the presence of a primary adrenal abnormality. Whereas mineralocorticoid therapy will correct the electrolyte abnormalities, the usual approach is to discontinue the offending agent.

B. Secondary (Hyporeninemic) Hypoaldosteronism

Other than rare cases of isolated hypoaldosteronism due to corticosterone methyloxidase deficiency, aldosterone deficiency occurring in the absence of glucocorticoid deficiency is most commonly the result of deficient secretion of renin by the renal juxtaglomerular apparatus (Fig. 7.3), usually associated with diffuse, histologically evident renal parenchymal disease that impairs glomerular filtration and renal tubule function (63–65). In these disorders, the proximate cause of aldosterone deficiency is presumed to be a subnormal circulating level of the major hormonal regulator of aldosterone secretion, angiotensin II, the intravascular generation of which from its precursor, angiotensin I, is initiated by renin-mediated enzymatic cleavage of the circulating angiotensin precursor, angiotensinogen, produced in the liver.

In addition to deficient renin secretion, hypoaldosteronism caused by angiotensin II insufficiency can occur as a result of impaired hepatic production of angiotensinogen (66). A syndrome of hypoaldosteronism possibly due to acquired insensitivity of the adrenal zona glomerulosa to angiotensin II has also been described (67).

1. HYPORENINEMIC HYPOALDOSTERONISM

a. Prevalence and Pathogenesis

In adults, mineralocorticoid deficiency occurs in numerous renal and extrarenal disorders that cause diminished renal secretion of renin (63–65,68–75). In affected patients, mineralocorticoid deficiency characteristically causes hyperkalemia and is commonly associated with hyperchloremic metabolic acidosis. Sodium depletion and renal sodium wasting are not invariably present, however, and in some patients, total body sodium and extracellular fluid volume are supernormal (70), raising the

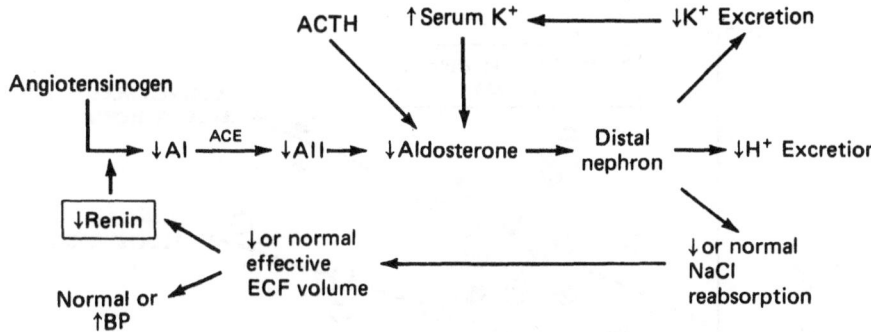

FIGURE 7.3. Pathophysiology of secondary (hyporeninemic) hypoaldosteronism. The proximate disorder is impaired renal secretion of renin (rectangle) with a consequent decrease in the normal trophic influence of angiotensin II on aldosterone secretion. Aldosterone secretion remains diminished despite normal levels of ACTH and increased levels of serum potassium owing to the critical role of angiotensin II in aldosterone biosynthesis. Frank volume depletion and hypotension do not occur invariably.

possibility that in such patients deficient renin secretion is a functional consequence of reduced renal clearance of sodium chloride.

The prevalence of aldosterone deficiency in hyperkalemic patients with diffuse renal parenchymal disease and chronic renal insufficiency was investigated in 31 such patients who had creatinine clearances ranging from 10 to 56 ml/min/1.73 m² and serum potassium concentrations ranging from 6.0 to 8.7 mEq/L when dietary potassium was unrestricted at the time of initial diagnosis (65). None of the patients had cortisol deficiency. By comparison with normal subjects, 14 of the 31 patients had frank hypoaldosteronism. Furthermore, when the effects of concomitant hyperkalemia on aldosterone production were assessed by comparing the rate of urine aldosterone excretion to the serum potassium concentration, 23 of the 31 patients were judged to have hypoaldosteronism. Thus, hypoaldosteronism occurs commonly in patients with chronic renal insufficiency who have hyperkalemia.

The results of this study also indicated that aldosterone deficiency contributed to the pathogenesis of hyperkalemia in these patients. A significantly greater degree of hyperkalemia was observed in the patients with hypoaldosteronism than in those without overt hypoaldosteronism in relation to the net dietary load of potassium (Fig. 7.4). It is conceivable that the presence of hyperkalemia in the second subgroup, although of lesser severity than in those with overt hypoaldosteronism, was related to a lesser degree of hypoaldosteronism in these patients. Viewed in another way, these data indicate that renal clearance of potassium is impaired in hyperkalemic patients with chronic renal insufficiency, and that the

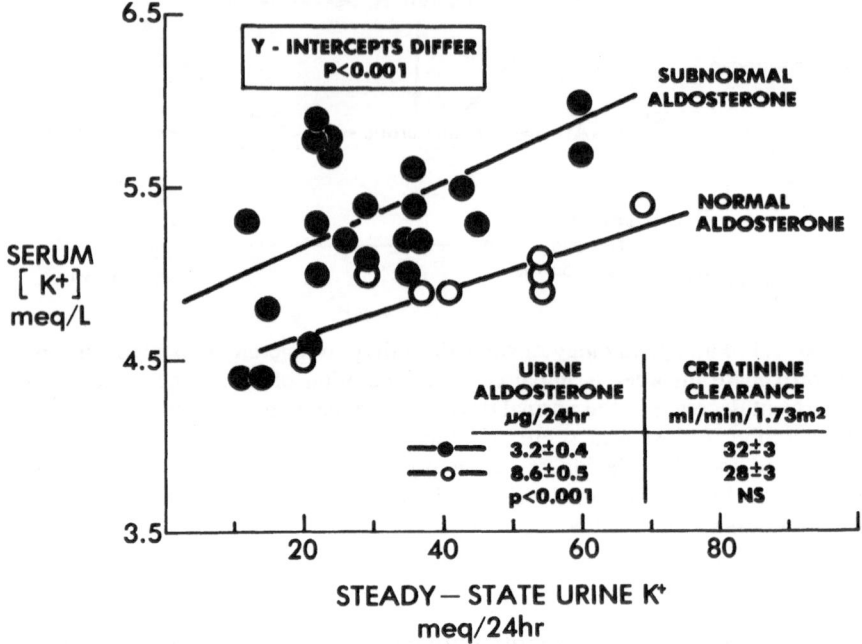

FIGURE 7.4. Relationship between serum potassium concentration and steady-state urinary potassium excretion in patients with isolated hypoaldosteronism (solid circles) and patients with chronic renal insufficiency without hypoaldosteronism (open circles). The findings may be interpreted as reflecting the relationship between serum potassium concentration and dietary potassium intake inasmuch as in the steady state, urinary potassium excretion reflects dietary potassium intake. For any given net dietary potassium intake, serum potassium levels are higher in the patients with hypoaldosteronism. (Reprinted from *Kidney International* (65) with permission.)

magnitude of this impairment is related to the extent to which aldosterone secretion is subnormal.

Hyporeninemia was present in more than 80% of these patients with isolated hypoaldosteronism and appeared to be quantitatively sufficient to account for the degree of aldosterone deficiency. The pathogenesis of hyporeninemia in such individuals is probably multifactorial, including such possibilities as structural damage to the juxtaglomerular apparatus (72), autonomic neuropathy (73), and failure to convert the inactive renin precursor (prorenin) to the active form of the molecule (74). In some patients, particularly those with diabetes mellitus, it has been suggested that a defect in the final step of aldosterone biosynthesis in the zona glomerulosa may also be present (74,75).

To study the function of the zona glomerulosa in such patients we measured levels of plasma aldosterone and its immediate biosynthetic precursor, 18-hydroxycorticosterone (18-OHB), in patients with isolated hypoaldosteronism and compared them to those in patients with a variety of adrenal and extraadrenal disorders (76). A disproportionate increase in 18-OHB relative to aldosterone was not observed in the group with isolated hypoaldosteronism (Table 7.2). Because concomitant hyporeninemia may have masked the expression of a biosynthetic block, we infused des-Asp'-angiotensin II (angiotensin III) as a specific stimulus of zona glomerulosa production (77). In the group with isolated hypoaldosteronism, the response of both aldosterone and 18-OHB to angiotensin III was reduced significantly in comparison to normal controls and to those with chronic renal insufficiency without hypoaldosteronism. The impaired aldosterone and 18-OHB secretory responses in patients with isolated hypoaldosteronism were similar to those in patients who had recently undergone unilateral adrenalectomy for an aldosterone-producing adenoma; in such patients, the prolonged hypermineralocorticoid state that had been present preoperatively resulted in suppression of the function of the contralateral adrenal gland and a transient state of hypoaldosteronism postoperatively (78). The impaired responses of both 18-OHB and aldosterone to angiotensin III indicate a generalized reduction in zona glomerulosa function rather than an enzymatic defect in adrenal biosynthesis as the cause of hypoaldosteronism in patients with isolated hypoaldosteronism. Such an impairment may result from prolonged hyporeninemia in this syndrome.

Other authors have suggested that mechanisms other than hyporeninemia may account for hypoaldosteronism in some patients with this syndrome, since aldosterone levels remain low despite hyperkalemia, a known and potent stimulus to aldosterone secretion (79). To evaluate the relative roles of potassium and angiotensin in the control of aldosterone secretion in patients with chronic renal insufficiency, we administered the angiotensin-converting enzyme inhibitor captopril to patients from the subgroup without hypoaldosteronism (80). This resulted in a prompt increase in plasma potassium to levels comparable to those seen in the subgroup with hyporeninemic hypoaldosteronism. Despite the hyperkalemia, urinary aldosterone excretion decreased to levels not significantly different from those in patients with hyporeninemic hypoaldosteronism, presumably as a result of the reduction in angiotensin II levels. Normal subjects studied in a similar fashion had no significant change in plasma potassium or urinary aldosterone excretion; presumably the reduction in angiotensin II levels was not as great, owing to the much greater increases in plasma renin activity that occurred in the normal subjects. Thus, treatment with an angiotensin-converting enzyme inhibitor can unmask an impaired renin secretory response mechanism in patients with chronic renal insufficiency, resulting in hypoaldosteronism, renal potassium retention, and hyperkalemia. Furthermore, these studies

TABLE 7.2. Evaluation of corticosterone methyloxidase activity as reflected by the 18-OHB/aldosterone ratio in disorders affecting the renal-adrenal axis.

Disorder	N	Plasma 18-OHB (ng/dl)	Plasma aldosterone (ng/dl)	18-OHB:aldo ratio	PRC[/PRA] (ng/ml/h)	Plasma K$^+$ (mEq/L)
Aldosterone-producing adenoma	31	163.1 ± 16.3***	58.3 ± 6.0***	2.8 ± 0.2	0.5 ± 0.1***	2.8 ± 0.1***
Idiopathic hyperaldosteronism	15	31.4 ± 2.9*	15.2 ± 1.2***	2.1 ± 0.2	0.9 ± 0.2***	3.3 ± 0.1***
Salt-losing nephropathy	3	485 ± 208***	197 ± 119***	3.1 ± 0.5	[6.3 ± 1.9]***	5.4 ± 0.5***
Bartter's syndrome	9	43.5 ± 7.2**	15.3 ± 2.7**	2.9 ± 0.2	[11.1 ± 1.4]	2.7 ± 0.1***
Isolated hypoaldosteronism	13	15.4 ± 2.4**	4.9 ± 0.5***	3.2 ± 0.4	1.4 ± 0.3***	5.2 ± 0.1***
Normal controls	15	23.3 ± 2.2	8.9 ± 0.5	2.6 ± 0.2	[1.6 ± 0.2] 5.1 ± 0.1	4.1 ± 0.1

Values are mean ± SEM. Comparison between patient groups and normal control subjects was done using Student's t-test for unpaired variables.
* $P < 0.05$.
** $P < 0.01$.
*** $P < 0.001$.
Reprinted with permission from Kater CE, Biglieri EG, Rost CR, Schambelan M, Hirai J, Chang BCF, Brust N. The constant plasma 18-hydroxycorticosterone to aldosterone ratio: An expression of the efficacy of corticosterone methyloxidase type II activity in disorders with variable aldosterone production. J Clin Endocrinol Metab 60:225–228, © by The Endocrine Society (1985).

provide additional support for the concept that angiotensin II is required for a normal adrenal secretory response to potassium (81).

b. Treatment

Inasmuch as patients with this syndrome are usually asymptomatic, it may be argued that treatment is not always necessary. Nevertheless, a life-threatening arrhythmia was the presenting finding in the index case (82) and has been noted in several of our patients who had markedly elevated serum potassium levels. The long-term consequences of untreated metabolic acidosis on bone mineralization and other cellular functions must also be considered.

Administration of fludrocortisone generally results in a prompt increase in renal potassium and hydrogen ion excretion and amelioration of hyperkalemia and metabolic acidosis (71). The doses of fludrocortisone required for optimal effects generally range from 100 to 300 μg per day, levels that had previously been considered "superphysiologic" (71) but that now appear to be more consistent with the range of doses that are required to normalize renin levels in patients with Addison's disease (43). Some degree of hyporesponsiveness to the renal effect of mineralocorticoids in patients with hyporeninemic hypoaldosteronism may reflect their underlying renal insufficiency.

In addition to treatment with exogenous mineralocorticoid hormone, hyperkalemia and acidosis can be ameliorated in patients with hyporeninemic hypoaldosteronism by the oral administration of sodium bicarbonate or of sodium polystyrene sulfonate, a resin that binds potassium and releases sodium in the lumen of the gastrointestinal tract. Amelioration of hyperkalemia is mediated by the movement of potassium into cellular compartments secondary to increased plasma bicarbonate concentration (during sodium bicarbonate therapy) or by increased excretion of potassium in the feces (during resin therapy). Despite impairment in renal bicarbonate resorption, administration of 1.5–2.0 mEq sodium bicarbonate per kilogram body weight per day may be sufficient to sustain correction of acidosis, and such therapy usually mitigates the severity of hyperkalemia. Often, the serum potassium concentration can be maintained at less than 5.0 mEq/L with combined alkali therapy and modest dietary potassium restriction.

Treatment of hyperkalemia and acidosis with these agents is not always effective and safe, however, particularly in patients with hypertension or a demonstrated increase in extracellular fluid volume and total exchangeable sodium before treatment (70). Treatment with fludrocortisone increases renal tubular reabsorption of sodium chloride and might exacerbate hypertension and lead to other deleterious consequences of extracellular fluid volume expansion. Administration of sodium bicarbonate entails an increased systemic load of sodium and does not usually fully correct hyperkalemia or reliably protect against acute endogenous or exogenous potassium loads. Chronic treatment with sodium polystyrene sulfonate also entails an increased sodium load, is poorly accepted, and can lead to fecal impaction.

Alternative therapeutic measures that can be considered in patients with hyporeninemic hypoaldosteronism include restriction of dietary potassium (83) and administration of a loop diuretic such as furosemide. Chronic administration of furosemide is known to result in an increase in net acid and potassium excretion in dogs (84) and to inhibit sodium chloride resorption in the loop of Henle, an effect that results in extracellular fluid volume contraction and secondary hyperaldosteronism. The kaliuretic response to furosemide can be attributed in part to inhibition of potassium reabsorption in the loop of Henle. In addition, both the kaliuretic and acid excretory response can be attributed to increased secretion of potassium and hydrogen ion in the distal nephron secondary to the combined effect of increased distal delivery of sodium and increased circulating levels of aldosterone.

We have recently demonstrated that administration of furosemide, in doses of 40–120 mg daily, increases potassium and net acid excretion and ameliorates substantially hyperkalemia and metabolic acidosis in such patients (85). In the patients studied, the magnitude of the acid excretory and kaliuretic effect of furosemide was directly correlated with the level of endogenous aldosterone. In those patients with the most severe degree of hypoaldosteronism, the beneficial effect of furosemide was greatly enhanced by pretreatment with small doses of fludrocortisone. The combination of fludrocortisone and furosemide offers the advantages of mutual potentiation of kaliuretic and acid-excretory effect and countervailing natiuretic and chloruretic effect, providing a mechanism for reducing hyperkalemia and acidosis while controlling body content of sodium chloride in relation to dietary sodium chloride level by the proper adjustment of the relative doses of the two agents.

C. Pseudohypoaldosteronism

Hyperkalemia and hyperchloremic metabolic acidosis occur in several clinical disorders referred to by the term "pseudohypoaldosteronism," because signs and symptoms suggestive of aldosterone deficiency are present, but aldosterone levels are, in fact, normal or supernormal, and the administration of exogenous mineralocorticoid, even in large amounts, is ineffective. Two pathophysiologically distinct forms of pseudohypoaldosteronism have been described.

1. TYPE I PSEUDOHYPOALDOSTERONISM

Classic (type I) pseudohypoaldosteronism is an apparently congenital and, in some cases, familial disorder characterized by failure to thrive, dehydration and hyponatremia due to renal salt wasting, hyperkalemia due to renal potassium retention, and renal tubular acidosis (86–90). In affected infants, plasma renin activity and plasma and urinary aldosterone concentrations are markedly elevated (Fig. 7.5), and neither glucocorticoid deficiency nor excess is present. Refractoriness of renal sodium wasting and potassium retention to mineralocorticoid administration is evident even during prolonged periods of parenteral administration of large amounts of deoxycorticosterone and aldosterone and oral administration of fludrocortisone. Supplementation of the diet with large amounts of sodium chloride can greatly ameliorate hyponatremia and hyperkalemia, relieve symptoms, and permit normal or improved growth. Characteristically, the severity of renal sodium wasting and potassium retention diminishes after the period of infancy, permitting discontinuation of sodium chloride supplements without recurrence of overt hyponatremia or hyperkalemia. Disordered renal handling of sodium and potassium persists, however, evidenced by recurrence of hyponatremia and hyperkalemia in response to restriction of dietary sodium chloride (89).

The pathophysiologic manifestations of pseudohypoaldosteronism of infancy are consistent with a cellular defect that interferes with the action of aldosterone in those segments of the renal tubule normally responsive to aldosterone, specifically the collecting tubules and ducts. Generalized glomerular and tubular dysfunction is not present in affected patients, and examination of biopsied renal tissue usually reveals only a mild hyperplasia of the juxtaglomerular apparatus. These patients are thus distinguishable from patients with type I pseudohypoaldosteronism associated with

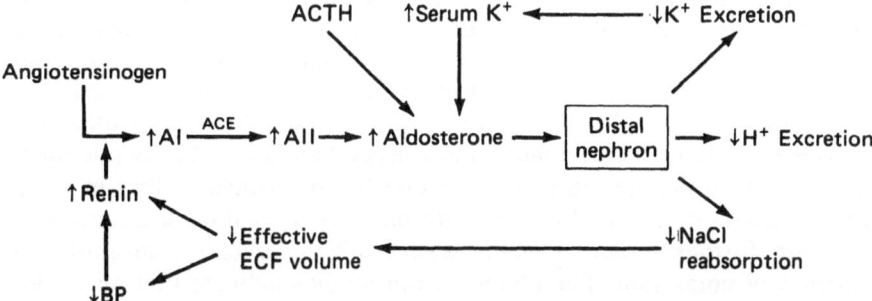

FIGURE 7.5. Pathophysiology of type I (classic) pseudohypoaldosteronism. The proximate abnormality is in the distal nephron, where the response to mineralocorticoids is impaired as a consequence of a diminished number of specific mineralocorticoid receptors or because of structural damage. Impaired sodium chloride reabsorption and retention of potassium and hydrogen ion occur as in primary hypoaldosteronism. Since the adrenal cortex is intact, the augmented activity of the renin-angiotensin system and hyperkalemia result in markedly increased levels of aldosterone.

acquired generalized renal parenchymal destruction (e.g., salt-losing nephritis caused by methicillin nephritis; medullary cystic disease) (91,92).

Armanini et al. (93) recently demonstrated a decreased number of type I mineralocorticoid-binding sites in monocytes obtained from 3 patients with pseudohypoaldosteronism of infancy, including the index case reported by Cheek and Perry (86). Because the steroid specificity of these receptors is indistinguishable from the renal minerolocorticoid receptor, it seems quite likely that a similar deficiency of type I receptors is present in the kidney. It is not certain that the same fundamental pathogenetic disturbance underlies classic pseudohypoaldosteronism of infancy in all reported cases, however, nor is it certain that a genetic abnormality is present in all cases. The familial occurrence of the disorder has been documented in only a small fraction of reported cases.

2. Type II Pseudohypoaldosteronism

Renal tubular responsiveness to the action of aldosterone also appears to be impaired in a rare syndrome characterized by hyperkalemia, hyperchloremic metabolic acidosis, hypertension, hyporeninemia, and abnormally reduced aldosterone production (94–101). Glomerular filtration rate is in the normal range. Mineralocorticoid resistance is revealed by the persistence of hyperkalemia and a markedly reduced kaliuretic response to large amounts of exogenously administered mineralocorticoid steroids. The antinatriuretic and antichloruretic responses to mineralocorticoid steroids may be intact. A primary defect in renal potassium secretion has been proposed (94–98).

In 1 patient with type II pseudohypoaldosteronism, we found that fractional renal potassium excretion was subnormal and increased only minimally during administration of large amounts of mineralocorticoid steroids (100). However, distal tubule potassium secretion increased greatly when distal sodium delivery was increased with anions other than chloride (sulfate or bicarbonate). Thus, hyperkalemia and mineralocorticoid resistance did not appear to be due to an intrinsic defect in renal potassium secretion. The finding of normal salivary and fecal secretion of potassium further mitigated against a generalized defect in transepithelial transport of potassium. The results of our studies indicate that mineralocorticoid-resistant renal hyperkalemia and acidosis in type II pseudohypoaldosteronism cannot be attributed to the absence of a renal potassium secretory mechanism but instead may depend on the amount of chloride available for reabsorption in the distal nephron. The findings suggest that the primary abnormality in this syndrome is a defect of the distal nephron that increases the reabsorptive avidity of the distal nephron for chloride, which (a) limits the sodium and mineralocorticoid-dependent voltage driving force for potassium and hydrogen secretion, resulting in hyper-

kalemia and acidosis, and (b) augments distal sodium chloride reabsorption, resulting in hyperchloremia, volume expansion, and hypertension. Such a "chloride shunt" might arise as a result of an abnormal
increase in the permeability of the distal nephron to chloride. Consistent
with the presence of such a chloride shunt, restriction of dietary sodium
chloride or administration of a chloruretic diuretic (furosemide, thiazide)
ameliorates hyperkalemia and acidosis (96,97,99–101).

Acknowledgments. These studies were supported by U.S. Public Health
Service grants DK06415 from the National Institute of Diabetes, Diges-
tive, and Kidney Diseases; HL11046 from the National Heart, Lung, and
Blood Institute; and RR79 and RR83 from the Division of Research
Resources.

References

1. Garrod O, Davies SA, Cahill G Jr (1955) The action of cortisone and
   desoxycorticosterone acetate on glomerular filtration rate and sodium and
   water exchange in the adrenalectomized dog. J Clin Invest 34:761–776.
2. Vander AJ, Malvin RL, Wilde WS, Lapides J, Sullivan LP, McMurray VM
   (1958) Effects of adrenalectomy and aldosterone on proximal and distal
   tubular sodium reabsorption. Proc Soc Exp Biol Med 99:323–332.
3. Hierholzer K, Wiederholt M, Stolte H (1966) Hemmung der Natriumresorp-
   tion in Proximalen und Distalen Konvolut adrenalektomierter Ratten.
   Pfluegers Arch 291:43–62.
4. Garg LC, Knepper MA, Burg MB (1981) Mineralocorticoid effects on
   Na-K-ATPase in individual nephron segments. Am J Physiol 240:F536–
   F544.
5. Gross JB, Imai M, Kokko JP (1975) A functional comparison of the cortical
   collecting tubule and the distal convoluted tubule. J Clin Invest 55:1284–
   1294.
6. Stone DK, Seldin DW, Kokko JP, Jacobson HR (1983) Mineralocorticoid
   modification of rabbit medullary collecting duct acidification: A sodium-
   independent effect. J Clin Invest 72:77–83.
7. Uhlich E, Baldamus CA, Ullrich KJ (1969) Einflu von Aldosteron auf den
   Natriumtransport in den Sammelrohren der Saugetierniere. Pfluegers Arch
   308:111–126.
8. Tisher CC (1981) Anatomy of the kidney. In Brenner BM, Rector FC Jr
   (eds.): The Kidney. Saunders, Philadelphia, pp 3–75.
9. Feldman D, Funder JW, Edelman IS (1972) Subcellular mechanisms in the
   action of adrenal steroids. Am J Med 53:545–560.
10. Funder JW, Feldman D, Edelman IS (1973) The role of plasma binding and
    receptor specificity in the mineralocorticoid action of aldosterone. Endocri-
    nology 92:994–1004.
11. Baxter JD, Schambelan M, Matulich DT, Spindler BJ, Taylor AA, Bartter
    FC (1976) Aldosterone receptors and the evaluation of plasma mineralocorti-
    coid activity in normal and hypertensive states. J Clin Invest 58:579–589.
12. Samuels HH, Tomkins GM (1970) Relation of steroid structure to enzyme
    induction in hepatoma tissue culture cells. J Mol Biol 52:57–74.

13. Crabbe J, DeWeer P (1964) Action of aldosterone on the bladder and skin of the toad. Nature 202:298–299.
14. Edelman IS, Bogoroch R, Porter GA (1963) On the mechanism of action of aldosterone on sodium transport: The role of protein synthesis. Proc Natl Acad Sci USA 50:1169–1177.
15. Fanestil DD (1968) Mode of spironolactone action: Competitive inhibition of aldosterone binding to mineralocorticoid receptors. Biochem Pharmacol 17:2240–2242.
16. Vandewalle A, Farman N, Bencsath P, Bonvalet JP (1981) Aldosterone binding along the rabbit nephron: An autoradiographic study on isolated tubules. Am J Physiol 240:F172–F179.
17. Crabbe J (1961) Stimulation of active sodium transport by the isolated toad bladder with aldosterone in vitro. J Clin Invest 40:2103–2110.
18. Stokes JB, Ingram MJ, Williams AD, Ingram D (1981) Heterogeneity of the rabbit collecting tubule: Localization of mineralocorticoid hormone action to the cortical portion. Kidney Int 20:340–347.
19. Al-Awqati Q, Norby LH, Mueller A, Steinmetz PR (1976) Characteristics of stimulation of H transport by aldosterone in turtle urinary bladder. J Clin Invest 58:351–358.
20. Koeppen BM, Helman SI (1982) Acidification of luminal fluid by the rabbit cortical collecting tubule perfused in vitro. Am J Physiol 242:F521–F531.
21. Hanley MJ, Kokko JP (1978) Study of chloride transport across the rabbit cortical collecting tubule. J Clin Invest 62:39–44.
22. Handler JS, Preston AS, Orloff J (1969) Effect of adrenal steroid hormones on the response of the toad's urinary bladder to vasopressin. J Clin Invest 48:823–833.
23. Koeppen BM, Biagi BA, Giebisch G (1983) Intracellular microelectrode characterization of the rabbit cortical collecting duct. Am J Physiol 244:F35–F47.
24. Stokes JB (1985) Mineralocorticoid effect on K permeability of the rabbit cortical collecting tubule. Kidney Int 28:640–645.
25. Katz AI, Epstein FH (1967) The role of sodium-potassium activated adenosine triphosphatase in the reabsorption of sodium by the kidney. J Clin Invest 46:1999–2011.
26. Petty KJ, Kokko JP, Marver D (1981) Secondary effect of aldosterone on Na-K ATPase activity in the rabbit cortical collecting tubule. J Clin Invest 68:1514–1521.
27. Geering K, Girardet M, Bron C, Krachenbuhe J-P, Rossier BC (1982) Hormonal regulation of (Na,K,-ATPase biosynthesis on the toad bladder: Effect of aldosterone and 3,5,3'-triiodo-L-thyronine. J Biol Chem 257:10338–10343.
28. Edelman IS, Fanestil DD (1970) Mineralocorticoids. In Litwack G (ed): Biochemical Actions of Hormones. Academic Press, New York, pp 321–364.
29. Stanton BA, Biemesderfer D, Wade JB, Giebisch G (1981) Structural and functional study of the rat distal nephron: Effects of potassium adaptation and depletion. Kidney Int 19:36–48.
30. Stetson DL, Wade JB, Giebisch G (1980) Morphologic alterations in the rat medullary collecting duct following potassium depletion. Kidney Int 17:45–56.
31. Stokes JB (1981) Potassium secretion by cortical collecting tubule: Relation to sodium absorption, luminal sodium concentration, and transepithelial voltage. Am J Physiol 241:F395–F402.
32. Ludens JH, Fanestil DD (1974) Aldosterone stimulation of acidification of

urine by isolated urinary bladder of the Colombian toad. Am J Physiol 226:1321–1326.

33. Al-Awqati Q (1978) H transport in urinary epithelia. Am J Physiol 235:F77–F88.

34. Dixon TE, Al-Awqati Q (1979) Urinary acidification in turtle bladder is due to a reversible proton-translocating ATPase. Proc Natl Acad Sci USA 76:3135–3138.

35. Stone DK, Seldin DW, Kokko JP, Jacobson HR (1983) Anion dependence of rabbit medullary collecting duct acidification. J Clin Invest 71:1505–1508.

36. Lombard WE, Kokko JP, Jacobson HR (1979) Bicarbonate transport in cortical and outer medullary collecting tubules. Am J Physiol 244:F289–F296.

37. Blizzard RM, Kyle M (1963). Studies of the adrenal antigens and antibodies in Addison's disease. J Clin Invest 42:1653–1660.

38. Bondy PK (1985) Disorders of the adrenal cortex. In Wilson JD, Foster DW (eds): Williams Textbook of Endocrinology, 7th Ed. Saunders, Philadelphia, pp 816–893.

39. Brown JJ, Fraser R, Lever AF, Robertson JIS, James VHT, McCusker J, Wynn V (1968): Renin, angiotensin, corticosteroids, and electrolyte balance in Addison's disease. Q J Med 37:97–118.

40. Haack D, Mohring J, Mohring B, Petri M, Hackenthal E (1977) Comparative study on development of corticosterone and DOCA hypertension in rats. Am J Physiol 233:F403–F411.

41. Boykin J, DeTorrente A, Erickson A, Robertson G, Schrier RW (1978) Role of plasma vasopressin in impaired water excretion of glucocorticoid deficiency. J Clin Invest 62:738–744.

42. Sebastian A, Sutton JM, Hulter HN, Schambelan M, Poler SM (1980) Effect of mineralocorticoid replacement therapy on renal acid-base homeostasis in adrenalectomized patients. Kidney Int 18:762–773.

43. Smith SG, Markandu ND, Banks RA, Dorrington-Ward P, MacGregor GA, Bayless J, Prentice MG, Wise P (1984) Evidence that patients with Addison's disease are undertreated with fludrocortisone. Lancet 1:11–14.

44. Thompson DG (1979) Mineralocorticoid replacement in Addison's disease. Clin Endocrinol (Oxf) 10:499–506.

45. Miller PD, Waterhouse C, Owens R, Cohen E (1975) The effect of potassium loading on sodium excretion and plasma renin activity in Addisonian man. J Clin Invest 56:346–353.

46. Baxter JD, Tyrell JB (1981) The adrenal cortex. In Felig P, Baxter JD, Broadus AE, Frohman LA (eds): Endocrinology and Metabolism. McGraw-Hill, New York, pp 385–510.

47. Finkelstein M, Shaefer JM (1979) Inborn errors of steroid biosynthesis. Physiol Rev 59:353–406.

48. Keenan BS, Holcombe JH, Kirkland RT, Potts VE, Clayton GW (1979) Sodium homeostasis and aldosterone secretion in salt-losing congenital adrenal hyperplasia. J Clin Endocrinol Metab 48:430–436.

49. Horner JM, Hintz RL, Luetscher JA (1979) The role of renin and angiotensin in salt-losing 21-hydroxylase-deficient congenital adrenal hyperplasia. J Clin Endocrinol Metab 48:776–783.

50. Landau RL, Lugibihl K (1958) Inhibition of the sodium retaining influence of aldosterone by progesterone. J Clin Endocrinol Metab 18:1237–1245.

51. Jacobs DR, Van der Poll J, Gabrilove JL, Soffer LJ (1961) 17α-hydroxyprogesterone—a salt-losing steroid: Relation to congenital adrenal hyperplasia. J Clin Endocrinol Metab 21:909–922.

52. Ulick S (1976) Diagnosis and nomenclature of the disorders of the terminal

portion of the aldosterone biosynthetic pathway. J Clin Endocrinol Metab 43:92–96.

53. Rosler A, Rabinowitz D, Theodor R, Ramirez LC, Ulick S (1977) The nature of the defect in a salt-wasting disorder in Jews of Iran. J Clin Endocrinol Metab 44:279–281.
54. Veldhuis JD, Kulin HE, Santen RJ, Wilson TE, Melby JC (1980) Inborn error in the terminal step of aldosterone biosynthesis: Corticosterone methyl oxidase type II deficiency in a North American pedigree. N Engl J Med 303:117–121.
55. Williams FA Jr, Schambelan M, Biglieri EG, Carey RM (1983) Acquired primary hypoaldosteronism due to an isolated zona glomerulosa defect. N Engl J Med 309:1623–1627.
56. Carey RW, Schambelan M, Biglieri EG, Bright GM (1984) Primary hypoaldosteronism due to zona glomerulosa defect. (Letter to the editor.) N Engl J Med 310:1394–1395.
57. Marieb NJ, Melby JC, Lyall SS (1974) Isolated hypoaldosteronism associated with idiopathic hypoparathyroidism. Arch Intern Med 134:424–429.
58. Saenger P, Levine LS, Irvine WJ, Gottesdiener K, Rauh W, Sonino N, Chow D, New MI (1982) Progressive adrenal failure in polyglandular autoimmune disease. J Clin Endocrinol Metab 54:863–868.
59. Zipser RD, Davenport MW, Martin KL, Tuck ML, Warner NE, Swinney RR, Davis CL, Horton R (1981) Hyperreninemic hypoaldosteronism in the critically ill: A new entity. J Clin Endocrinol Metab 53:867–873.
60. Stern N, Beck FWJ, Sowers JR, Tuck M, Hsueh WA, Zipser RD (1983) Plasma corticosteroids in hyperreninemic hypoaldosteronism: Evidence for diffuse impairment of the zona glomerulosa. J Clin Endocrinol Metab 57:217–220.
61. O'Kelly R, Magee F, McKenna TJ (1983) Routine heparin therapy inhibits adrenal aldosterone production. J Clin Endocrinol Metab 56:108–112.
62. Warren SE, O'Connor DT (1980) Hyperkalemia resulting from captopril administration. JAMA 244:2551–2552.
63. DeFronzo RA (1980) Hyperkalemia and hyporeninemic hypoaldosteronism. Kidney Int 17:118–134.
64. Phelps KR, Lieberman RL, Oh MS, Caroll HJ (1980) Pathophysiology of the syndrome of hyporeninemic hypoaldosteronism. Metabolism 29:186–199.
65. Schambelan M, Sebastian A, Biglieri EG (1980) Prevalence, pathogenesis, and functional significance of aldosterone deficiency in hyperkalemic patients with chronic renal insufficiency. Kidney Int 17:89–101.
66. Landier F, Guyene TT, Boutignon H, Nahoul K, Corvol P, Job J-C (1984) Hyporeninemic hypoaldosteronism in infancy: A familial disease. J Clin Endocrinol Metab 58:143–148.
67. Morimoto S, Kim KS, Yamamoto I, Uchida K, Takeda R, Kornel L (1979) Selective hypoaldosteronism with hyperreninemia in a diabetic patient. J Clin Endocrinol Metab 49:742–747.
68. Schambelan M, Stockigt JR, Biglieri EG (1972) Isolated hypoaldosteronism in adults: A renin deficiency syndrome. N Engl J Med 287:573–578.
69. Weidmann P, Reinhart R, Maxwell MH, Rowe P, Coburn JW, Massry SG (1973) Syndrome of hyporeninemic hypoaldosteronism and hyperkalemia in renal disease. J Clin Endocrinol Metab 36:965–977.
70. Oh MS, Carroll HJ, Clemmons JE, Vagnucci AH, Levison SP, Whang ESM (1974) A mechanism for hyporeninemic hypoaldosteronism in chronic renal disease. Metabolism 23:1157–1165.

71. Sebastian A, Schambelan M, Lindenfeld S, Morris RC Jr (1977) Amelioration of metabolic acidosis with fludrocortisone therapy in hyporeninemic hypoaldosteronism. N Engl J Med 297:576–583.
72. Sparagana M (1975) Hyporeninemic hypoaldosteronism with diabetic glomerulosclerosis. Biochem Med 14:93–103.
73. Tuck ML, Sambhi MP, Levin L (1979) Hyporeninemic hypoaldosteronism in diabetes mellitus: Studies of the autonomic nervous system's control of renin release. Diabetes 28:237–241.
74. DeLeiva A, Christlieb AR, Melby JC, Graham CA, Day RP, Luetscher JA, Zager PG (1976) Big renin and biosynthetic defect of aldosterone in diabetes mellitus N Engl J Med 295:639–643.
75. Tuck ML, Mayes DM (1980) Mineralocorticoid biosynthesis in patients with hyporeninemic hypoaldosteronism. J Clin Endocrinol Metab 50:341–347.
76. Kater CE, Biglieri EG, Rost CR, Schambelan M, Hirai J, Chang BCF, Brust N (1985) The constant plasma 18-hydroxycorticosterone to aldosterone ratio: An expression of the efficacy of corticosterone methyloxidase type II activity in disorders with variable aldosterone production. J Clin Endocrinol Metab 60:225–228.
77. Schambelan M, Kater CE, Biglieri EG, Sebastian A (1982) Response of plasma 18-hydroxycorticosterone and aldosterone to infusion of des-Asp^1-angiotensin II demonstrates a generalized reduction of adrenal zona glomerulosa function in isolated hypoaldosteronism. Program 64th Annual Meeting Endocrine Society, p 189.
78. Kater CE, Biglieri EG (1982) Zona fasciculata origin of 18-hydroxycorticosterone in the chronically suppressed zona glomerulosa. J Clin Endocrinol Metab 55:628–633.
79. Dluhy RG, Axelrod L, Underwood RH, Williams GH (1972) Studies of the control of plasma aldosterone concentration in normal man. II. Effect of dietary potassium and acute potassium infusion. J Clin Invest 51:1950–1957.
80. Peters W, Schambelan M, Sebastian A, Biglieri EG (1983) Aldosterone ameliorable hyperkalemia induced by angiotensin converting enzyme inhibitor. Kidney Int 23:131.
81. Pratt JH (1982) Role of angiotensin II in potassium-mediated stimulation of aldosterone secretion in the dog. J Clin Invest 70:667–672.
82. Hudson JB, Chobanian AV, Relman AS (1957) Hypoaldosteronism. A clinical study of a patient with an isolated adrenal mineralocorticoid deficiency, resulting in hyperkalemia and Stokes-Adams attacks. N Engl J Med 257:529–536.
83. Maher T, Schambelan M, Kurtz I, Hulter HN, Jones JW, Sebastian A (1984) Amelioration of metabolic acidosis by dietary potassium restriction in hyperkalemic patients with chronic renal insufficiency. J Lab Clin Med 103:432–445.
84. Bosch JP, Goldstein MH, Levitt MF, Kahn T (1977) Effect of chronic furosemide administration on hydrogen and sodium excretion in the dog. Am J Physiol 232:F397–F404.
85. Sebastian A, Schambelan M, Sutton JM (1984) Amelioration of hyperchloremic acidosis and hyperkalemia with furosemide therapy in patients with chronic renal insufficiency and type 4 renal tubular acidosis. Am J Nephrol 4:287–300.
86. Cheek DB, Perry JW (1958) A salt wasting syndrome in infancy. Arch Dis Child 33:252–256.
87. Donnell GN, Litman N, Roldan M (1959) Pseudohypoadrenalocorticism. Am J Dis Child 97:813–828.
88. Raine DN, Roy J (1962) A salt-losing syndrome in infancy. Pseudohypoadrenocorticalism. Arch Dis Child 37:548–556.
89. Postel-Vinay M C, Alberti GM, Ricour C, Limal J-M, Rappaport R, Royer P

(1974) Pseudohypoaldosteronism: Persistence of hyperaldosteronism and evidence for renal tubular and intestinal responsiveness to endogenous aldosterone. J Clin Endocrinol Metab 39:1038–1044.

90. Rosler A, Theodor R, Boichis H, Gerty R, Ulick S, Alagem M, Tabachnik E, Cohen B, Rabinowitz D (1977) Metabolic responses to the administration of angiotensin II, K and ACTH in two salt-wasting syndromes. J Clin Endocrinol Metab 44:292–301.

91. Cogan MG, Arieff AI (1978) Sodium wasting, acidosis and hyperkalemia induced by methicillin interstitial nephritis: Evidence for selective distal tubular dysfunction. Am J Med 64:500–507.

92. Stanbury SW, Mahler RF (1959) Salt-wasting renal disease: Metabolic observations on a patient with salt-losing nephritis. Q J Med 28:425–447.

93. Armanini D, Kuhnle U, Strasser T, Dorr H, Butenandt I, Weber PC, Stockigt JR, Pearce P, Funder JW (1985) Aldosterone receptor deficiency in pseudohypoaldosteronism. N Engl J Med 313:1178–1181.

94. Arnold JE, Healy JK (1969) Hyperkalemia, hypertension and systemic acidosis without renal failure associated with a tubular defect in potassium excretion. Am J Med 47:461–472.

95. Gordon RD, Geddes RA, Pawsey CGK, O'Halloran MW (1970) Hypertension and severe hyperkalaemia associated with suppression of renin and aldosterone and completely reversed by dietary sodium restriction. Aust Ann Med 4:287–294.

96. Spitzer A, Edelmann CM Jr, Goldberg LD, Henneman PH (1973) Short stature, hyperkalemia and acidosis: A defect in renal transport of potassium. Kidney Int 3:251–257.

97. Weinstein SF, Allan DME, Mendoza SA (1974) Hyperkalemia, acidosis, and short stature associated with a defect in renal potassium excretion. J Pediatr 85:355–358.

98. Brautbar N, Levi J, Rosler A, Leitesdorf E, Djaldeti M, Epstein M, Kleeman CR (1978) Familial hyperkalemia, hypertension, and hyporeninemia with normal aldosterone levels: A tubular defect in potassium handling. Arch Intern Med 138:607–610.

99. Lee MR, Ball SG, Thomas TH, Morgan DB (1979) Hypertension and hyperkalaemia responding to bendrofluazide. Q J Med 48:245–258.

100. Schambelan M, Sebastian A, Rector FC, Jr (1981) Mineralocorticoid-resistant renal hyperkalemia without salt wasting (type II pseudohypoaldosteronism): Role of increased renal chloride reabsorption. Kidney Int 19:716–727.

101. Sanjad SA, Keenan BS, Hill LL (1983) Renal hypoprostaglandism, hypertension, and type IV renal tubular acidosis reversed by furosemide. Ann Intern Med 99:624–627.

8
Bartter's Syndrome

MARK D. OKUSA and MARGARET JOHNSON BIA

Our understanding of the pathogenesis of Bartter's syndrome, although not complete, has evolved greatly since the first description of the disorder in 1962 (1). This chapter reviews the salient features of the syndrome, discussing previous and current concepts of its etiology and treatment, as well as the pathogenesis of the multiple hormonal derangements that accompany Bartter's syndrome. To illustrate the confusion that can arise concerning the basic definition of Bartter's syndrome and the essential defects required for proper diagnosis, we begin with clinical summaries of 2 patients with primary renal potassium wasting who demonstrated some but not all of the classic features of Bartter's syndrome. These cases will be analyzed at the conclusion of the chapter in light of the discourse that follows.

I. Case Summaries

J.W., a 21-year-old man, was admitted to the Clinical Research Center of Yale–New Haven Hospital for evaluation of hypokalemia. He had been in excellent health all his life until 4 weeks earlier, when he developed muscle weakness progressing to limb paralysis following the ingestion of 16 mg of dexamethasone which he received after a dental extraction. On admission to another hospital, blood pressure was found to be 125/65, and physical examination revealed no abnormalities except for an almost complete paralysis of arms and legs and diminished arm reflexes. There was documented severe hypokalemia (serum potassium concentration 0.9 mEq/L) with mild metabolic alkalosis (blood pH 7.46; serum bicarbonate concentration 34 mEq/L). Urine potassium concentration was inappropriately high (32 mEq/L), and the remaining chemistry values were normal except for a depressed serum magnesium level (1.1 mEq/L). Normal strength in arms and legs returned following potassium replacement. However, serum potassium continued to remain low (2.6–3.0 mEq/L), despite supplements of 120–200 mEq of potassium per day. Urine potassium concentrations ranged between 20 and 100 mEq/L, documenting inappropriate renal potassium excretion. The patient was then referred to the Clinical Research Center at Yale for evaluation.

The patient's parents had normal serum electrolytes, but his 23-year-old sister, M.W., had significant hypokalemia (serum potassium 2.5 mEq/L) along with a normal blood pressure, an elevated serum bicarbonate level (33 mEq/L), and a depressed serum magnesium level. The patient and his sister were therefore admitted for evaluation of the cause of their hypokalemia.

Pertinent laboratory data for both patients are listed in Table 8.1. In addition to the serum electrolyte abnormalities, they had renal potassium wasting documented during a balance study in which urine and stools were collected daily on a constant potassium intake. On 100 mEq of potassium intake per day, stool potassium was negligible (<10 mEq/day), and urinary potassium exceeded intake (Table 8.1). Multiple urine and stool analyses for diuretic and laxative screening were negative in both patients. On a sodium-restricted diet (10 mEq Na/day), urinary sodium excretion by M.W. was <10 mEq/day by the second day of restriction. However, for J.W., urinary Na excretion was still >20 mEq day on the fourth day. Plasma renin activity was markedly elevated in both patients, and plasma aldosterone concentration was at the upper limits of normal (Table 8.1). Following infusion with 2 L of normal saline over 4 h, a test used to exclude primary aldosteronism (2), plasma aldosterone concentrations was suppressed to below 5 ng/dl in both patients. Fractional chloride reabsorption in the loop of Henle was at the lower limits of normal (3–7) in both patients during water diuresis (Table 8.1).

Because primary renal potassium wasting was associated with hyperreninemia and slightly elevated plasma aldosterone concentration, the diagnosis of Bartter's syndrome or a variant of Bartter's syndrome was made in both patients. J.W. refused further therapy, since he had been asymptomatic despite his hypokalemia until he received dexamethasone. M.W. was treated during hospitalization with a short course of verapamil which had no effect on urinary potassium excretion. She was then discharged on 60 mEq of potassium per day with serum potassium remaining <3 mEq/L.

II. Historical Perspective

A. Original Description of the Syndrome

The original report by Frederick Bartter in 1962 described 2 patients with hypokalemia and metabolic alkalosis, hyperaldosteronism, and elevated plasma angiotensin II levels associated with marked resistance to the pressor action of infused angiotensin II and norepinephrine and hyperplasia of the juxtaglomerular apparatus on kidney biopsy (1). Subsequent reports described other biochemical abnormalities including defects in

TABLE 8.1. Laboratory data.

	Patient	
	M.W.	J.W.
Serum sodium (mEq/L)	141	138
Serum potassium (mEq/L)	2.4–2.6	2.2–2.5
Serum bicarbonate (mEq/L)	32	30
Serum chloride (mEq/L)	96	95
Urine potassium (mEq/L)	40–60	50–60
Serum uric acid (mg/dl)	4.2	5.3
Serum magnesium (mg/dl)	1.1	1.5
Supine plasma renin activity (ng/ml/min)[a]	13	22
Supine plasma aldosterone (ng/dl)[b]	17	16
Daily urinary potassium excretion (mEq)[c]	110	135
Glomerular filtration rate (ml/min)	141	113
Loop of Henle fractional chloride reabsorption (%)[d]	73	70

[a] Normal, 0.2–2.0 ng/ml/min.
[b] Normal, 1–16 ng/dl.
[c] During balance study on an intake of 100 mEq potassium per day.
[d] Following intravenous water loading. Lower limit of normal, 70–85% (3–7).

erythrocyte sodium transport (8–13), erythrocytosis (14,15), gout with elevated uric acid levels (16), and defects in platelet aggregation (17,18). Depressed serum magnesium levels have been found in some cases (19–30), and elevated urinary calcium excretion with nephrocalcinosis in others (31–33). Since not all patients with renal potassium wasting manifest all of these abnormalities, it is possible that Bartter's is only one of several different syndromes sharing similar or different etiologies leading to excessive urinary losses of potassium (34).

Muscle weakness has been the main physical sign of the disorder (1), although some patients, especially children, also have nocturia because of an associated defect in urinary concentrating ability (1,35,36). Children can also exhibit growth and mental retardation (35,37), features not usually seen in adult cases.

B. Genetic Transmission

A genetic basis for the syndrome is suggested by its familial occurrence (38,39). Furthermore, certain associated abnormalities, such as the defect in erythrocyte sodium transport, have been demonstrated in normokalemic family members of patients with the syndrome (9). Autosomal-recessive transmission is considered likely because of the equal incidence of the syndrome in males and females (39) and because of its occurrence in the offspring of consanguineous parents (40). Although a characteristic HLA typing pattern was observed in one family whose members had a form of renal potassium wasting (41), no significant HLA linkage pattern was found in a family with true Bartter's syndrome (42). Sporadic cases also occur, especially in adults.

C. Evolution of Concepts Regarding Pathogenesis

Initially, the potassium wasting in patients with Bartter's syndrome was thought to be the result of resistance to the pressor action of angiotensin II, resulting in compensatory overproduction of renin and aldosterone (1). After it was realized that the subnormal response to angiotensin was nonspecific, investigators speculated that a derangement in macula densa function led to the hyperreninemia and hyperaldosteronism (35). It was obvious that the hyperaldosteronism was secondary, since it was suppressible with volume expansion (43–46). A defect in chloride reabsorption in the loop of Henle was next postulated as the proximate cause of the potassium wasting (5–7). In fact, Dr. Bartter himself believed that this defect was the most important biochemical feature characteristic of the syndrome (47). Subsequent studies documented this defect in some (5–7,20–22,47) but not all patients (3,4,22,49–50) with renal potassium wasting. More recently, further doubt has been cast on the validity of the hypothesis invoking an impairment in chloride reabsorption, since its presence depends in part on the method used to detect it (49).

Another theory that gained widespread acceptance in the mid-1970s maintained that renal prostaglandin overproduction might be primarily responsible for the potassium wasting by stimulating excess renin secretion (53–55). However, it has subsequently been demonstrated that, although prostaglandin levels are elevated in most patients with Bartter's syndrome, suppression of their production rarely reduces renal potassium excretion to normal (30,54,56–61).

In the subsequent sections of this chapter, we will discuss these theories in more detail and explain why many of the abnormalities originally thought to be the primary cause of potassium wasting in Bartter's syndrome may actually be the result of potassium depletion. We will also discuss several of the metabolic derangements present in patients with Bartter's syndrome and review their contribution to the potassium wasting in these patients.

III. Etiology and Pathophysiology

A. Renin-Angiotensin-Aldosterone

1. VASCULAR RESISTANCE TO ANGIOTENSIN II

In his original description, Bartter postulated that the primary cause of the syndrome was vascular insensitivity to the pressor effect of angiotensin II (1). This would lead to decreased vascular resistance, secondary stimulation of the renin-angiotensin-aldosterone system, and subsequent hypokalemia from hyperaldosteronism. Since then, several lines of evidence have made this hypothesis untenable.

First, pressor resistance to angiotensin II is not unique to Bartter's syndrome but is found in other states of potassium depletion such as surreptitious vomiting (62,63,81), normal pregnancy (64,65), cirrhosis (66), nephrotic syndrome (67), and adrenal insufficiency (68). Second, clinical manipulations such as volume expansion (43–46) or the administration of drugs such as inhibitors of prostaglandin synthesis (53,54,69,70) or propranolol (54) have been shown to improve or normalize the vascular responsiveness to angiotensin II in patients with Bartter's syndrome even though the hypokalemia is not corrected. Thus, it is evident that although vascular unresponsiveness to angiotensin II is present in patients with Bartter's syndrome, it is not the primary cause of this disorder.

Possible mechanisms that could explain the vascular resistance to angiotensin II infusion in patients with Bartter's syndrome include the hypokalemia itself, tachyphylaxis from high endogenous levels of angiotensin II, and elevation in the levels of vasodilatory hormones. Hypokalemia alone has been shown to be vasodepressive in the dog (71) and may have a similar effect in human subjects. Increased angiotensin II, which has been demonstrated in patients with Bartter's syndrome (69,72), could induce tachyphylaxis to infused angiotensin by decreasing the absolute number of angiotensin II receptor sites (74) or by causing a relative decrease in receptor sites through occupancy by endogenous angiotensin (73). Increased vasodilatory hormones, such as prostaglandins, may modulate the vascular responsiveness to angiotensin II; in support of this are the findings that inhibitors of prostaglandin synthesis can reverse the resistance to angiotensin II infusion (53,54,56,61,69,70,75) and that plasma bradykinin, another potent vasodilator, is elevated in patients with Bartter's syndrome (57).

2. HYPERALDOSTERONISM

Initially, hyperaldosteronism was believed to be the major cause of hypokalemia in Bartter's syndrome. However, as is evident from the 2 patients described above as well as from previous reports (7,43,76,77), high serum or urinary levels of aldosterone are not always present. In fact, since potassium depletion inhibits aldosterone secretion, the degree of hyperaldosteronism depends, in part, on the degree of the hypokalemia (43). Thus patients with Bartter's syndrome and severe hypokalemia tend to have the lowest plasma aldosterone levels. Perhaps more important with respect to the role of aldosterone in the genesis of hypokalemia is the fact that potassium excretion is not decreased to normal following medical adrenalectomy with aminoglutethamide (43), surgical adrenalectomy (78,79), or treatment with spironolactone (1,45). Thus it is evident that hyperaldosteronism is not the primary cause of potassium wasting in patients with Bartter's syndrome, although it may aggravate potassium losses in some patients.

3. HYPERRENINEMIA

Hyperreninemia in Bartter's syndrome seems to be largely determined by the degree of potassium depletion. This interpretation stems from several lines of evidence. First, several other hypokalemic disorders, such as vomiting and the abuse of diuretics, are also associated with hyperreninemia (29,41,62,80–82). Second, experimentally induced potassium depletion in animals leads to hyperreninemia (71). The precise mechanism whereby potassium depletion stimulates renin release is still not completely understood, but studies suggest that prostaglandins may play an important modulating role. For example, prostaglandins can stimulate renin release (53–55,61,71,83,84), and hyperreninemia can be completely corrected following administration of inhibitors of prostaglandin synthesis (54,57,85). Hyperreninemia in Bartter's syndrome may also result from volume contraction in those patients who have associated salt wasting (1). Additionally, the defect in chloride reabsorption in the loop of Henle may contribute to the hyperreninemia (86). This chloride reabsorption in the thick ascending limb of the loop of Henle, which contains the macula densa, is a direct determinant of renin release independent of changes in arterial pressure or plasma volume (87). A defect in chloride reabsorption could stimulate renin release in patients with Bartter's syndrome in a manner similar to that which occurs following the administration of loop diuretics to normal individuals.

B. Defect in Cell Membrane Transport

Evidence suggesting abnormal function in platelets (18), erythrocytes (8–13), muscle cells (88), and renal tubular cells (5–7) has prompted the postulate that Bartter's syndrome arises from a generalized membrane defect. Garrick et al. recently revived interest in this hypothesis (50). They proposed that a generalized membrane defect could explain not only the defective chloride reabsorption in the loop of Henle but also the vascular resistance to infusion with pressor agents. Because of the obvious difficulties in investigating transport abnormalities in patients with Bartter's syndrome and because there is no animal model of this disorder, there is only a handful of studies on this subject. In most of these, investigators have observed an elevation in the intracellular sodium concentration in erythrocytes or muscle cells (8–13,88). The defect in sodium transport responsible for this finding has been more difficult to define, probably owing to technical variables. Nevertheless, several studies have demonstrated increased erythrocyte sodium efflux (8–12).

Recent evidence, however, suggests that a generalized cell membrane defect is not the primary cause of potassium wasting but is the result of the hypokalemia. Korff et al. evaluated erythrocyte transport in 9 patients with Bartter's syndrome before and after correcting hypokalemia with potassium chloride supplements (13). Although they found increased sodium concentration in erythrocytes and increased ouabain-sensitive and furosemide-sensitive sodium efflux, correction of hypokalemia normalized these transport processes. Thus, as is the case with many of the other abnormalities found in Bartter's syndrome, the abnormality in red cell sodium and the increased sodium efflux seem to be the result of potassium depletion rather than a sign of a generalized or specific membrane defect.

C. Renal Prostaglandin Overproduction

Prostaglandin overproduction has been demonstrated in several patients with Bartter's syndrome, and inhibitors of prostaglandin synthesis have been successfully used in their treatment (53–55,59). In 1 patient studied by Fichman et al., administration of indomethacin caused a marked decrease in PGA concentration in addition to a decrease in plasma renin activity and plasma levels of aldosterone and angiotensin II (54). Verberckmoes et al. also reported the successful treatment with indomethacin of 1 patient (53); although prostaglandin levels were not provided, histologic examination of a renal biopsy specimen revealed massive hyperplasia of the interstitial cells of the medulla, which are cells that synthesize prostaglandins when grown in tissue culture (89,90). Similar results have been reported with ibuprofen (55) and aspirin (59).

Since these early reports, there have been numerous clinical studies evaluating the contribution of prostaglandins to metabolic and clinical features of Bartter's syndrome. These studies confirmed that inhibitors of prostaglandin synthesis decrease the elevated urinary excretion of the prostaglandins PGA (54), PGE_2 (30,55–57,59,76), PGD_2 (51), 6-keto PGF_1 (51), and PGF_2 (55,59,76) and improve some of the metabolic abnormalities associated with the syndrome (30,51,54–57,76). Although this evidence suggests an important role for prostaglandins, other data indicate that elevated levels of prostaglandins do not account for all of the metabolic abnormalities. First, if prostaglandin overproduction were the primary cause, all patients with Bartter's syndrome should have elevated plasma and/or urinary prostaglandin levels; such is not the case (22,51,91). Second, if prostaglandins were responsible for the potassium wasting, treatment with inhibitors of prostaglandin synthesis should correct this metabolic abnormality. As shown in Fig. 8.1 and as described elsewhere (30,54,56–61,85), treatment with inhibitors of prostaglandin synthesis may decrease urinary prostaglandins and plasma renin activity yet fail to completely correct the hypokalemia. Third, although it has been

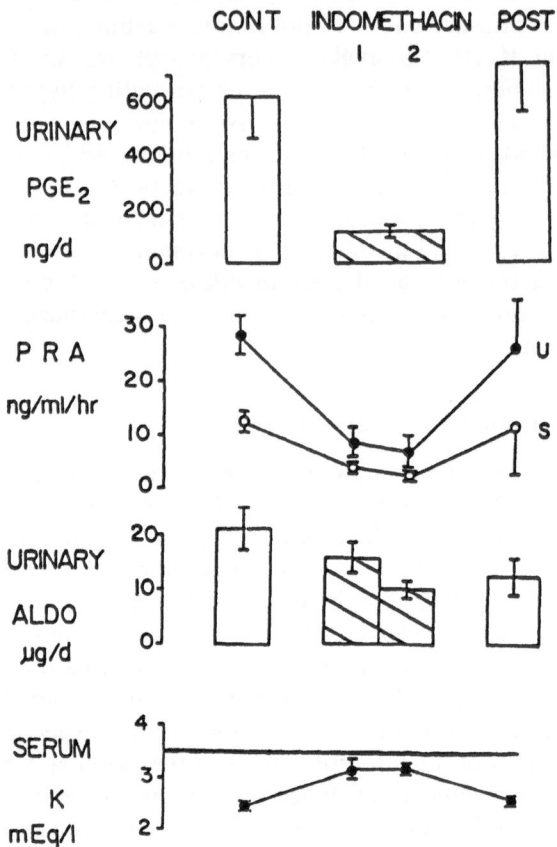

FIGURE 8.1. Effects of treatment with indomethocin on urinary prostaglandin excretion (PGE$_2$); plasma renin activity (PRA), supine (S) and upright (U); urinary aldosterone (aldo) excretion; and serum K concentration in 5 women with Bartter's syndrome. Indomethacin corrected the hormonal derangements without restoration of serum potassium concentration to normal. (Adapted from reference 85 with permission.)

postulated that prostaglandins are responsible for defective chloride transport in the loop of Henle in patients with Bartter's syndrome, it has been clearly demonstrated that inhibition of prostaglandin synthesis lowers prostaglandin levels without improving chloride transport in patients with this defect (21,47).

Data suggesting that elevated prostaglandin production is not the primary cause of Bartter's syndrome have also emerged from studies of other hypokalemic states. For example, patients with potassium depletion from psychogenic vomiting and diuretic abuse who exhibit metabolic abnormalities (hyperreninemia, hypokalemia, resistance to angiotensin II infusion, metabolic alkalosis) similar to those of patients who have true Bartter's syndrome have increased urinary PGE_2 excretion (63,81,92,93). These findings suggest that prostaglandin overproduction is a secondary rather than a primary phenomenon in patients with potassium depletion.

D. Defect in Renal Tubular Sodium Chloride Transport

Although hypokalemia, metabolic alkalosis, and renal potassium wasting are constant characteristics of patients with Bartter's syndrome, defects in the renal tubular handling of sodium chloride have not been consistently observed. In patients with such abnormalities, impaired handling of chloride in the loop of Henle has been observed most often (5–7,20–22,47), although defective sodium chloride transport in the proximal (6,46,94) or distal tubule (4) has occasionally been described.

Utilizing clearance techniques under conditions of a water diuresis, defects in chloride transport in the loop of Henle can be defined (95). Fractional chloride transport has been shown to be lower in patients with Bartter's syndrome than in control subjects or in patients with similar degrees of potassium depletion from other causes (5–7,20,22,47). Some investigators have interpreted this defect in chloride transport to be the primary cause of Bartter's syndrome (5–7,47). However, this impairment has not been a universal finding in all patients (3,22,44,46), as shown in Table 8.2, which summarizes published results of clearance studies evaluating segmental renal tubular function in response to a water diuresis in patients with Bartter's syndrome. The methods used to produce the water diuresis (i.e., intravenous or oral water or hypotonic saline) have varied, and the pitfalls of these clearance techniques must be considered. First, the results of such studies will vary with the degree of suppression of antidiuretic hormone and the magnitude of tubular flow rate. Second, the validity of the methodology itself using free water clearance to detect defects in tubular transport has been challenged (95). Finally, recent evidence indicates that the method of producing a water diuresis can affect the results.

Hernandez et al. performed oral water and intravenous hypotonic saline loading in 4 patients with Bartter's syndrome and in 5 normal subjects (49). They found that the diluting ability was similar in these groups following oral water loading. However, intravenous hypotonic saline infusion caused a greater decrease in fractional solute reabsorption in patients with Bartter's syndrome than in normal subjects, possibly owing to the greater degree of solute delivery to the loop with intravenous compared to oral water loading. This finding has been confirmed by others (50).

As can be seen in Table 8.2, most of the studies in which a defect in chloride transport has been reported have been performed using intravenous hypotonic saline or dextrose and water loading, whereas defects

TABLE 8.2. Results of tests for fractional chloride reabsorption in patients with renal potassium wasting.

Method[b]	Investigators	Fractional chloride reabsorption[a]	
		Defect	No defect
Intravenous dextrose and water	Stein (3)		+
	Gill and Bartter (47)	+	
	Bartter et al. (6)	+	
	Favre et al. (21)	+	
	Solomon et al. (22)	+	
	Ogihara et al. (98)	+	
Intravenous hypotonic saline	Tomko et al. (46)		+
	Chaimovitz et al. (5)	+	
	Fujita et al. (7)	+	
	Baehler et al. (20)	+	
	Uribarri et al. (4)		+
	Hernandez et al. (49)	+	
	Garrick et al. (50)[c]	+	+
Oral water	Norby et al. (44)		+
	Solomon et al. (22)[c]	+	+
	Ogihara et al. (98)		+
	Hernandez et al. (49)		+
	Garrick et al. (50)		+

[a] Criteria for defect based on values for fractional chloride reabsorption in normal subjects from Stein (3).
[b] Refers to method by which water diuresis was induced.
[c] Some patients with a defect; others without.

have seldom been diagnosed with oral water loading. These data support the notion that results of loop chloride reabsorption depend on the method used to demonstrate this function, and they underscore the difficulty in evaluating the results of previous studies.

In cases where a defect in chloride transport in the loop of Henle was demonstrated, it was initially speculated that this impairment could arise from prostaglandin overproduction, since animal studies have demonstrated that prostaglandins inhibit chloride transport in the thick ascending limb of the loop of Henle (96,97). However, as mentioned above, recent evidence clearly indicates that these two features are independent. Treatment with indomethacin decreases prostaglandin production without correcting the defect in fractional chloride absorption (47).

E. Primary Potassium Wasting

Several patients with potassium wasting but with completely normal renal tubular function have been reported (3,22,44,98). These patients manifest all the features of Bartter's syndrome with the exception of the defect in sodium chloride handling. The defect in potassium transport in such patients appears to be localized to the late distal tubule or early collecting duct (3). It is difficult to conceptualize a primary tubular defect leading to potassium wasting, since urinary potassium excretion results from tubular secretion. Most reported defects in the renal tubular handling of potassium, which as those that occur in patients with chronic renal failure or with tubulointerstitial renal diseases, lead to decreased and not increased renal potassium secretion (109). Thus, most defects in renal tubular potassium excretion cause hyperkalemia.

F. Kallikrein-Kinin System

Several studies have reported the presence of an activated kallikrein-kinin system in patients with Bartter's syndrome, evidenced by elevated urinary excretion of kallikrein (22,57,61,70) and plasma bradykinin (57). It has been postulated that bradykinin, a potent vasodilator, in conjunction with elevated levels of prostaglandins, offsets the vasoconstrictor effects of angiotensin II.

The physiologic regulation and activation of kallikrein is complex and has recently been reviewed by Fuller and Funder (99). With respect to Bartter's syndrome, the two aspects of particular interest are the production of kallikrein and its interaction with the prostaglandins and with the renin-angiotensin-aldosterone system. In patients with Bartter's syndrome, inhibition of prostaglandin synthesis normalizes elevated levels of urinary kallikrein (22,57,61) and plasma bradykinin (57). These results suggest that prostaglandins mediate the observed elevation in bradykinin, which in turn contributes to the hyporesponsiveness to infusions with pressor agents.

Further evidence supporting the contribution of the kallikrein-kinin system in the altered vascular reactivity of Bartter's syndrome is provided in a recent investigation utilizing aprotinin, an inhibitor of the proteolytic activity of glandular kallikrein (70). Following aprotinin administration in 4 patients with Bartter's syndrome, the amount of angiotensin II required to raise diastolic blood pressure 20 mm Hg was reduced by 30–55% (70). This occurred without significant changes in plasma renin activity or urinary prostaglandin excretion. In the same patients, treatment with indomethacin, which inhibits plasma renin activity, urinary PGE excretion, and urinary kallikrein, caused a greater decrease in the amount of angiotensin II required for blood pressure reduction than did use of aprotinin. These results suggest that both prostaglandin production and the activated kallikrein-kinin system contribute to the vascular hyporeactivity in Bartter's syndrome.

There is also considerable evidence to suggest that the renin-angiotensin-aldosterone system is involved in the regulation of the kallikrein-kinin system. In particular, aldosterone seems to modulate kallikrein synthesis (99). For example, high urinary kallikrein levels are found in primary aldosteronism (100–102). In addition, release of kallikrein from renal

cortical cells is increased with aldosterone and decreased with spironolactone treatment (103). In human subjects, low-sodium and high-
potassium diets are associated with elevated aldosterone and urinary
kallikrein, and these increases are blocked by spironolactone (104). These
results suggest that aldosterone is an important regulator of the kallikrein-
kinin system.

From the evidence cited above, it is apparent that both an activated
kallikrein-kinin system and elevated prostaglandin production contribute
to the vascular resistance to infused pressor agents that occurs in patients
with Bartter's syndrome. Regulation of the kallikrein-kinin system is
complex but appears influenced, in part, by aldosterone. Thus the two
sets of vasodilatory hormones, prostaglandin and the kallikrein-kinin
system, have independent as well as interrelated features in their regulation and action (99).

G. Hypomagnesemia

Hypomagnesemia is a common finding in patients with Bartter's
syndrome, occurring in 20–50% of reported cases (19–30). Although it
is known that magnesium deficiency can lead to renal potassium
wasting (105), the mechanism underlying this association is not well
defined. Prominent potassium wasting in association with hypomagnesemia and magnesium wasting has been reported in children other than
those with true Bartter's syndrome (23,24,27,28). Absence of a tubular
defect in chloride reabsorption or of juxtaglomerular hyperplasia in
renal biopsy distinguishes such children from those in the latter
category (34). In addition, growth and development are usually normal
in children with "magnesium-losing tubulopathy" (see Table 8.5)
(23,24). As discussed below, more information is needed before
deciding whether this syndrome is actually a variant of Bartter's
syndrome or represents a separate entity arising from primary
magnesium wasting.

Studies examining the efficacy of magnesium replacement on potassium wasting in patients with potassium-losing syndrome have yielded
conflicting results. Using balance studies, Baehler et al. found that
intravenous magnesium replacement in a patient with Bartter's syndrome dramatically reduced urinary potassium excretion and increased
serum potassium and magnesium concentrations (20). Other investigators have reported similar findings (25,29). In contrast, some investigators have found that magnesium replacement does not correct renal
potassium wasting (24,26). Why some of these patients improve with
magnesium supplements and others do not is not entirely clear.
Different amounts of magnesium replacement or different routes of
administration in patients with disparate etiologies for the potassium
wasting may have contributed to the varying results.

H. Defect in Calcium Excretion

Both increases and decreases in urinary calcium excretion have been reported in patients with Bartter's syndrome or its variants. Whereas most studies have described a decrease in urinary calcium excretion (2,29,106), there are reports of patients with hypercalciuria and nephrocalcinosis, in association with hyperreninemia, hyperaldosteronism, and potassium wasting (31–33). It is likely that these children, who are usually born prematurely and frequently suffer from growth retardation, represent a separate entity which has been called "calcium-losing tubulopathy" (34), especially since the presence of nephrocalcinosis and histologic evidence of tubular damage are lacking in patients with Bartter's syndrome. Mechanisms for the calcium wasting in these patients could include a defect in calcium and chloride reabsorption in the loop of Henle or elevated prostaglandin excretion causing a renal leak of calcium (107). Treatment of patients with calcium-losing tubulopathy can lead to dramatic improvement in growth and development in association with decreases in hypercalciuria, hyperreninemia, hyperaldosteronism, and increases in serum potassium (108).

I. Overview

The foregoing discussion reviews the physiologic abnormalities that characterize patients with Bartter's syndrome, as well as various pathophysiologic concepts that have evolved in recent years regarding the etiology of this syndrome. Many of the abnormalities once thought to be the primary cause of Bartter's syndrome are now felt to be the consequence of hypokalemia. Figure 8.2 summarizes the major defects in Bartter's syndrome and their interrelationship. The principal abnormality is renal potassium wasting, which can occur as a consequence of several defects. Some patients may have isolated potassium wasting as the sole tubular defect. Others have a defect in chloride reabsorption in the loop of Henle which may augment sodium and tubular flow to distal potassium secretory sites, leading to renal potassium wasting. Still others have hypomagnesemia, which can produce renal potassium wasting. Regardless of the primary mechanism, renal wasting results in total body potassium depletion and leads to a variety of secondary metabolic and hormonal responses. In addition to causing a defect in cell membrane transport processes, potassium depletion has three other major effects: (a) increased synthesis of renal prostaglandins, which stimulate aldosterone synthesis both directly and via increased renin and angiotensin II production; (b) increased production of vasodilatory prostaglandins; and (c) inhibition of aldosterone secretion. It should be noted that whereas potassium depletion depresses aldosterone secretion, elevated angiotensin levels enhance it. Thus, in patients with Bartter's syndrome, the net

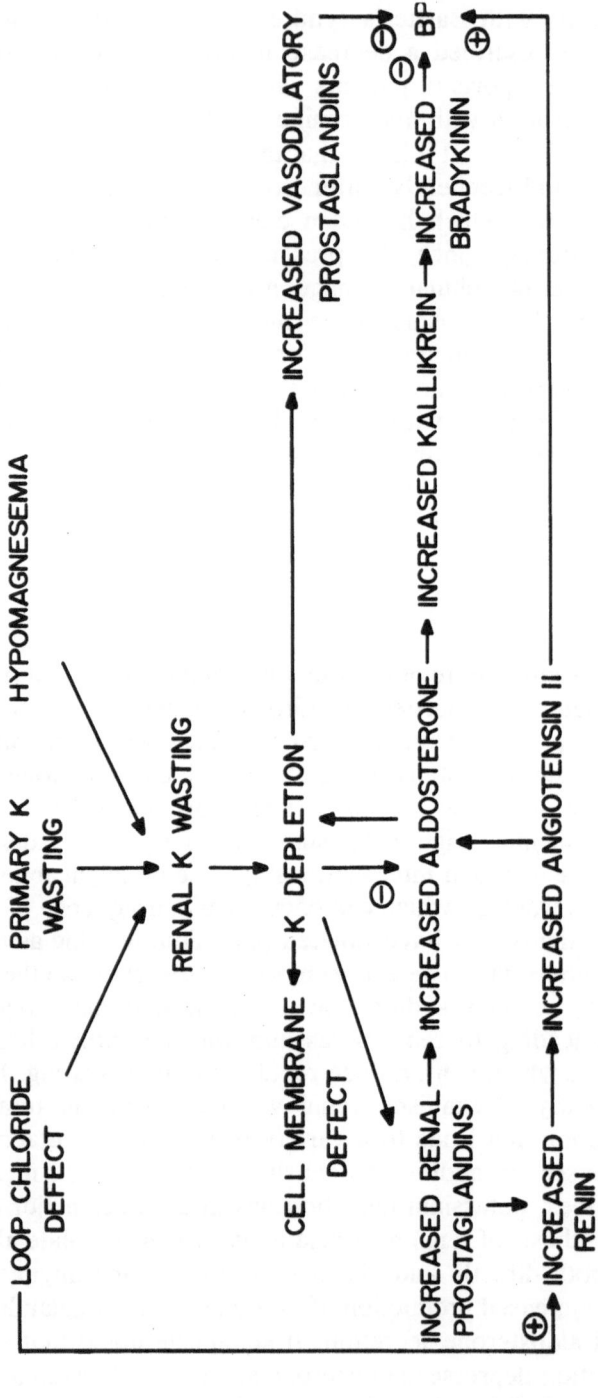

FIGURE 8.2. Metabolic derangements in patients with Bartter's syndrome. The major defect, renal K wasting, could be a primary event or could result from a defect in chloride reabsorption or from hypomagnesemia. The resultant K depletion leads to the remaining hormonal abnormalities described in patients with the syndrome.

effect tends to be a mildly elevated aldosterone level, which further exacerbates renal potassium loss and activates kallikrein activity, thereby increasing plasma bradykinin levels. Enhanced activity of the kallikrein system, in conjunction with elevated levels of vasodilatory prostaglandins, offsets the vasoconstrictive action of increased angiotensin II, resulting in a normal blood pressure in these patients.

IV. Differential Diagnosis

Persistent hypokalemia may be due to excess potassium losses in urine, stool, or skin or to abnormal shifts of potassium into cellular compartments. An in-depth discussion of defects in renal and extrarenal potassium regulation can be found elsewhere (109). Hypokalemia due to potassium losses can occur with or without hypertension (Table 8.3). Hypokalemia associated with hypertension is usually associated with mineralocorticoid excess (Table 8.3). Hypokalemia without hypertension also may be associated with hyperaldosteronism, but in this situation the increased aldosterone is usually secondary to the volume contraction caused by an underlying process. The major causes of normotensive hypokalemia include surreptitious vomiting, diuretic or laxative abuse, and congenital syndromes associated with renal potassium wasting. Primary hyperaldosteronism without hypertension is a rare cause of potassium wasting (116–118). Although renal tubular acidosis can also induce potassium losses, the presence of acidosis distinguishes this disorder from the others, which are associated with a metabolic alkalosis.

A. Abuse of Diuretics and Laxatives and Surreptitious Vomiting

Patients who abuse diuretics or laxatives or who suffer from bulimia may exhibit the same constellation of clinical signs observed in a patient with Bartter's syndrome (Table 8.4) (62,63,81,82,110–113). Such patients have

TABLE 8.3. Etiology of renal K wasting.

With hypertension	Without hypertension
Primary hyperaldosteronism (adrenal adenoma, hyperplasia)	Diuretic abuse
Secondary hypertension (malignant hypertension, renin tumor)	Laxative abuse
Hypermineralocorticoidism (licorice, desoxycorticosterone)	Vomiting (bulimia)
Cushing's syndrome (Cushing's disease, ectopic ACTH, adrenal Ca)	Bartter's syndrome and others
Other (Liddle's syndrome)	Mg deficiency
	Renal tubular[a] acidosis

[a] The only disorder associated with acidosis, not alkalosis.

TABLE 8.4. Causes of hypokalemia without hypertension.[a]

	Serum aldosterone	PRA[b]	Renal prostaglandin production	Juxtaglomerular hyperplasia	Urine chloride
Surreptitious vomiting	↑	↑	↑	+[c]	↓
Occult diuretic abuse	↑	↑	↑	+	↑
Occult laxative abuse	↑	↑	↑	+	↓
Bartter's syndrome and other primary K-wasting syndromes	↑	↑	↑	+	↑

[a] All have hypokalemia and metabolic alkalosis.
[b] Plasma renin activity.
[c] Present.

hypokalemia and metabolic alkalosis secondary to the sodium, potassium, and chloride loss. The consequent volume contraction activates the renin-angiotensin system, resulting in hyperaldosteronism and metabolic alkalosis that in turn aggravate renal potassium losses. Vascular resistance to infusion with angiotensin II, elevated renal prostaglandin production secondary to the potassium depletion (62,81), and hyperplasia of the juxtaglomerular region on renal biopsy have also been reported in these cases (62,82). As summarized in Table 8.4, the hormonal profile characteristic of a patient with Bartter's syndrome is also observed in patients with these disorders of abuse. In fact, with increasing emphasis in our society on maintenance of low body weight and the growing incidence of anorexia nervosa and bulimia in adolescents, abuse of diuretics and laxatives and surreptitious vomiting, rather than a congenital defect, are more common causes of these metabolic abnormalities.

These patients have been labeled as having "pseudo-Bartter's syndrome" and are notorious for denying their abusive habits (82,112,113). One patient even allowed an AV fistula to be placed in her arm for chronic intravenous potassium infusion rather than admit that she was surreptitiously ingesting diuretics (112). Although patients with these disorders display many of the abnormalities common to those of Bartter's syndrome, there are ways to distinguish them. Urine chloride concentration or excretion is low in patients with gastrointestinal losses (laxative abuse and vomiting) but high in patients with Bartter's syndrome and diuretic abuse (Table 8.4). On the other hand, only the presence of diuretics in the urine can definitively confirm diuretic abuse. It is now recommended that multiple urinary screens for diuretics be carried out before the diagnosis of Bartter's syndrome is made in an adult (82).

B. Congenital Disorders

Several familial disorders associated with normotensive renal potassium wasting have been described (Table 8.5). Whether these syndromes

TABLE 8.5. Congenital disorders associated with renal potassium wasting.[a]

	Growth and development	Renal concentrating ability	Fractional chloride reabsorption	Juxtaglomerular hyperplasia	Urinary calcium excretion	Urinary magnesium excretion
True Bartter's syndrome	Impaired	Impaired	Impaired	Present	Decreased	Increased
Magnesium-losing tubulopathy	Normal	Normal	?Normal[b]	Absent	Decreased	Increased
Calcium-losing tubulopathy	Impaired	Impaired	?Normal[b]	Usually present	Increased	Normal
Familial hypokalemic tubulopathy[c]	Impaired	Not described	Normal	Normal	Not described	Not described

[a] All occurring with normal blood pressure, hyperreninemia, hyperaldosteronism, elevated renal prostaglandin production, and vascular resistance to angiotensin II. Reprinted from reference 34, pp. 121–131 by courtesy of Marcel Dekker, Inc.
[b] More subjects need to be evaluated before this finding is confirmed.
[c] Only one family reported to date (41).

should be designated as variants of Bartter's syndrome or as Bartter's syndrome with associated defects, or whether they represent truly separate entities, will be a matter of speculation until more is known about the specific etiologies of these disorders. These syndromes share the common features of renal potassium wasting, metabolic alkalosis, hyperreninemia, hyperaldosteronism, elevated renal prostaglandin production, and vascular resistance to angiotensin II. In addition, a defect in chloride transport in the thick ascending limb (in a patient not on diuretics) must be present before the diagnosis of true Bartter's syndrome can be made (47).

Familial magnesium-losing tubulopathy in children is associated with normal growth and development (34); in contrast, children with Bartter's syndrome commonly have growth and mental retardation (1,35). Furthermore, hyperplasia of the juxtaglomerular apparatus is absent in children with primary magnesium wasting (24,27,28). Impaired fractional chloride reabsorption appears to be normal in children with magnesium-losing tubulopathy, although more subjects need to be evaluated to confirm this finding (34). In children with calcium-losing tubulopathy, urinary calcium excretion is elevated; in contrast, it is depressed in most patients with true Bartter's syndrome (106).

The exact reason for the association between nephrocalcinosis and renal potassium wasting is not known. Although growth, development, and renal concentrating ability are impaired, as they are in subjects with calcium-losing tubulopathy, fractional chloride reabsorption was normal in the one family in whom it was evaluated. Another congenital potassium-losing tubulopathy has recently been described in 3 out of 4 siblings in a single family (41). This disorder is distinguishable from true Bartter's syndrome mainly by the absence of juxtaglomerular hyperplasia, normal fractional chloride reabsorption, and a distinct HLA linkage (41,42). Whether these congenital syndromes represent distinct entities or just variants of true Bartter's syndrome is of more than just academic interest, since these patients respond differently to different therapeutic modalities (108).

C. Primary Hyperaldosteronism

Most patients with primary hyperaldosteronism secondary to adrenal hyperplasia, adenoma, or tumor will have hypertension (blood pressure ≥ 140/90) at the time they present with hypokalemia (115). However, there are at least 3 case reports of normotensive patients with this diagnosis. In all 3 cases, adrenal adenoma (116,117) or hyperplasia (118) was found at surgery. Furthermore, the hypokalemia was corrected after surgical intervention in at least 2 of the cases (116,117). No uniform explanation has been offered for the absence of hypertension in these cases.

V. Clinical Evaluation

When a patient presents with normal blood pressure and persistent hypokalemia clearly not due to drugs, the first step that should be taken to establish the diagnosis is to measure the potassium concentration in a random urine sample (Table 8.6). If the urine potassium concentration is greater than 20 mEq/L when the serum potassium level is ≤ 3.5 mEq/L, renal potassium wasting is established. In the absence of hypertension, diuretic or laxative abuse, surreptitious vomiting, or a congenital syndrome associated with renal potassium wasting, Bartter's syndrome is most likely.

A low chloride concentration in a random urine sample will rule out the diagnosis of Bartter's syndrome and provide evidence for laxative abuse or surreptitious vomiting. In contrast, patients with diuretic abuse may exhibit high urinary chloride excretion if they are taking such drugs at the time of collection. Therefore, it is important to analyze several urine samples for diuretics, since drug ingestion may be sporadic, and several samples may be negative (82). Furthermore, it should be realized that most "diuretic screens" test mainly for furosemide and the thiazide derivatives, making them unreliable in a patient surreptitiously taking a different kind of diuretic.

Once the disorders of occult diuretic or laxative use and vomiting have been excluded by the urine tests described above, additional tests should be performed to define further the defect in renal potassium wasting. Measurement of plasma aldosterone and renin levels with the patient in the supine position before and after volume expansion will differentiate the rare case of normotensive primary hyperaldosteronism from Bartter's syndrome. Basal plasma renin activity is low, and the plasma aldosterone concentration cannot be suppressed with saline in those with primary aldosteronism (2,115), whereas plasma renin activity is high and plasma aldosterone suppresses with volume expansion in patients with Bartter's syndrome (43–46).

With the diagnosis of primary aldosteronism excluded, urinary levels of magnesium and calcium can be measured to define the type of potassium-losing tubulopathy. Fractional chloride reabsorption under conditions of water diuresis can also be evaluated, although, as described above, results of this test depend in part on how the study is performed (49). Measurement of urinary prostaglandins and of resistance to angiotensin II infusion can also be performed, but abnormalities in these parameters, if present, will not be specific for Bartter's syndrome. Similarly, a renal biopsy to demonstrate hyperplasia of the juxtaglomerular apparatus may be performed, but this will not clearly differentiate Bartter's syndrome from other disorders associated with potassium depletion (62,81,82). Diagnostic tests required to evaluate patients for Bartter's syndrome are listed in Table 8.6.

TABLE 8.6. Diagnostic evaluation for Bartter's syndrome.[a]

1. Measure urine potassium concentration
2. Measure urine chloride concentration
3. Screen urine for diuretics
4. Measure baseline plasma aldosterone and renin levels—before and after volume expansion
5. Measure fractional chloride reabsorption
6. ?Measure urinary prostaglandins[b]
7. ?Document vascular resistance to angiotensin II infusion[b]
8. ?Document hyperplasia of juxtaglomerular apparatus with a renal biopsy[b]

[a] In a patient with persistent hypokalemia and normal blood pressure.
[b] Not specific for Bartter's syndrome.

VI. Treatment

A. Should Hypokalemia Be Treated?

Before discussing the forms of treatment available, it is important to consider the essential question of whether or not treatment of hypokalemia is necessary. Careful analysis of current data raises some question about this matter, especially when the hypokalemia is mild (119,120).

The major organs affected by hypokalemia are the heart, kidneys, and neuromuscular system. Although there is little doubt that hypokalemia may exacerbate digitalis toxicity, considerable controversy exists as to whether hypokalemia itself increases the risk of ventricular arrhythmia (119–124). Some reports have documented an association between hypokalemia itself and ventricular ectopy or sudden death (121,122), but others have not (123,124). However, even in the studies demonstrating an increased frequency of ectopy or sudden death in patients with hypokalemia, other causes for the problem, such as magnesium deficiency in patients on diuretics, have not been excluded.

Neuromuscular manifestations of hypokalemia may involve skeletal muscle (weakness, paralysis, and rhabdomyolysis) or smooth-muscle function (ileus, gastric distension, and bladder retention). In the absence of other metabolic derangements, these effects do not occur unless hypokalemia is severe (<2.5 mEq/L). However, many patients, especially those in whom the condition is chronic, tolerate severe depressions in serum potassium concentration with no symptoms at all. The major renal manifestation of chronic hypokalemia is a concentrating defect which, if severe, can cause symptomatic polyuria and polydipsia (125). Although reversible decreases in glomerular filtration rate can occur in patients with severe hypokalemia (126), permanent renal damage has been more difficult to document. Chronic renal failure and biopsy-proven interstitial fibrosis have been reported in a few children with Bartter's syndrome and in patients with hypokalemia secondary to diuretic abuse (127,128). However, in these cases, it is not clear whether it is the hypokalemia per se or some other metabolic derangement that is responsible for the renal lesion. Indeed, hypokalemia does not appear to be a frequent cause of chronic renal damage, raising serious questions about its role in this disorder.

Hypokalemia may also effect development in children. Physical and mental retardation were cardinal features described in the original report of children with Bartter's syndrome (1,35). Although these abnormalities have always been attributed to the hypokalemia, potassium supplementation has had little effect on reversing them (108). Furthermore, children with potassium depletion associated with magnesium-losing tubulopathy have normal growth and development (34). Thus, the issue of whether the growth defects are actually caused by, or just associated with, potassium depletion is unresolved.

The question of whether or not to treat is of more than academic interest since, especially in patients with congenital syndromes associated with potassium depletion, potassium supplementation in large amounts or drugs with significant side effects are required to restore potassium balance to normal. Thus, the following guidelines for treatment are suggested. Certainly, patients on digitalis preparations, those with heart disease, and those with signs or symptoms of hypokalemia should receive treatment. Additionally, attempts to restore serum potassium levels toward normal should be made in children for their potential influence on growth and development. It also seems prudent to treat patients with significant hypokalemia to prevent the occurrence of dangerous depressions in serum potassium levels should another metabolic event occur. Whether all patients with hypokalemia should be treated, especially those with milder degrees of potassium depletion, is still an intensely debated issue (119,120).

Patients with Bartter's syndrome and other forms of potassium-losing nephropathies usually have significant degrees of hypokalemia and probably should be treated. Standard drugs available for treatment have included potassium supplements, magnesium supplements, potassium-sparing diuretics, and inhibitors of prostaglandin synthesis. More recently, inhibitors of angiotensin-converting enzyme and calcium channel blockers have been used on an experimental basis.

B. Treatment Regimens (Fig. 8.3)

Regardless of the treatment employed, only a partial correction of hypokalemia has been achieved in most cases of Bartter's syndrome. Historically, the therapy has been directed toward correcting the secondary hyperaldosteronism with agents such as spironolactone (1,45) and aminoglutethamide (43), or with bilateral adrenalectomy (78,79), and

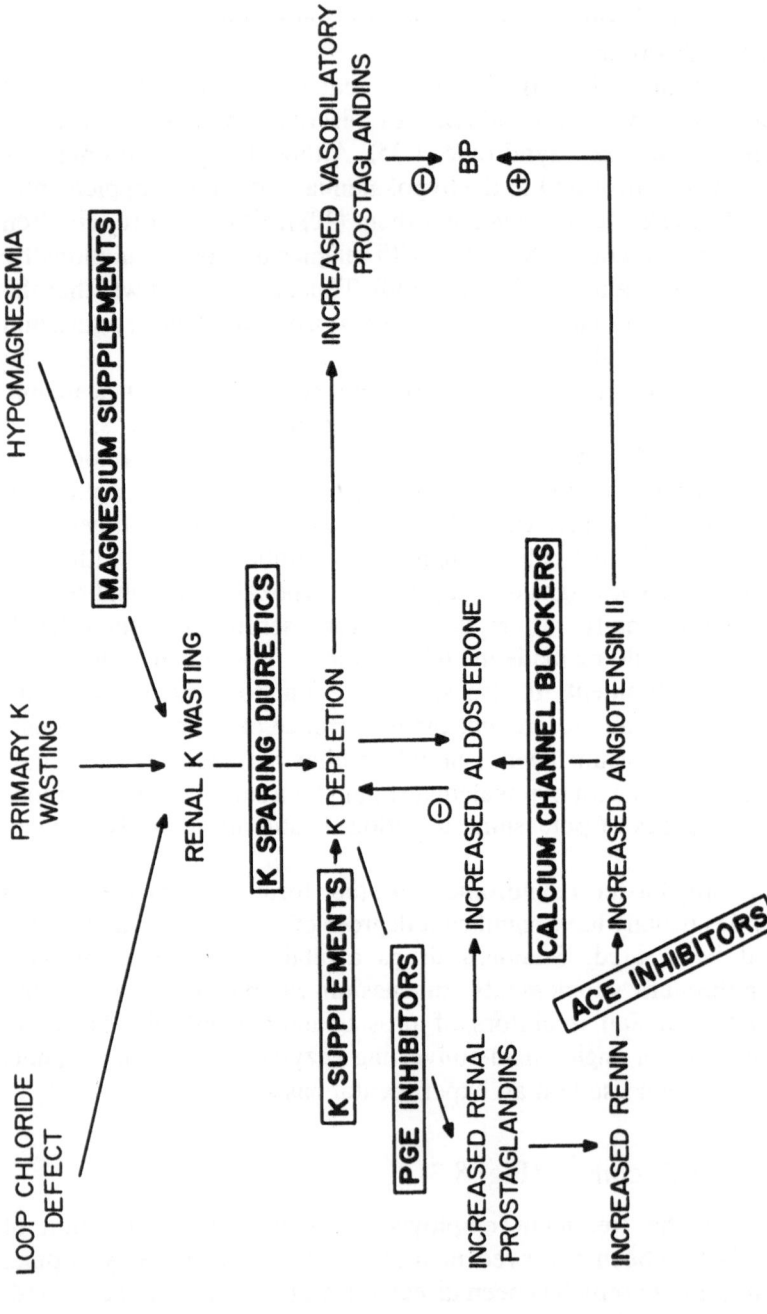

FIGURE 8.3. Types of treatment used to treat patients with Bartter's syndrome (shown in boxes) according to where in the pathogenesis of the disorder they operate to ameliorate the potassium wasting.

correcting the elevated prostaglandin levels with inhibitors of prostaglandin synthesis. Despite these attempts and the current availability of a variety of new pharmacologic agents, such as inhibitors of angiotensin-converting enzyme and calcium channel blockers, reports of complete reversal of the renal potassium wasting are still few.

1. POTASSIUM SUPPLEMENT

The cornerstone in the treatment of Bartter's syndrome is potassium supplementation. Large doses, ranging from 120 to 400 mEq/day, are usually required (108). Symptomatic improvement is frequently observed despite only partial correction of the hypokalemia. Compliance is a major problem, as frequent doses are required, and many patients tolerate oral potassium poorly primarily because of gastrointestinal irritation. A variety of potassium supplements are now available, and the choice should be a matter of patient preference. As a general rule, however, potassium chloride is the preferred salt, since most of these patients have an associated metabolic alkalosis. If liquid potassium chloride is not tolerated, potassium chloride in a wax matrix may produce fewer gastrointestinal side effects. Even with large amounts of supplemental potassium chloride, it is usually inevitable that most patients will require other forms of treatment.

2. MAGNESIUM SUPPLEMENT

Magnesium replacement in patients with potassium-losing syndromes has yielded conflicting results. Although some investigators have reported a decrease in potassium excretion (20,25,29), others have not (24,26). In cases where potassium excretion has decreased with magnesium replacement, the changes occurred without a decrease in plasma renin activity, plasma aldosterone concentration, or renal prostaglandin production (20,25). Although metabolic abnormalities may not be corrected with magnesium replacement, some patients have responded with dramatic improvement in muscle strength (19). Thus magnesium replacement should be initiated in hypokalemic patients with concomitant hypomagnesemia. Since the degree of hypomagnesemia is often mild, oral treatment (20–60 mEq/day) should be sufficient (108). Large doses should be avoided, since they frequently promote diarrhea. Such gastrointestinal side effects of magnesium preparations limit patient compliance with this therapy.

3. INHIBITORS OF PROSTAGLANDIN SYNTHESIS

Many studies have shown that treatment with inhibitors of prostaglandin synthesis in patients with Bartter's syndrome leads to increases in serum potassium levels and improvement in the associated metabolic abnormalities. However, few studies have shown complete correction of the hypokalemia (30,54,56–61) (Fig. 8.1). Inhibition of prostaglandin synthesis may raise serum potassium by either decreasing renal perfusion, thereby producing decreased flow and sodium delivery to the distal tubule with subsequent decrease in renal potassium secretion, or by decreasing aldosterone production (130). Although inhibition of prostaglandin synthesis does not usually restore serum potassium concentration to normal, it significantly decreases the hyperreninemia and hyperaldosteronism associated with Bartter's syndrome (34,57,85) and restores vascular responsiveness to pressor agents (53,54,69,70). Inhibition of prostaglandin production also decreases renal calcium excretion (107). In the subset of children with hypokalemia and calcium-losing tubulopathy, use of these agents may produce dramatic improvement in the metabolic abnormalities, increase serum potassium concentration, and improve growth and development (107,108).

Common side effects of inhibitors of prostaglandin synthesis include gastrointestinal distress, bleeding, and salt and water retention. Such effects often limit the use of these agents, especially for a long time. Because of these side effects, it is recommended that treatment with potassium and magnesium supplements alone or with potassium-sparing diuretics be initiated before starting treatment with inhibitors of prostaglandin synthesis (108).

4. POTASSIUM-SPARING DIURETICS

Several potassium-sparing diuretics have been used to correct the hypokalemia in Bartter's syndrome. Spironolactone, a steroidal competitive antagonist of aldosterone, can partially correct the hypokalemia (1,45). However, its prolonged use may be limited by the occurrence of gynecomastia, androgenlike effects, menstrual irregularities, and minor gastrointestinal upset. Triamterene, a nonsteroidal potassium-sparing diuretic, may also decrease renal potassium losses. Its mechanism of action is not completely understood but appears to be related to the tubular transport of potassium that is independent of aldosterone (131). Amiloride, a more recently available potassium-sparing diuretic, has also been used in the treatment of Bartter's syndrome. Although its effectiveness has not been fully evaluated, some evidence suggests that it is more effective than spironolactone in raising serum potassium (132). The mechanism by which amiloride inhibits potassium secretion is believed to be secondary to interference of sodium entry into the distal tubule and collecting duct cells, thereby reducing negative potential difference in the lumen (133). Although experience with amiloride is limited in the U.S., its use in Europe indicates that it is better tolerated and has fewer side effects than spironolactone. In our opinion, such data make amiloride the potassium-sparing diuretic of choice. Doses of 5–10 mg once or twice daily in conjunction with potassium supplement are recommended.

5. ANGIOTENSIN-CONVERTING ENZYME INHIBITOR

Because the renin-angiotensin-aldosterone system is activated in patients with Bartter's syndrome, several investigators have evaluated the effects of captopril (72,134–137) and enalapril (138) in these patients. Results demonstrate that treatment with these angiotensin-converting enzyme inhibitors leads to a decrease in aldosterone. However, significant improvement was noted after 1 month of therapy, but not after only 5 days (134–138). Therefore, treatment with angiotensin-converting enzyme inhibitor for at least 1 month may be necessary for a significant response to occur.

Caution must be exercised when angiotensin-converting enzyme inhibitors are given to patients with Bartter's syndrome, because they may cause serious hypotension. Most investigators report mild to moderate hypotension following acute treatment with captopril (134–136), whereas blood pressure following long-term treatment is better tolerated (136,138). Nevertheless, a recent report described severe hypotension requiring treatment with norepinephrine following captopril treatment in a patient with Bartter's syndrome (72). This result is not unexpected, considering the evidence that levels of vasodilatory prostaglandins and bradykinin are elevated in patients with the syndrome, and that the effect of such increases are usually offset by high circulating levels of angiotensin II. Thus, when angiotensin II production is decreased, significant hypotension may occur (72). Hemodynamic stability may be better served by combined use of inhibitors of both prostaglandin synthesis and angiotensin-converting enzyme. However, the expense and the side effects of such a regimen may limit its use.

6. CALCIUM CHANNEL BLOCKERS

Calcium channel blockers are another class of pharmacologic agents with potential value in the treatment of Bartter's syndrome. In isolated adrenal cells, angiotensin II increases calcium entry which then acts as a second messenger to stimulate aldosterone production, an effect that is blocked by verapamil (139,140). Indeed, Nadler et al. reported that the treatment of primary aldosteronism with nifedipine for 4 weeks normalized blood pressure and potassium and reduced serum aldosterone levels (52). Since patients with Bartter's syndrome are characterized by secondary hyperaldosteronism, calcium channel blockade may similarly reduce levels of aldosterone, thus decreasing the degree of renal potassium wasting. However, since there are no published reports in which calcium channel blockers have been used in the treatment of Bartter's syndrome, and since hypotension is a potential hazard, such treatment should be considered experimental and undertaken with caution.

We treated our patient, M.W., with verapamil for 5 days. Blood pressure remained stable, but plasma aldosterone levels decreased minimally (16–12 ng/dl), and serum potassium level remained unchanged (2.5. mEq/L). Thus, in this case, verapamil did not ameliorate the potassium wasting.

VII. Analysis of Cases

In view of the above discussion, several points about our cases, J.W. and M.W., can be made. First, it is likely that they have a congenital abnormality, since both siblings are affected by the same disorder. Both have hypokalemia due to potassium wasting, metabolic alkalosis, normal blood pressure, hyperreninemia, and minimal hyperaldosteronism. Both have normal glomerular filtration rates, and neither has significant salt wasting, although J.W. was unable to reduce sodium below 10 mEq/L over 4 days on a low-sodium diet. When loop of Henle tubular function was evaluated by loading with intravenous dextrose and water, fractional chloride reabsorption (73% and 70%, respectively) was at or below the lower limits of normal, depending on the standard used (3–7). If one considers their fractional chloride reabsorption to be impaired, then they can be said to have true Bartter's syndrome by the strictest criteria (47). On the other hand, their normal growth and development and their significant hypomagnesemia, as well as the near normal fractional chloride reabsorption, would also justify the diagnosis of magnesium-losing tubulopathy. The overlap of these syndromes is obvious, making it difficult to separate all affected individuals into discrete categories. Regardless of this dilemma, initiation of treatment with potassium and magnesium supplements would seem most advisable no matter what label is used to describe the potassium-losing syndrome.

It is interesting that our index case, J.W., developed severe symptomatic hypokalemia (serum K < 1 mEq/L) only after receiving a short course of dexamethasone. It is likely that he had chronic hypokalemia, acutely exacerbated by additional potassium losses induced by dexamethasone. Acute administration of glucocorticoids has been shown to induce kaliuresis in human subjects (48) and in animals (114,129). The effect is distinguished from the kaliuresis caused by mineralocorticoid administration, since it is associated with an increase in urine sodium excretion and urine flow rate (114,129). Thus glucocorticoids should be used with caution in patients with an underlying potassium-wasting syndrome.

Acknowledgments. Supported in part by a grant from the Adult Clinical Research Center.

References

1. Bartter FC, Pronove P, Gill JR, MacCardle RC (1962) Hyperplasia of the juxtaglomerular complex with hyperaldosteronism and hypokalemic alkalosis. Am J Med 33:811–828.
2. Kem D, Weinberger M, Mayes D, Nugent C (1971) Saline suppression of plasma aldosterone in hypertension. Arch Intern Med 128:380–386.
3. Stein JH (1985) The pathogenic spectrum of Bartter's syndrome. Kidney Int 28:85–93.
4. Uribarri J, Alveranga D, Oh MS, Kuhar NM, Del Monte ML, Caroll JH (1985) Bartter's syndrome due to a defect in salt reabsorption in the distal convoluted tubule. Nephron 40:52–56.
5. Chaimovitz C, Levi J, Better OS, Oslander L, Benderli A (1973) Studies on the site of renal salt loss in a patient with Bartter's syndrome. Pediatr Res 7:89–94.
6. Bartter FC, Delea CS, Kawasaki T, Gill JR (1974) The adrenal cortex and the kidney. Kidney Int 6:272–280.
7. Fujita T, Sakaguchi H, Shibagaki M, Fukui T, Nomura M, Sekiguchi S (1977) The pathogenesis of Bartter's syndrome. Functional and histologic studies. Am J Med 63:467–474.
8. Gall G, Vaitukaitis J, Haddow JE, Klein R (1971) Erythrocyte Na flux in a patient with Bartter's syndrome. J Clin Endocr 32:562–567.
9. Gardner JD, Simopoulos AP, Lapey A, Shibolet S (1972) Altered membrane sodium transport in Bartter's syndrome. J Clin Invest 51:1565–1571.
10. Oliver JF, Delaney VB, Bourke E (1978) Increased erythrocyte sodium permeability in patients with Bartter's syndrome. Mineral Electrolyte Metab 1:225–230.
11. Cole CH, O'Regan S (1981) Effect of treatment with prostaglandin synthetase inhibitors on the erythrocyte sodium transport abnormality of Bartter's syndrome. Pediatr Res 15:926–929.
12. Mongeau J-G, Garay R, De Mendonca M, Broyer M, Meyer P (1983) Erythrocyte Na and K transport systems in children with Bartter's syndrome: Increase in passive sodium permeability. Kidney Int 23:530–535.
13. Korff JM, Siebens AW, Gill JR (1984) Correction of hypokalemia corrects the abnormalities in erythrocyte sodium transport in Bartter's syndrome. J Clin Invest 74:1724–1729.
14. Jepson J, McGarry EE (1968) Polycythemia and increased erythropoeitin production in a patient with Bartter's syndrome. Blood 32:370–375.
15. Erkelens DW, Statius van Eps LW (1973) Bartter's syndrome and erythrocytosis. Am J Med 55:711–719.
16. Meyer W, Gill J, Bartter F (1975) Gout as a complication of Bartter's syndrome. Am J Med 83:56–59.
17. Stoff JS, Stemerman M, Steer M, Salzman E, Brown RS (1980) Defect in platelet aggregation in Bartter's syndrome. Am J Med 68:171–180.
18. Güllner H-G, Kafka MS, Gill JR (1984) Insensitivity of platelet adenylate cyclase to prostaglandin E_1 in patients suffering from diseases of involving hypokalemia. Ann Intern Med 101:342–343.
19. Mace JW, Hambridge K, Gotlin RW, Dubois RS, Soloman CS, Katz FH (1973) Magnesium supplementation in Bartter's syndrome. Arch Dis Child 48:485–487.
20. Baehler RW, Work J, Kotchen TA, McMorrow G, Guthrie G (1980) Studies on the pathogenesis of Bartter's syndrome. Am J Med 69:933–938.

21. Favre L, Williams G, Favre H, Paunier L, Vallaton MB (1985) Relationship of renal prostaglandin to distal transport of sodium chloride in normokalemic and hypokalemic man. Mineral Electrolyte Metab 11:186–191.
22. Solomon LR, Bobinski H, Astley P, Golby FS, Mallick NP (1982) Bartter's syndrome—observations on the pathophysiology. Q J Med 203:251–270.
23. Spencer RW, Voyce MA (1976) Familial hypokalaemia and hypomagnesaemia. Acta Paediatr Scand 65:505–507.
24. McCredie DA, Blair-West JR, Scoggins, BA, Shipman R (1971) Potassium-losing nephropathy of childhood. Med J Aust 1:129–135.
25. Güllner HG, Gill JR, Bartter FC (1981) Correction of hypokalemia by magnesium repletion in familial hypokalemia alkalosis with tubulopathy. Am J Med 71:578–582.
26. Gitelman HJ, Graham JB, Welt LG (1966) A new familial disorder characterized by hypokalemia and hypomagnesemia. Trans Assoc Am Phys 79:221–235.
27. Evans RA, Carter JN, George CRP, Walls RS, Newland RC, McDonnell GD, Lawrence JR (1981) The congenital "magnesium-losing kidney." Q J Med 197:39–52.
28. Paunier L, Sizonenko PC (1976) Asymptomatic chronic hypomagnesemia and hypokalemia in a child: Cell membrane disease? J Pediatr 88:51–55.
29. Runeberg L, Collan Y, Jokinen EJ, Länhevirta J, Aro A (1975) Hypomagnesemia due to renal disease of unknown etiology. Am J Med 59:873–881.
30. Zipser RD, Rude RK, Zia PK, Fichman MP (1979) Regulation of urinary prostaglandins in Bartter's syndrome. Am J Med 67:263–267.
31. McCredie DA, Rotenberg E, Williams AC (1974) Hypercalciuria in potassium-losing nephropathy: A variant of Bartter's syndrome. Aust Pediatr J 10:286–295.
32. Dillon MJ, Shah V, Mitchell MD (1979) Bartter's syndrome: 10 cases in childhood. Q J Med 48:429–446.
33. Betend B, David L, Vincent M, Hernier M, Francois R (1979) Successful indomethacin treatment of two pediatric patients with severe tubulopathies. Helv Paediatr Acta 34:329–344.
34. Gill JR (1985) Potassium losing tubulopathies: Bartter's syndrome and similar disorders. In Welton PK, Welton AA, and Walker WG (eds): Potassium in Cardiovascular and Renal Medicine. Marcel Dekker, New York, pp 121–131.
35. Cannon PJ, Leeming JM, Sommers SC, Winters RW, Laragh JH (1968) Juxtaglomerular cell hyperplasia and secondary hyperaldosteronism (Bartter's syndrome): A reevaluation of the pathophysiology. Medicine 47:107–131.
36. Brackett NC, Koppel M, Randall RE, Nixon WP (1968) Hyperplasia of the juxtaglomerular complex with secondary aldosteronism without hypertension (Bartter's syndrome). Am J Med 44:803–818.
37. Simopoulos AP (1979) Growth characteristics in patients with Bartter's syndrome. Nephron 23:131–136.
38. Sutherland LE, Hartroft P, Balis U, Bailey JD, Lynch MJ (1970) Bartter's syndrome. Acta Paediatr Scand (Suppl) 201:1–24.
39. Delaney VB, Oliver JF, Simms M, Costello J, Bourke R (1981) Bartter's syndrome: Physiologic and pharmacologic studies. Q J Med 198:213–232.
40. Sotos JF (1970) Genetic Disorders of Man. Little, Brown, Boston, p 739.
41. Güllner H-G, Bartter FC, Gill JR, Dickman PS, Wilson CB, Tiwari JL (1983) A sibship with hypokalemic alkalosis and renal proximal tubulopathy. Arch Intern Med 143:1534–1540.
42. Delaney V, Watson AJ, Pollack M, Dupont B, Bourke E (1984) HLA typing in Bartter's syndrome. Am J Med Genet 19:1779–1782.

43. Goodman AD, Vagnucci AH, Hartroft PM (1969) Pathogenesis of Bartter's syndrome. N Engl J Med 281:1435–1439.
44. Norby L, Mark AL, Kaloyanides GJ (1976) On the pathogenesis of Bartter's syndrome: Report of studies in a patient with this disorder. Clin Nephrol 6:404–413.
45. White MG 91972) Bartter's syndrome, a manifestation of renal tubular defects. Arch Intern Med 129:41–47.
46. Tomko DJ, Yeh BPY, Falls WF (1976) Bartter's syndrome. Study of a 52 year old man with evidence for a defect in proximal tubular sodium reabsorption and comments on therapy. Am J Med 61:111–118.
47. Gill JR, Bartter FC (1978) Evidence for a prostaglandin-independent defect in chloride reabsorption in the loop of Henle as a proximal cause of Bartter's syndrome. Am J Med 65:766–772.
48. Yunis SL, Bergcovitch DD, Stein RM, Levitt MF, Goldstein MH (1964) Renal tubular effects of hydrocortisone and a aldosterone in normal hydropenic man: Comment in sites of action. J Clin Invest 43:1668–1669.
49. Hernandez R, Kurtz I, Schambelan M, Conran J, Rector FC, Morris RC, Sebastian A (1986) Can tests of renal diluting ability reveal the proximate cause of Bartter's syndrome? Kidney Int 29:190.
50. Garrick R, Ziyadeh FN, Jorkasky D, Golfarb S (1985) Bartter's syndrome: A unifying hypothesis. Am J Nephrol 5:379–384.
51. Stoff JS, MacIntyre DE, Brown RS, Salzman EW (1979) Prostacyclin overproduction in Bartter's syndrome. Lancet 2:1169–1170.
52. Nadler JL, Hsueh W, Horton R (1985) Therapeutic effect of calcium channel blockade in primary aldosteronism. J Clin Endocrinol Metab 60:896–899.
53. Verberckmoes RG, Van Damme B, Clement J, Amery A, Michielsen P (1976) Bartter's syndrome with hyperplasia of renomedullary cells: Successful treatment with indomethacin. Kidney Int 9:302–307.
54. Fichman MP, Telfer N, Zia P, Speckart P, Golub M, Rude R (1976) Role of prostaglandins in the pathogenesis of Bartter's syndrome. Am J Med 60:785–797.
55. Gill JR, Frölich JC, Bowden RE, Taylor AA, Keiser HR, Seyberth HW, Oates JA, Bartter FC (1976) Bartter's syndrome: A disorder characterized by high urinary prostaglandins and a dependence of hyperreninemia on prostaglandin synthesis. Am J Med 61:43–51.
56. Bowden RE, Gill JR, Radfar N, Taylor AA, Keiser HR (1978) Prostaglandin synthetase inhibitors in Bartter's syndrome. Effect of immunoreactive prostaglandin E excretion. JAMA 239:117–121.
57. Vinci JM, Gill JR, Bowden RE, Pisano JJ, Izzo JL, Radfar N, Taylor AA, Zusman RM, Bartter FC, Keiser HR (1978) The kallikrein kinin system in Bartter's syndrome and its response to prostaglandin synthetase inhibition. J Clin Invest 61:1671–1682.
58. Dunn MJ (1981) Prostaglandins and Bartter's syndrome. Kidney Int 19:86–102.
59. Norby L, Flamenbaum W, Lentz R, Ramwell P (1976) Prostaglandins and aspirin therapy in Bartter's syndrome. Lancet 2:604–606.
60. Rudin A, Aurell M, Hansson L, Westberg G (1979) Effect of sar^1-ala^8-angiotensin II on blood pressure and renin in Bartter's syndrome, before and after treatment with prostaglandin synthetase inhibitors. Scand J Clin Lab Invest 39:543–550.
61. Halushka PV, Wohltmann H, Privitera PJ, Hurwitz G, Margolius HS (1977) Bartter's syndrome: Urinary prostaglandin E–like material and kallikrein; indomethacin effects. Ann Intern Med 87:281–286.
62. Ramos E, Hall-Craggs M, Demers LM (1979) A case of surreptitious, habitual vomiting simulating Bartter's syndrome. JAMA 43:1070–1072.

63. Radfar N, Gill JR, Bartter FC, Bravo E, Taylor AA, Bowden RE (1978) Hypokalemia, in Bartter's syndrome and other disorders, produces resistance to vasopressors via prostaglandin overproduction. Proc Soc Exp Biol Med 158:502–507.

64. Gant NF, Daley GL, Chand S, Whalley PJ, MacDonald PC (1973) A study of angiotensin II pressor response throughout primigravid pregnancy. J Clin Invest 52:2682–2689.

65. Chelsey LC, Talledo E, Bohler CS, Zuspan FP (1965) Vascular reactivity to angiotensin II in pregnancy and nonpregnant women. Am J Obstet Gynecol 91:837–842.

66. Laragh JH, Cannon PJ, Bentzel CJ, Sicinski AM, Meltzer JI (1963) Angiotensin II, norepinephrine, and renal transport of electrolytes and water in normal man and in cirrhosis with ascites. J Clin Invest 42:1179–1192.

67. Johnson CI, Jose AD (1963) Reduced vascular response to angiotensin II in secondary hyperaldosteronism. J Clin Invest 42:1411–1420.

68. Kuchel O, Horky K, Pazourek M, Gregorova I (1964) Pressor hyporeactivity to angiotensin in Addison's disease. Lancet 2:1316–1317.

69. Fujita T, Ando K, Sato Y, Yamashita K, Nomura M, Fukui T (1982) Independent roles of prostaglandins and the renin-angiotensin system in abnormal vascular reactivity in Bartter's syndrome. Am J Med 73:71–76.

70. Rodriquez-Portales JA, Lopez-Moreno JM, Mahana D (1985) Inhibition of the kallikrein-kinin system and vascular reactivity in Bartter's syndrome. Hypertension 7:1017–1022.

71. Galvez OG, Bay WH, Roberts BW, Ferris TF (1977) The hemodynamic effects of potassium deficiency in the dog. Circ Res 40 (Suppl I):I11–I16.

72. Sasaki H, Okumura M, Kawasaki T (1981) Captopril and Bartter's syndrome. Nephron 41:303–304.

73. Thurston H (1976) Vascular angiotensin receptors and their role in blood pressure control. Am J Med 61:768–778.

74. Devynck MA, Meyer P (1976) Angiotensin receptors in vascular tissue. Am J Med 61:758–767.

75. Richards CJ, Mark AL, Van Orden DE, Kaloyanides GJ (1978) Effects of indomethacin on the vascular abnormalities of Bartter's syndrome. Circulation 58:544–549.

76. Senba S, Konishi K, Saruta T, Ozawa Y, Kato E, Amagasaki Y, Nakata I (1984) Hypokalemia and prostaglandin over production in Bartter's syndrome. Nephron 37:257–263.

77. Saruta T, Fujimaki M, Senba S, Saito I, Konishi K (1984) Aldosterone and other mineralocorticoids in Bartter's syndrome. J Lab Clin Med 103:848–853.

78. Trygstad CW, Mangos JA, Bloodworth JMB, Lobeck CC (1969) A sibship with Bartter's syndrome: Failure of total adrenalectomy to correct the potassium wasting. Pediatrics 44:234–242.

79. Takayasu H, Aso Y, Nakauchi K, Kawabe K (1971) A case of Bartter's syndrome with surgical treatment followed for four years. J Clin Endocrinol Metab 32:842–845.

80. Gill JR (1981) Prostaglandins in Bartter's syndrome and in potassium-deficient disorders that mimic it. Mineral Electrolyte Metab 6:76–81.

81. Veldhuis JD, Bardin CW, Demers LM (1979) Metabolic mimicry of Bartter's syndrome by covert vomiting. Am J Med 66:361–363.

82. Jamison R, Ross J, Kempson R, Sufit CR, Parker TE (1982) Surreptitious diuretic ingestion and pseudo-Bartter's syndrome. Am J Med 73:142–147.

83. Yun J, Kelly G, Bartter FC, Smith H (1977) Role of prostaglandins in the control of renin secretion in the dog. Circ Res 40:459–464.

84. Gerber JG, Branch RA, Nies AS, Gerkens JF, Shand DB, Hollifield J, Oates JA (1978) Prostaglandins and renin release. II. Assessment of renin secretion following infusion of PG1$_2$, E$_2$ and D$_2$ into the renal artery of anesthetized dogs. Prostaglandins 15:81–88.
85. Bartter FC, Gill JR, Frölich JC, et al. (1976) Prostaglandins are overproduced by the kidneys and mediate hyperreninemia in Bartter's syndrome. Trans Assoc Am Phys 89:77–91.
86. Gill JR (1985) Bartter's syndrome. In Gonick HC, Buckalew WM (eds): Renal Tubular Disorders: Pathophysiology, Diagnosis and Management. Marcel Dekker, New York, pp 457–473.
87. Burg MB, Bourdeau JE (1978) Function of the thick ascending limb of Henle's loop. In Cogel HG, Ullrich KJ (eds): Newer Aspects of Renal Function. Excerpta Medica, Amsterdam, pp 91–102.
88. Delaponte C, Stulzaft J, Lorait C, Broyer M (1978) Muscle electrolytes and fluid compartments in six children with Bartter's syndrome. Clin Sci 54:223–231.
89. Muirhead EE, Germain G, Leach BE, Pitcock JA, Stephenson P, Brooks B, Brosius WL, Daniels EG, Hinman JW (1972) Production of renomedullary prostaglandins by renomedullary interstitial cells grown in tissue culture. Circ Res (Suppl II):II161–II172.
90. Dunn MJ, Staley RS, Harrison M (1976) Characterization of prostaglandin production in tissue culture of rat medullary cells. Prostaglandins 12:37–49.
91. Dray F (1978) Bartter's syndrome: Contrasting patterns of prostaglandin excretion in children and adults. Clin Sci Mol Med 54:115–118.
92. Nivet H, Grenier B, Rolland JC, Lerbanchu Y, Dray F (1978) Raised urinary prostaglandins in patients without Bartter's syndrome. Lancet 1:333–334.
93. Nielsen I, Hesse B, Christensen P (1979) On the pathogenic role of prostaglandins in Bartter's syndrome. Acta Med Scand 625:135–140.
94. Chan JCM, Malekzadey MH, Anand SK (1975) Defect in renal tubular sodium reabsorption in a patient with Bartter's syndrome. Clin Proc Child Hosp Natl Med Ctr 31:67–71.
95. Schuster VL, Seldin DW (1985) Renal clearance. In Seldin DW, Giebisch G (eds): The Kidney: Physiology and Pathophysiology. Raven Press, New York, pp 365–369.
96. Higashihara E, Stokes JB, Kokko JP, Campbell WB, Dubose TD (1979) Cortical and papillary micropuncture examination of chloride transport in segments of the rat kidney during inhibition of prostaglandin production. J Clin Invest 64:1277–1287.
97. Stokes JB (1979) Effect of prostaglandin E$_2$ on chloride transport across the rabbit thick ascending limb of Henle. Selective inhibition of the medullary portion. J Clin Invest 64:495–502.
98. Ogihara T, Maruyama A, Nugent CA, Hata T, Mikami H, Kumahara Y (1982) Familial Bartter's syndrome. Arch Intern Med 142:906–908.
99. Fuller PJ, Funder JW (1986) The cellular physiology of glandular kallikrein. Kidney Int 29:953–964.
100. Margolius HS, Geller R, Pisano JJ, Sjoerdsma A (1971) Altered urinary kallikrein excretion in human hypertension. Lancet 2:1063–1065.
101. Margolius HS, Horowitz D, Pisano JJ, Keiser HR (1974) Urinary kallikrein excretion in hypertensive man. Relationships to sodium intake and sodium retaining steroids. Circ Res 35:820–825.
102. Lechi A, Covi G, Lechi C, Corgnati A, Arosio E, Zatti M, Scuro LA (1978) Urinary kallikrein excretion and plasma renin activity in patients with essential hypertension and primary aldosteronism. Clin Sci Mol Med 55:51–55.

103. Margolius HS, Chao J, Kaizu T (1976) The effects of aldosterone and spironolactone on renal kallikrein. Clin Sci Mol Med 51:2795–2825.
104. Margolius HS, Horwitz D, Geller RG, Alexander RW, Gill JR, Pisano JJ, Keiser HR (1974) Urinary kellikrein excretion in normal man. Relationship to sodium intake and sodium-retaining steroids. Circ Res 35:812–819.
105. Shils ME (1969) Experimental human magnesium depletion. Medicine 48:61–85.
106. Rudin A, Sjögren B, Aurell M (1984) Low urinary calcium excretion in Bartter's syndrome. N Engl J Med 310:1190.
107. Houser M, Zimmerman B, Davidman M, Smith C, Sinaiko A, Fish A (1984) Idiopathic hypercalciuria associated with hyperreninemia and high urinary prostaglandin E. Kidney Int 26:176–182.
108. Gill JR (1985) Bartter's syndrome. In Krieger DT, Bardin CW (eds): Current Therapy in Endocrinology and Metabolism. B.C. Deeker, Philadelphia, pp 134–136.
109. Smith JD, Bia MJ, DeFronzo RA (1985) Clinical disorders of potassium metabolism. In Arieff A, DeFronzo R (eds): Fluid, Electrolyte and Acid Base Disorders. Churchill Livingstone, New York, pp 413–509.
110. Katz F, Eckert RC, Gebott M (1972) Hypokalemia caused by surreptitious self administration of diuretics. Ann Intern Med 76:85–90.
111. Fleming BJ, Genuth SM, Gould AB, Kamionkowski MD (1975) Laxative induced hypokalemia, sodium depletion and hyperreninemia. Ann Intern Med 83:60–62.
112. Katz FH (1975) Self administration of diuretics. Ann Intern Med 83:575–576.
113. Gossian VV, Werk EE (1972) Surreptitious laxation and hypokalemia. Ann Intern Med 76:67.
114. Bia MJ, Tyler K, DeFronzo R (1982) The effect of dexamethasone on renal electrolyte excretion in the adrenalectomized rat. Endocrinology 111:882–888.
115. Bravo E, Terazi R, Duston H, Fouad F, Textor S, Figgord R, Viot D (1983) The changing clinical spectrum of primary aldosteronism. Am J Med 74:641–651.
116. Snow MH, Nicol P, Wilkinson R, Hall R, Johnston IDA, Hacking PM (1976) Normotensive primary aldosteronism. Br Med J 1:1125–1126.
117. Shiroto H, Ando H, Ebitani I, Hara M, Numazaw K, Kawamura S, Sasaki H (1980) Normotensive primary aldosteronism. Am J Med 69:603–606.
118. Zipser RD, Speckart PF (1978) Normotensive primary aldosteronism. Ann Intern Med 88:655–656.
119. Harrington JT, Isner JM, Kassirer JP (1982) Our national obsession with potassium. Am J Med 73:155–159.
120. Kaplan NM (1984) Our appropriate concern about hypokalemia. Am J Med 77:1–4.
121. Holland O, Nixon J, Kuhmert LV (1981) Diuretic-induced ventricular ectopic activity. Am J Med 70:762–768.
122. Duke M (1978) Thiazide-induced hypokalemia. Association with acute myocardial infarction and ventricular fibrillation. JAMA 239:43–45.
123. Medical Research Council (1983) Ventricular extra systoles during thiazide treatment: Substrate of NRC mild hypertension trial. Br Med J 287:1249–1253.
124. Papademetriou V, Fletcher R, Khatri I, Frees E (1983) Diuretic-induced hypokalemia in uncomplicated systemic hypertension: Effect of plasma potassium correction in cardiac arrhythmias. Am J Cardiol 52:1017–1022.
125. Relman A, Schwartz W (1985) The kidney in potassium depletion. Am J Med 24:764–773.

126. Relman A, Schwartz (1956) The nephropathy of potassium depletion. N Engl J Med 255:195–203.
127. Bock KD, Cremer W, Werner U (1978) Chronic hypokalemic nephropathy: A clinical study. Klin Wochenschr 56 (Suppl 1):91–96.
128. Riemenschneider T, Bohle A (1983) Morphologic aspects of low potassium and low sodium nephropathy. Clin Nephrol 19:271–279.
129. Bia MJ, Tyler K, DeFronzo R (1983) The effect of dexamethasone on renal potassium excretion and acute potassium tolerance. Endocrinology 113:1690–1696.
130. Saruta T, Kaplan NM (1972) Adrenal cortical steroidogenesis: The effects of prostaglandins. J Clin Invest 51:2246–2251.
131. Mudge GH (1980) Drugs affecting renal function and electrolyte metabolism. In Gilman AG, Goodman LS, Gilman A (eds): The Pharmacologic Basis of Therapeutics. Macmillan, New York, pp 885–915.
132. Griffing GT, Komanicky P, Aurecchia SA, Sindler BH, Melby JC (1982) Amiloride in Bartter's syndrome. Clin Pharmacol Ther 31:713–718.
133. Giebisch G, Malnic G, Berliner RW (1986) Renal transport and control of potassium excretion. In Brenner BM, Rector FC (eds): The Kidney. Saunders, Philadelphia, pp 177–205.
134. Mizuno K, Yamazaki M, Fukuchi S (1979) Hypotensive response to angiotensin I–converting enzyme inhibitor in Bartter's syndrome. N Engl J Med 300:1057.
135. Aurell M, Rudin A (1981) Effects of captopril in Bartter's syndrome. N Engl J Med 304:1609.
136. Hené RJ, Koomans HA, Boer P, Dorhout-Mees EJ (1982) Long term treatment of Bartter's syndrome with captopril. Br Med J 285:696.
137. Hené RJ, Koomans HA, Boer P, Dorhout-Mees EJ (1983) Effect of captopril in Bartter's syndrome. Nephron 35:275.
138. Hené RJ, Koomans HA, Stolpe AUD, Verhoef GEG, Dorhout-Mees EJ (1986) Effect of enalapril on serum K renin-aldosterone system and intrarenal Na-handling in patients with Bartter's syndrome. Kidney Int 29:190.
139. Fakunding JL, Catt KJ (1980) Dependence of aldosterone stimulation in adrenal glomerulosa cells on calcium uptake: Effects of lanthanum and verapamil. Endocrinology 107:1345–1353.
140. Foster R, Lebo MV, Rasmussen H, Marusic ET (1981) Calcium: Its role in the mechanism of action of angiotensin II and potassium in aldosterone production. Endocrinology 109:2196–2201.

Index